CW01429405

THE **CONSPIRACY** AGAINST **LITHIUM**

The Suppressed Essential Nutrient and Its Benefits for Mental Health

Michael Nehls, MD, PhD

Skyhorse Publishing

For Corvin

It goes without saying that I dedicate a book about Lithos, the "Philosopher's Stone," to a philosopher and friend in the shared and eternal struggle for truth.

Skyhorse Publishing books may be purchased in bulk at special discounts for sales promotion, corporate gifts, fund-raising, or educational purposes. Special editions can also be created to specifications. For details, contact the Special Sales Department, Skyhorse Publishing, 307 West 36th Street, 11th Floor, New York, NY 10018 or info@skyhorsepublishing.com.

Skyhorse® and Skyhorse Publishing® are registered trademarks of Skyhorse Publishing, Inc.®, a Delaware corporation.

Visit our website at www.skyhorsepublishing.com.

Please follow our publisher Tony Lyons on Instagram @tonylyonsisuncertain.

10 9 8 7 6 5 4 3 2 1

Library of Congress Cataloging-in-Publication Data is available on file.

Print ISBN: 978-1-5107-8403-1
eBook ISBN: 978-1-5107-8405-5

Jacket design by David Ter-Avanesyan

Printed in the United States of America

On the Pronunciation of Lithium

The word lithium is derived from the Greek *líthos*. Therefore, chemists assume that it is pronounced as—depicted here using the International Phonetic Alphabet (IPA)—['lɪθiəm.] Also permissible is ['lɪθiʌm.] It's a similar situation to that of the element strontium, which chemists pronounce ['strɑn(t)siəm.].

As to not irritate the listener, I chose to use the colloquially more common pronunciation of ['lɪθiəm.] The chemical community may forgive me, especially since we don't really know how words in ancient Greece were actually pronounced.

General Information

The generic masculine is used, as typical, in the gender-neutral sense.

For simplicity's sake, I usually only mention lithium and not lithium ions, even though, scientifically-speaking, lithium in nature is always an ion or a saltlike ionic compound. It is therefore also so in biological contexts.

Disclaimer

The publisher and author have verified to the best of their knowledge and belief that the information in this book is correct and complete for proper interpretation. However, neither the publisher nor the author assumes any liability for the content of the work or for any errors. The information in this book is based on publicly available documents and studies, accessible via the website links provided when possible, except where explicitly identified as circumstantial speculation or logical deduction. The author does not adopt the contents of the websites cited as his own. However, since the internet is forgetful and subject to increasing censorship, reference is made to the status at the time of initial publication. The source information is usually accompanied by the date the respective website was last accessed.

For legal consumer protection reasons, no promises may be made about the effects of medical devices unsupported by studies. This also applies to food (supplements). Therefore, all information in this book regarding lithium's effects should always be viewed with reservation: unless directly evident from the studies cited, it is based on logical conclusions and ultimately represents a pertinent free expression of opinion. Neither the author nor publisher assume any liability for this. This also applies to quantities. These usually refer to an average adult's presumed requirements. Variations in dosage may result, among other things, from differences in lifestyle, gender, age, body size, and any preexisting medical conditions. The information in this book is expressly not to be understood as medical advice. You should always supplement responsibly based on your needs and in consultation with a trusted doctor—this book does not replace medical advice. This disclaimer also applies to possible interactions with other active ingredients (more on this in Chapter 3).

Lithium is not approved as a food supplement in most of the world's countries, nor throughout the EU, even in essential microdoses. By law, it may only be manufactured and sold in pharmacies with a doctor's prescription. As a doctor licensed to practice in Germany, the author follows the legal advice not to recommend alternative lithium-containing dietary supplements. The author cannot make any statements about their quality—but this does not necessarily mean that they are of lower quality than, for example, pharmacy products from Germany, Austria, and Switzerland. However, due to EU legislation, pharmacy products must be GMP-certified. GMP stands for *good manufacturing practice*, which is intended to ensure consistent quality standards in producing and testing medicinal products or active ingredients.

CONTENTS

A Guide to Health (1921) Mahatma Gandhi (1869–1948)

~~~

Ignorance is one of the root-causes of disease.
Our ignorance of the most elementary laws of health
leads us to adopt wrong remedies or drives us into
the hands of the veriest quacks.

There is nothing so closely connected with us as our body,
but there is also nothing perhaps of which our ignorance
is so profound, or our indifference so complete.
It is the duty of every one of us to get over this indifference.

Everyone should regard it as his bounden duty
to know something of the fundamental facts
concerning his body.
This kind of instruction should indeed
be made compulsory in our schools.

Indeed, if we consider the depth
of our ignorance in such matters, we shall
have to hang down our heads in shame.

To assert that the average man cannot be expected
to know these things is simply absurd.
....

The following pages are aimed at those who are willing to learn.
~~~

ESSENTIAL OPENING REMARKS

And I'm looking for the silver lining,
silver lining in the clouds
And I'm searching for,
searching for the philosopher's stone.
—Van Morrison ("The Philosopher's Stone")

In the spring of 2024, while I was busy exploring the fundamental role of lithium in living organisms, I received an invitation to the Greek island of Kos. There I had the honor of participating as a guest speaker at a naturopathy conference—in the homeland of Hippocrates (c. 460–370 BC), arguably the most famous physician in antiquity. The conference culminated in a solemn ceremony at the world-famous Asclepieion. At this historic site, where medical treatments based on Hippocrates's teachings were already being performed in the fourth century BC, the conference participants were to receive their certificates of attendance between the still-preserved columns and statues, just as, over two thousand years ago, physicians received their diplomas upon completing their training as holistic practitioners.

When I entered the historic site that day for the first time in my life, I was immediately gripped by an overwhelming, physical sensation that had been completely unknown to me until then and that can hardly be put into words. Past experiences came to mind. I saw myself again on the edge of the mighty Grand Canyon and the mystical-looking summit of Kilimanjaro appeared before my mind's eye. These had undoubtedly been great, moving experiences. But the feeling that the Asclepieion evoked in me was, in a way, quite extraordinary. I felt a deep connection with the history of this place, which continues to have a powerful impact on the present. An incredible energy filled me and I felt an overwhelming need to be part of the revival of this ancient medical knowledge that this special place represents. I realized that lithium, which I was very interested in, could be an effective means to this end and possibly the basis for systemic medicine's long overdue renaissance. My thoughts had come full circle. I felt that I had recognized the meaning of the German doctor Gerlinde Nyncke's (1925–2007) poetic quote: "The philosopher's stone cannot be found with reason."[1] My earlier, purely logical efforts to discover the all-important factor for a long and healthy life through analytical research had led me astray. Lithium now seemed to me to be, if only symbolically, the "philosopher's stone" sought for ages—"symbolic" because I knew from my previous work that there could not, in principle, be one single factor guaranteeing a healthy and long life. But in that moment of emotional breakthrough, the mysterious descriptions of this philosopher's stone, completely a mystery to me until then, suddenly took on a whole new, practical meaning: lithium may be the sought-after philosopher's stone, but only because this trace element is exactly what most people still lack. I also found the coincidence remarkable that the name lithium is derived from the Greek word *líthos*, meaning stone. Indeed, for many generations, many people have been suffering, becoming ill, and dying from conditions of often severe and undetected lithium deficiency. Therefore, for countless people, lithium could indeed have effects expected from a philosopher's stone, granting, among other things, unprecedented health, clarity, resilience, curiosity, or the courage for self-reflection—and, as I will show in this book, even a longer life and even greater opportunities for wisdom!

However, due to its interdependence with the body, the mind can only develop optimally if the body doesn't suffer from a single elementary deficiency. This applies to lithium as well. Lithium alone cannot magically compensate for other essential deficiencies. However, if one wants to find a real-world equivalent of the philosopher's stone to keep with this symbol, it loses its concrete form. According to legend, it serves to cure and prevent diseases and should embody the principle that is proven to underlie true health: the elimination of all kinds of deficiencies. Seen in this light, this multifaceted "philosopher's stone" can also be said to have a rejuvenating effect—after all, we age more quickly due to a chronic lack of essential vital substances, as well as a lack of intangible substances, for example when our soul lacks human warmth. In this interpretation, the growing wisdom that the stone bestows consists of recognizing our deficiencies at any given moment and then finding ways to remedy the respective deficiency. What we lack, and is therefore particularly beneficial to our health, when balanced, always makes our previously unattained health potential attainable for us.

An early and, in this respect, less enigmatic description of this mystical stone, remarkably compatible with the interpretation presented here, is attributed to the Greek alchemist Zosimos of Panopolis (460–520 AD):

> *There is a stone which is not a stone, a precious thing which has not value,*
> *a thing of many shapes which has no shape, this unknown thing which is known to all.*[2]

It should not surprise us that certain concepts that arise from a deep intuitive understanding of natural relationships and human hopes, and which have existed in the form of terms or symbols since the beginning of cultural records, are continually being reinterpreted based on new scientific findings. The same applies here: from a systems biology perspective—according to which biological life requires both various essential vital substances and nonmaterial factors for optimal health—it is logical that there cannot be a single "philosopher's stone" that is always the same ("there is a stone which is not a stone"). As a metaphor, this can only ever stand for the scarcest essential resource ("a thing of many shapes which has no shape"). The lack of an essential resource means that our physical and, above all, mental development lags behind our potential. Only when we encounter our own personal "philosopher's stone" and correct all deficiencies can we heal our body and free our mind.

The parallels presented between lithium (chemical symbol Li) and the legendary philosopher's stone have led me to a profound realization about true health. But astonishing similarities in this symbolism go even further and symbolically underline further fundamental insights about lithium: the atomic nucleus of lithium—one of the oldest elements in the universe along with hydrogen and helium—consists of seven nucleons (three protons and four neutrons). If attempting a stylized symmetrical representation, one can easily arrive at the basic structure of the mythical symbol of the "Flower of Life," the so-called "Seed of Life," which could thus also become a symbol for lithium. (See the front cover and Fig. 1.) Interestingly, in many spiritual teachings and beliefs, seven stands for perfection, enlightenment, and the attainment of higher levels of consciousness.

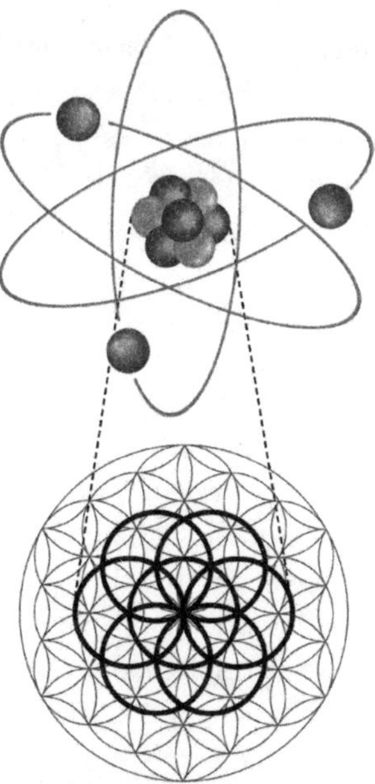

Figure 1

As part of the "sacred geometry," the Flower of Life embodies vitality, a key to enlightenment, and, among other things, universal harmony. Applied to biological processes, harmony is known as homeostasis. The parallel we can draw: as part of an open, dynamic system in constant exchange with the environment, a healthy balance of our physiological body functions simply cannot be achieved if there's a deficit of essential lithium. In particular, the development and maintenance of our "mental immune system" is impaired by chronic lithium deficiency. Our social adaptability and learning ability, our natural curiosity, our mental resilience and ability to think, which is also necessary for planning and implementing highly complex tasks—in other words, everything that makes us human, as well as our ability to understand ourselves and the world—is hindered by a lack of lithium and, conversely, represents potential that can be unlocked by compensating for a lithium deficiency. As fantastic as the rediscovery of essential lithium as the missing "keystone" for our mental health may be, it's the unnatural deficiency that gives this trace element its power. Even if it's not a miracle cure, essential lithium can act like a rediscovered philosopher's stone for many. That is, it is a missing key element for previously undreamt-of natural physical and mental health because we are currently short on it.

From the moment I came to this new realization, there was no stopping me. With the same enthusiasm with which I had spent decades searching in vain for the philosopher's stone through purely analytical or experimental work, i.e. by continually breaking down molecular biological processes and functions into ever smaller fragments or puzzle pieces until they were finally solved (ancient Greek: *análysis*), I made an epistemological U-turn—and this was absolutely necessary. Although analytical research provides an infinite amount of individual information, it doesn't provide an understanding of the life processes by which nature has succeeded in assembling inanimate matter into a living entity. In other words, no isolated gene, protein, or vital substance represents life itself. None of these puzzle pieces provide a coherent picture. But the fragmentation of nature into ever smaller puzzle pieces had been the essential feature of my research so far. In order to better understand nature, I had to start thinking not only analytically, but above all, systematically. This meant building on existing research and reassembling the puzzle pieces that already existed in abundance—the ongoing discovery and disassembly of which is constantly being driven forward into ever smaller pieces by the scientific community at large—and linking them together to form a coherent overall picture of life (ancient Greek: *sýnthesis*). Even though the result will always be an image and never life itself, I hoped that reversing the analytical process of data generation would create a synthetic process of knowledge. However, this was not the first time I took this path. My scientifically published theory on the development, prevention, and treatment of Alzheimer's disease[3] is based on the same approach, but it was only on Kos that I truly realized how crucial such an epistemological shift in direction is for our lives.

Not all the puzzle pieces I found fit the picture. In some cases, I had the impression, without intentionally suggesting, that this was due either to inadequate data collection or incorrect interpretation thereof. But I didn't just ignore those pieces; I decided to repeatedly point out which pieces of the puzzle had to be removed and the reasons why. This was necessary not only to refute the criticism of lithium's essentiality, which is based on these questionable puzzle pieces or studies, but also to give you a deeper insight into the nature of the molecular miracles lithium enables. As we all know, we learn the most from mistakes.

Every scientific argument is a journey into the unknown; the best equipment for this is a basic understanding of what is known so far to optimally master the journey into new territory. These foundations are laid out in Chapter 1. In order to systematically classify the diverse functions of the essential trace element lithium, it establishes a common terminology. This chapter covers everything from the cultural history and aberrations of medicine to the problematic discovery of essential nutrients and the Laws of the Minimum and Maximum, which serve to maintain the vital homeostasis of all biological systems. Their evolutionary biological purpose is reproduction, which must be protected by nature. To this end, the physical immune system, which every lifeform—from single-celled organisms to multicellular ones—possess in some form, first developed. Later, as a "higher" or superior function, so did the mental immune system. The latter distinguishes us as humans and evolved, like two sides of a coin, together with our extraordinary longevity. But it is not just a protective system, which is what the term "immune" implies. Through the associated ability to put ourselves in other people's shoes and recognize their intentions, the mental immune system also gives us the opportunity to imagine our own future, make plans and act consciously. As I will show, every biological system since life began has evolved in the presence of lithium and incorporated its properties. But no system, apart from longevity and reproduction, is as susceptible to a deficiency of this essential trace element as our mental immune system. As an essential human characteristic, this provides a highly intriguing starting point for the exciting evidence of lithium's essentiality.

I provide this evidence in Chapter 2 based on six arguments. In my opinion, each of these arguments taken alone would suffice to demonstrate lithium's essentiality. However, all six arguments together not only supply a comprehensive picture of the essentiality of lithium but also enable a deeper understanding of human biology and personhood. You'll see that between reason and madness there often lies only a trace of an element. At the same time, we deepen our knowledge of the broad spectrum of diseases associated with mental immunodeficiency syndrome, in the development of which essential lithium or lithium deficiency plays a decisive role.

In Chapter 3, the essential lithium requirements are discussed. Here, too, we enter new territory, which we'll approach cautiously. Due to its use as a high-dose drug with numerous side effects, it is important to understand how therapeutic lithium differs from essential lithium.

In Chapter 4, the interaction between the physical and mental immune systems is examined practically using several highly medically relevant topics. The concept of neuroinflammation as a pathological link between the two systems will help us to gain a deeper understanding of lithium's role in the prevention and causal restoration of biological balance, i.e., homeostasis, in disease. This covers everything from autism, which helps us develop a basic systemic understanding to many other neurological and mental illnesses, right up to neurodegenerative diseases. Another issue is chronic pain, also usually an expression of a dysfunctional or misguided endogenous immune system. Here too, essential lithium could help break this painful vicious cycle. Lastly, based on these findings and the fundamental laws of life (Chapter 1), this chapter provides a systemic and thus revolutionary approach for a new medicine for chronic diseases—both for their prevention as well as their causal therapy.

Chapter 5 deals with the topic of cytokine storm as an extreme example of a dysregulated and hyperactive immune system due to lithium deficiency, along with the life-threatening consequences in traumatic, infectious, and autoimmune diseases. Here we will also discuss lithium's diverse effects against viral pathogens, providing insights into the general functioning of the body's immune system.

In Chapter 6, we discuss longevity and the diverse physical effects of lithium deficiency. The text clarifies, for example, the apparent paradox that renal insufficiency is one of the most problematic side effects of high-dose lithium therapy while lithium in essential doses is relevant for maintaining kidney health. Lithium is also important for the functioning and maintaining of an efficient cardiovascular system; the loss of such functioning most often limits life expectancy in our modern world. Last but not least, the topics of cancer development and therapy and the possible supportive function of essential lithium are addressed.

Finally, in Chapter 7, I will try to find answers to some fundamental questions this book raises: Who presumes to place themselves above nature and deny the essentiality of a vital substance that has long since demonstrated all the requirements for an essential trace element in numerous studies, and has consistently produced astonishing, corroborating successes in countless applications in practice for decades? Why is it even banned as a dietary supplement? What could be the reasons for this? And what can we do to counter this?

Lithium could become the keystone in constructing a naturopathic or species-specific humane medicine. However, in addition, through its effect on the human mental immune system, as an "element of rational peacefulness," so to speak, it could also trigger a transformation of all of humanity toward greater social cohesion. It would thus also be the beginning of an era of mentally and psychologically healthier people, who can no longer be (so easily) incited to hatred and war by external influences, but who can independently develop peaceful solutions to conflicts. We could all then live in a more stable, healthier, and more livable world. For example, in the scientific publication of a clinical study, showing how the lives of seriously ill COVID-19 patients could be saved by lithium, the authors write wonderfully prosaically:

> *"From the 'Big Bang' to its presence in all living things, to its use in medicine, and in the storage of electricity, the third element of the Periodic Table may be paramount in life, which cannot be spelled without Li."*[4]

What still stands in the way of a new era of a mentally healthy, psychologically stable, spiritually open and thus peaceful humanity is the final proof of the trace element lithium's essentiality for human life, which you now hold in your hands. However, in order for lithium to receive the essential trace element status that it naturally deserves, it's especially important to win over experts to support my efforts to educate humanity. This is of enormous importance because it can be assumed that there are influential groups who, for reasons that you'll understand by reading the book, have a perfidious interest in ensuring that lithium is not recognized as an essential trace element.

This book therefore strikes a balance between educating laypeople interested in medicine and health policy and convincing scientists to help expand on the new findings compiled here so that doctors may implement them in their daily practice for their patients' benefit. Nevertheless, I have tried to present the facts in as general a manner as possible. Oversimplification—without presenting the essential arguments in detail—would make it more difficult for interested readers to fully understand how certain conclusions are drawn and reduce the chance of convincing critical scientists or doctors. For this reason, as in several earlier books which also aimed to expand the boundaries of general knowledge, I have decided to label further explanations and content that I consider worth knowing but which deviates from the main text as "Additional Information," displayed in gray. You can skip these at will and follow the main or continuous text, but it is certainly not a mistake to engage with them in order to delve a little deeper into the subject matter.

Evolution of Knowledge: This book is just the beginning. Proving that lithium is an essential trace element will change the world. I expect that once all humanity, or at least a large part of it, knows how important lithium is for mental health and performance, there will be even more research and testimonials. Even if the authorities may be reluctant for a time, many people will discover the benefits for themselves. I hope that many more scientists, inspired by the evidence presented here, will take a close look at lithium's function in essential quantities, which could lead to a flood of new discoveries. I will try to publish the most important findings from my point of view as soon as possible on both my German and English websites in a generally understandable form. From time to time, these new findings will be incorporated into new editions.

CHAPTER 1

THE FUNDAMENTAL LAWS OF LIFE AND THE MENTAL IMMUNE SYSTEM

I searched on all the beaches of my world,
looked along every path, to discover it at the edge.
I knocked around in the stone quarries,
even examined the stones that were thrown at me.
Almost every stone I pick up, I consider from all sides.
Meanwhile, the wise ones laugh at me.[1]
—Kristiane Allert-Wybranietz (1955–2017)

Science Gone Awry

In medical practice, one often hears the question "What's the matter with you?" It is asked in the sense of "What health problems do you have?" Even if we perceive these questions as synonymous, only the first question points to the empirical understanding that health is not possible to have if something essential to life is missing. The question is about the primary cause of the ailment, not just the symptomatic reason for visiting the doctor. The medical insight that basic human needs such as nutrition, sleep, physical and social activity are crucial for health goes back to the teachings of Hippocrates of Kos. He emphasized the importance of a way of life in harmony with (one's own) nature. Around 2,500 years ago, he pursued a primarily preventive, systemic, and deeply nature-oriented medical approach based on this. The founder of the Hippocratic Oath, which bears his name, recognized that lifestyle has an enormous influence on human well-being—for better or for worse. Deficits in basic human needs have been recognized as a major cause of developmental disorders and disease; the question of what we might be lacking arises from such an understanding. In this way, Hippocrates came astonishingly close to both the natural needs of humans and the causes of many diseases, even from today's perspective.[2] Together with his students, he laid the foundation for nature-oriented medicine as an empirical science: health or illness were not random or fateful developments; rather, they had natural causes that, once understood, could also be influenced naturally. Man was not at the mercy of divine whims, but, like all living things, part of nature and subject to its laws. With this revolutionary view, he undermined the previously unrestricted power of the priestly word in matters of illness and liberated medicine from faith: it was not gods, as believed, on Olympus or elsewhere that caused illness and suffering, but a way of life not in harmony with natural needs.

However, over time, the disastrous belief that health is a matter of luck, and illness our genetic destiny over which our lifestyle has virtually no influence slowly regained dominance in medical matters. It is conceivable that this momentous development was favored by certain interpretations of the creation myth from the Book of Genesis, which was translated into Greek (Septuagint) by Jewish scholars in the third century BC. The Roman Catholic Church, which incorporated this creation myth into Biblical canon and the doctrine of which also shapes secular Western culture, continues to promote such interpretations—mostly unconsciously—to this day. The concept of God creating humanity in His own image can give rise to the belief that people are outside of nature, over which man is set to rule on his Creator's behalf. Even though few people believe this literally today, this idea has shaped an essential part of our collective self-image for more than two millennia. It may have contributed subtly to the misconception that these natural principles of "biological conditioning," as described by Albrecht von Haller (1903–2000) in his highly recommended book *The Power and Mystery of Food*, apply only to the animal world and not to humans.[3] Von Haller writes that "the chemists concerned with the problems of practical agriculture were far ahead of their colleagues entrusted with human nutrition." In particular, "a psychological factor" did not play an "insignificant role," which in my opinion can likely be traced back to this culturally formative narrative.[4] In fact, for a long time I too felt a certain discomfort in applying the term "species-appropriate" to humans—after all, it was usually reserved exclusively for concepts of animal husbandry.

"Every farmer knew," says von Haller, "that his cattle's growth and health depended on the quality of the feed. But people were reluctant to allow this experience to apply to humans as well." In contrast, at the end of the nineteenth century, when the metabolic energy of our food was discovered, it was believed that only calories counted and that "a pleasant but subordinate process like food intake" should have no influence "on the development of a spiritual being" like us. "This assumption," says von Haller, "allowed us to conclude what we ate was more or less irrelevant." Consequently, the cause of diet-related diseases was sought everywhere save for diets. This idea still influences many doctors today, particularly in the field of mental illness; these are often due to a systemic deficiency, making the attempt to cure them exclusively through talk therapy absurd. It is of course equally absurd to support this misguided approach by administering medication. Also discovered toward the end of the nineteenth century by physician, microbiologist, and later Nobel Prize winner Robert Koch (1843–1910) was the world of pathogenic microbes—including the causes of plague, cholera, and tuberculosis, among others—leading most researchers astray. Building on this, the adverse health effects of nutrient deficiencies such as scurvy (vitamin C deficiency), beriberi (vitamin B1 deficiency) and, ultimately, almost all other vitamin deficiency conditions were erroneously considered the result of infections by yet unknown pathogens. According to von Haller, the belief in the special status of humans as well as "scientific prejudices" had "blocked the door to progress" and delayed the discovery of vital substances for many decades, costing the lives of countless millions of people.[5]

The graveness of this misconception can be seen from the fact that the spread and deadly course of many infectious diseases, like the Plague in the Middle Ages or the Spanish Flu at the end of World War I, are most likely due to nutritional deficiencies that weaken the immune system: "The microbe is nothing, the milieu is everything!" suspected French physician, chemist, and pharmacist Antoine Béchamp (1816–1908) back then.[6] Béchamp was a bitter adversary of French chemist, physicist. and biochemist Louis Pasteur (1822–1895), who, like his German colleague Robert Koch, insisted on the pathogen as the main cause of disease. Béchamp was thus not only right with regard to infectious diseases, which are based on malnutrition and the malfunctioning of the immune system (see my book *Herd Health*),[7] he also correctly identified the cause of all modern diseases of civilization early on. Even their designation is misleading, because even under civilized conditions, these can be prevented with a species-appropriate lifestyle.

The Law of the Minimum

In this case, time is fortunately on truth's side; it was inevitable that the true causes would sooner or later become apparent. Ultimately, no pathogen could be found that could've caused scurvy, for example. However, a deficiency of a certain nutrient (termed vitamin C) was easily proven to be the cause. This could soon no longer be denied, even though countless human lives had been needlessly sacrificed to this disease due to dogmatists' ignorant doctrine.

At the beginning of the twentieth century, the first vitamins and many essential trace elements were discovered in a relatively short time. Nevertheless, the pharmaceutical industry ultimately had the power to either banish these findings from the curricula or to minimize their significance. It is very likely attributable to the standardization and structuring of medical curricula, intended to serve the pharmaceutical industry's comprehensive interests, that during my medical studies in the 1980s, I wasn't taught the fundamental importance of a species-appropriate lifestyle, which must include a diet sufficient in essential vital substances. Nothing has changed in this deplorable state of affairs (to date, as of spring 2025). In my medical studies, I learned nothing about the Law of the Minimum, which governs all life on our planet, and which, since its publication by agricultural scientist Carl Sprengel (1787–1859) in 1828, every farmer must know if he wants his plants to thrive optimally.[8] Sprengel himself formulated it very vividly:

> *If a plant requires twelve elements for its development, it will never germinate if even one of them is missing, and it will always grow poorly if any one of them is not present in the quantity that the plant's nature demands.*

Of course, different plant species and ultimately all other living beings have individual needs that reflect their evolutionary development or genetic adaptation to the living conditions of their origin. The Law of the Minimum remains unaffected: individual needs always limit possibilities for development and thus fundamentally determine one's well-being. Accordingly, we too—as part of nature—can only develop and remain healthy to the extent that the scarcest resources which are part of our species-specific needs allow. That means any deficiency of vital vitamins and trace elements inevitably leads to developmental disorders and diseases. Our mental performance and mental health in general usually suffer the most from such deficiencies. This may be because our brain is more complex in structure and function than any other organ. Deficiencies in its functionality very quickly become apparent in the socially complex world of human life.

Our mental (and of course physical) performance is at its highest when the metaphorical barrel shown in Fig. 2 to illustrate the Law of the Minimum is completely full, i.e. when we are not lacking anything fundamental. In this example, each stave (each length of wood in the barrel) represents a different essential resource: from the meaning of life to physical and social activities to essential vital substances that flow into this barrel. The height of each stave is determined by the composition of the drops; if something is missing, it is shorter. If the defect is corrected, the stave magically grows upward. Under species-appropriate living conditions, the barrel is completely intact as we have sufficient access to everything needed to develop and maintain our mental performance—then and only then do we fully exploit our potential for mental development and health. In today's world, however, this metaphorical (magic) barrel has a problem: many staves do not reach the top rim. Depending on the individual deficits resulting from a lifestyle that is common today and mostly alien to our species, the barrel of our mental (and any other) performance is not completely filled.

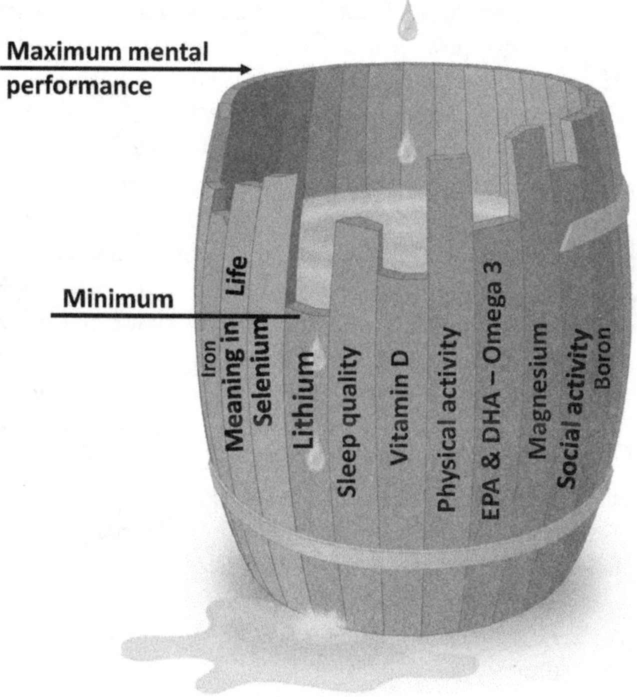

Figure 2

According to the Law of the Minimum, the contents filling the barrel only reach as high as the shortest stave. In this example, the shortest stave would be lithium. So, is lithium the most important trace element? In this example it would be, though not in every case. For other people, completely different deficiencies may be the primary limiting factors for filling the metaphorical barrel of their mental performance. And in this example, too, as soon as the lithium intake increases, e.g. via a dietary supplement (which extends the corresponding stave upward), the lack of vitamin D (i.e., now the shortest stave in this example) continues to limit the development and maintenance of mental performance in comparison to the theoretical potential. Perhaps the level would be higher, but even with sufficient lithium intake, the next performance-limiting resource in this case would be the vitamin D deficiency. It seems logical that solely supplying lithium to correct multiple deficiencies won't unlock the full potential of mental (and physical) performance if the other deficiencies aren't completely remedied. One can assume that supplementing up to the requirement limit is sensible under all circumstances and has positive health effects, even if other nutrient deficiencies continue to exist. It is important to understand, however, that lithium is not a miracle cure! Today, it can only be the "cornerstone" of a species-appropriate lifestyle because its essential importance is generally not known or is completely ignored, and its consumption eliminates even the last existing deficiency.

So, there is no single most important element or vital behavior (social or physical activity, adequate sleep, etc.) because all are necessary. Furthermore, it becomes clear that the lack of one essential factor can never be compensated for by another. Each stave represents an essential factor. Only all the factors together ensure full mental performance. It's obvious that a plant threatened by nitrogen deficiency cannot be kept alive by adding more potassium, phosphate, or magnesium. And only water helps against thirst. You don't need to have studied medicine or molecular genetics; common sense is enough to be aware. This simple, yet fundamental insight also means, however, that a drug can never cure a disease caused by an essential factor's deficiency (which applies to most, if not all, diseases of civilization). A drug can only alleviate the symptoms of the ongoing disease; even if the drug typically has many side effects and is nevertheless—or perhaps precisely because of this—highly lucrative (after all, the side effects must also be alleviated with medication). This may be the reason why pharmaceutical-funded universities try very hard to avoid pointing out this fundamental law of nature to medical students. Naturally, this critique of medication doesn't apply to compensatory interventions, where an absent bodily function is replaced by an appropriate external supply. Examples include insulin for type 1 diabetes and thyroid hormones for organic hypothyroidism.

Vitamin D Deficiency—A Deadly Example of Ignoring the Law of the Minimum. According to Hippocrates: "Whoever wishes to investigate medicine properly, should proceed thus: in the first place to consider the seasons of the year, and what effects each of them produces for they are not at all alike, but differ much from themselves in regard to their changes." Contrary to these wise words, pharmaceutically oriented conventional medicine ignores the sun-related winter drop in the 25-hydroxyvitamin D level (simply called vitamin D level) of much of the population of northern latitudes as the actual cause of the seasonal occurrence of respiratory diseases. Nor is such medicine interested in the countless studies that show that virtually no one would die from respiratory infections like COVID-19 or influenza if the winter vitamin D level, which is usually too low, were raised to values of 110 to 140 nmol/l (as measured in primitive peoples of East Africa, therefore can be considered natural levels).[9, 10] Recognizing the need for a natural vitamin D level would be tantamount to accepting Sprengel's Law of the Minimum (which cannot be true) and vitamin D supplementation as a cost-effective and natural solution to the problem (which must not be true) despite all evidence.[11] Instead, in the case of COVID-19, a highly lucrative yet also life-threatening genetic engineering intervention via mRNA injections was propagated, perfidiously declared a "vaccination." To this day, the fact that the modified viral genetic material, which encodes the life-threatening, genetically modified SARS-CoV-2 spike protein, and paralyzes and damages the immune system, appears to leave people unconcerned.[12] Corresponding injection campaigns continue, with catastrophic consequences for the entire world population. Particularly affected is the mental immune system, which has been damaged in many ways by the COVID-19 program, as I demonstrate in detail in my book *The Indoctrinated Brain*.[13] But declining birth rates[14] and rising excess mortality[15] are two more of the many consequences of ignoring the Law of the Minimum and the effects of an inhuman ideology with which a few seek to dominate all of humanity.

In today's conventional medicine, which is mostly shaped by the pharmaceutical industry's interests, prevention based on the Law of the Minimum plays only a minor role in teaching and practice. There is virtually no causal therapy for chronic diseases, which is what makes many of them chronic diseases. Following the revenue-oriented business model, we patients and future customers are told: a species-appropriate lifestyle and eliminating lifestyle factors that cause illness and nutritional deficiencies cannot protect us from suffering from lifestyle diseases and dying prematurely. The only solution to this, according to the seductive promise, is pharmaceutical innovations (more on this in Chapter 7). When it comes to medications, people often talk of so-called "magic bullets." However, these "magic bullets" are merely conceptually high-precision pharmaceuticals that act precisely and exclusively on the predetermined target proteins, as shown in Figure 3. They "shoot," so to speak, at the body's own "targets" that were previously declared to be defective or that regulate naturally malfunctioning signaling pathways, thereby deceiving us into believing that these are a causal therapy.

In order to properly contextualize this presumptuous claim, one must know the following: *Homo sapiens*, the wise man, has about 19,500 genes—surprisingly, only a few hundred genes more than, for example, a roundworm about one millimeter in size called *Caenorhabditis elegans*.[16] This raises a problematic question: How can the development of the highly complex human brain, consisting of billions upon billions of nerve cells, be encoded in our genome, which is itself, so to speak, "spare," with only four base pairs? After all, humans have just a few hundred more genes than this tiny worm, whose life is governed by a mere 302 nerve cells. Consider, too, the array of instincts vital for human survival, i.e. innate behavioral patterns encoded in the genetic makeup that humans can apply over the course of their lives in an environment

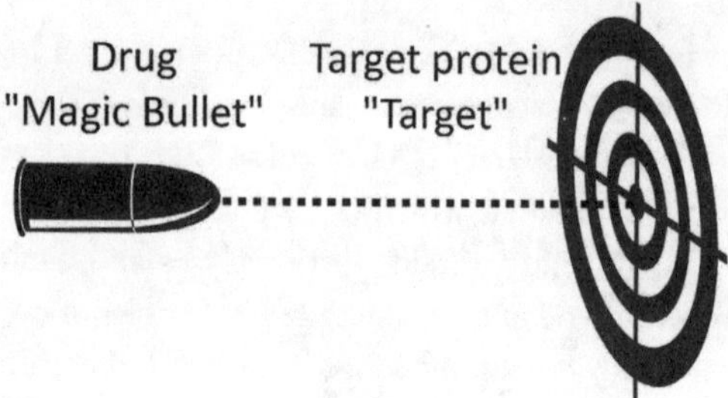

Figure 3

far more complex than that of a worm. The solution to the puzzle lies in the fact that all our proteins produced in the body according to our genes' instructions don't just have one single function. Their interactions in combination with a host of other proteins enable them to regulate a vast spectrum of biological functions. This includes the functional diversity of individual proteins, including their demand-based, temporal, and cell-specific activation. This is where epigenetics comes into play, i.e. the influence of the environment, our lifestyle and, finally, our thoughts and feelings, on our genes' regulation. In this way, our relatively small genetic makeup maintains our health—a true miracle, the result of an evolutionary process lasting billions of years, the creative power of which exceeds our imagination.

In applying this knowledge to pharmaceutical magic bullets, these would have to develop truly magical powers to actually have preventative and healing effects. They would have to influence their target protein in such a way that the interaction of thousands and thousands of other proteins, which it directly and indirectly controls, is changed so cleverly that diseases are prevented or even cured, even if one continues to maintain an alien lifestyle that disregards the Law of the Minimum and thus causes illness. However, as the term suggests, this is magical, wishful thinking and impossible in reality, which is why there can be no pharmacological "philosopher's stone" in the future either. Or, put a bit more simply: it cannot provide a highly precise and controllable drug effect that eliminates the disease's actual causes; after all, drugs cannot compensate for an insufficient vital nutrient. As we all know, no pill helps against thirst either. And so even the purely symptomatic effects (e.g., reducing pain) are accompanied by a long list of undesirable effects, commonly referred to as side effects. Ignorance of the Law of the Minimum, and, therefore, of a truly causal prevention and therapy strategy leads to chronic, purely medicinal symptom treatment with chronic side effects and, as a result, permanently increasing sales. In keeping with the motto "A patient cured is a customer lost," almost everything that serves to educate people about natural disease prevention is not only ignored but even attacked or ridiculed, such as orthomolecular medicine. The term "orthomolecular" is derived from the ancient Greek word *orthós* for "correct," and therefore optimal, supply of all essential molecular active ingredients, and from the Latin word *molecula* for "small mass." It refers to the adequate supply of essential molecules such as vitamins and trace elements. However, this knowledge is not part of medical training. To this day, doctors who think orthomolecularly and act causally, logically, preventively, and therapeutically based on the Law of the Minimum are the exception. First, one must personally acquire the foundational medical knowledge required; second, it requires a degree of courage to confront the powerful establishment for patients' benefit. This means questioning medical guideline content that's often unhelpful or, in many cases, even life-threatening. This is associated with certain hurdles, because for the doctor, compliance with the guidelines also means protection: as long as they adhere to them correctly and there is no gross negligence, they are generally protected from legal consequences should the treatment actually prove harmful to the patient (or even if that's the impression). In fact, these guidelines often primarily serve pharmaceutical interests rather than the patient's best possible well-being (even though patients are often led to believe, through corresponding propaganda, that getting a prescription is enough to experience relief without taking any responsibility themselves). Last but not least, treatment based on natural, system-relevant needs is hindered by the fact that the doctor must then expect lower income. This is because prevention—or even a cure through eliminating the disease's actual cause—is not provided for in the guidelines and is therefore generally not reimbursed by health insurance companies citing said guidelines. A typical example is the causal prevention of seasonal respiratory infectious diseases by raising the 25-hydroxyvitamin D level, which drops in winter. For example, the renowned *New England Journal of Medicine* in July 2022 issued "A Decisive Verdict on Vitamin D Supplementation." According to this study, "vitamin supplements do not have important health benefits in the general population of older adults, even in those with low 25-hydroxyvitamin D levels." Therefore, health-care providers should "stop screening for 25-hydroxyvitamin D levels or recommending vitamin D supplements, and people should stop taking vitamin D supplements to prevent major diseases or extend life."[17] At minimum, one can palpably sense the life-threatening ignorance of systems biology principles concerning human nature—if not even a contemptuous greed for profit—that drives such pronouncements.

Dietary Supplement versus Medication. Even with a wholesome and nutrient-rich diet, most people suffer from a lack of aquatic omega-3 fatty acids as well as iodine, zinc, selenium, and—at least in winter—vitamin D3 as well. As I will show here, this applies particularly to the trace element lithium, the essentiality of which has always been denied. Dietary supplementation is the rational way to compensate for deficiencies in vital nutrients that cannot be adequately supplied through food. Even if you don't necessarily feel sick due to the lack of vital nutrients (you've grown accustomed to this state), you are not as physically and mentally fit as you could be. A deficiency can entail both acute (see above: *Vitamin D Deficiency—A Deadly Example of Ignoring the Law of the Minimum*) as well as chronic health problems (preventable diseases of civilization such as Alzheimer's disease).[18] Therefore, a dietary supplement, such as iodized table salt, usually brings only advantages and no disadvantages. By contrast, in most cases, medications are only artificial molecules that do not occur in the body under natural conditions. They interfere unnaturally with biological processes yet cannot correct their malfunction (itself often due to a lack of essential nutrients): in lieu of remedying a deficiency, it purely treats symptoms. In clinical trials, protocols usually provide for testing the drug against a dummy drug (placebo). The goal is to determine whether an actual substance-related effect occurs. The hope is to demonstrate a symptomatic benefit. In return, symptomatic disadvantages or undesirable effects are accepted as possible side effects—up to a certain extent which is considered tolerable. (It's not uncommon, however, for the study design to be adapted so that the occurrence of intolerable side effects can be largely excluded, simply overlooked, or sugarcoated by statistical methodology until the product is launched on the market). The same procedure is also used in clinical studies with vital substances, e.g. to find out whether the intake of vital substances actually prevents diseases that their insufficiency might cause. While this may sound reasonable at first, the design actually has a serious conceptual flaw: as in the general population, most of the subjects in such studies, which are concerned with the treatment or prevention of multicausal lifestyle diseases, have multiple vital substance and often other lifestyle-related deficiencies, all of which individually increase the probability of disease (not to exclude the Law of the Maximum, see below). However, the Law of the Minimum requires a multicausal approach to achieve a complete health effect or, depending on the problem, any health effect at all. Remedying deficiencies that promote disease is therefore something completely different from artificially treating symptoms—it seems as if this is mostly deliberately ignored, although it is so obvious. Studies that examine the health effects of simultaneously correcting several disease-causing deficiencies are therefore, lamentably, the exception. Yet, when put into practice, the combined individual measures exhibit a synergistic effect, typically resulting in significant success ("typically" because the approach often doesn't resolve all deficiencies).[19]

The Law of the Maximum

Commercial interests ignore the Law of the Maximum as well as the Law of the Minimum, to the detriment of our general health. The Law of the Maximum states that excess is harmful for achieving and maintaining a healthy homeostasis—again, you don't need to have studied medicine to understand this; common sense wholly suffices. Chronic distress is just as unhealthy for us as certain types and intensities of radiation or too much fine dust in the air we breathe, etc. The Law of the Maximum naturally applies to all poisons, often harmful even in small quantities. However, harmful excesses include even vital substances known to tend toward overdose. According to the Swiss-Austrian physician Theophrastus Bombast von Hohenheim (1493–1541), who became known under the name Paracelsus:

> *All things are poison, and nothing is without poison;*
> *the dosage alone makes it so a thing is not poison.*

Accordingly, when using any medication—since these are foreign products that interfere with our bodily functions unnaturally—the question of toxicity is like the proverbial elephant in the room. Serious evidence, analyses, and studies exist which show that prescription drugs, the supposed saviors, are now the most common cause of death worldwide—and this doesn't just apply to genetic engineering since COVID-19.[20] In addition, there's a steady increase in chronic, supposedly environmental diseases due to increasingly precarious health care, which is becoming more apparent as a result of ignoring the Law of the Minimum. The associated increase in medication consumption, for example in the USA, leads to the frightening prediction that people born in 2019 will take prescription medications for about half their lives.[21] Insidiously, their side effects lead to the prescription of further medications, causing further side effects. These further fuel the vicious cycle of polypharmacy, i.e. the simultaneous prescription of many medications to one person. Of course, true health is never achieved this way; rather, the body only becomes increasingly unbalanced. However, restoring the patient's physical, mental, and emotional balance should always underscore a genuine interest in healing.

Homeostasis—Dynamic Stability Between Extremes

Homeostasis is derived from the Ancient Greek words *homoios* for "equal" and *stásis* for "standstill" and means "equality." However, this state only seems still. In nature there is nothing static. Rather, a dynamic balance of forces has emerged, which is why "balance" is the better and established translation. As with bicycling, a stable state is only achieved through movement. The Greek philosopher Heraclitus (520–460 BC) is credited with the insight *panta rhei*, ancient Greek for "everything flows," which also corresponds to our current physical understanding of the universe. In particular, every form of life exists only in constant exchange with the environment; without dynamism there is no life, but also without balance there is no life. We therefore live in the tension between the Law of the Minimum and the Law of the Maximum, as shown in Figure 4 below.

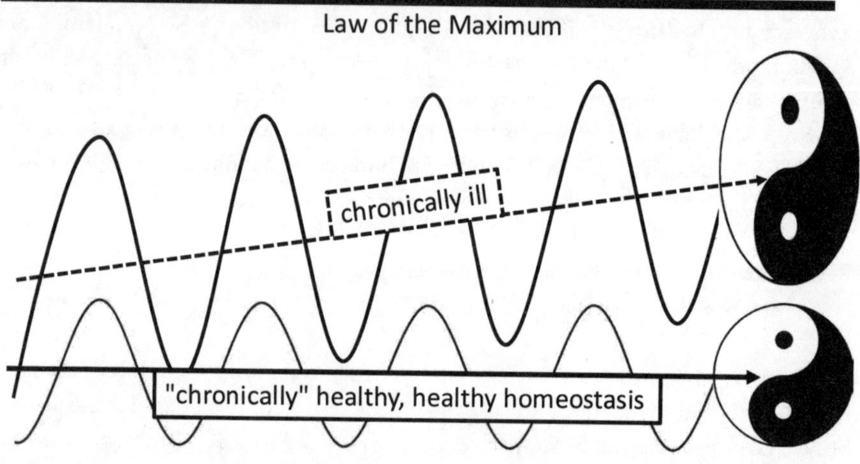

Figure 4

Ideally, the "Barrel of the Minimum" is full and our systems needn't strive in vain to compensate for deficiencies. As a result, exposure to toxic influences would be contained, enabling the interplay of forces, illustrated in Figure 4 as wavelike dynamics, to oscillate around a stable horizontal line. Feedback mechanisms in the body are responsible for such a stabilizing dynamic. These mechanisms consist of networks of inhibitory and activating signals and messenger substances, receptors, and intracellular information processing networks (see also Additional Information: *Somatic Cells Are Biological Information Processors* in Chapter 2, Argument 2). If you remove the temporal dimension from the graphic, this wave becomes a yin and yang symbol—two terms from Chinese philosophy that stand for polar opposite yet complementary forces.

A lack of vital resources (violating the Law of the Minimum) as well as an excess of stress (violating the Law of the Maximum) that can no longer be compensated for change the dynamics. Our systems, which use complex feedback mechanisms, will continue to try to achieve homeostasis, as shown by the dashed, ascending line in Figure 4: In the yin and yang symbol, the balancing forces have been symbolically preserved, but the conditions have changed and the system is, in a sense, "out of control." It is obvious from this graphic that such a lifestyle cannot work in the long run and must end in chronic illnesses. These days, these diseases are often labeled as "diseases of civilization" or "lifestyle diseases," as if disease is a logical consequence of the fortunate circumstance that more and more people are reaching old age. However, compared to indigenous peoples far removed from medical progress, a higher percentage of newborns today reach a higher age, not because they are born healthier and then live healthier lives, but because they are more likely to survive childhood thanks to better hygiene (see below: *Evolution of the Grandmother*). However, in comparison, they endure longer illnesses and, even after strokes, heart attacks, or similar lifestyle-related events, remain tethered to the pharmaceutical industry for the duration of their artificially extended lives, all thanks to highly efficient acute medicine and surgery. This only ends when they die either from the underlying disease or from the side effects of their medication, which in many cases is difficult to distinguish. The problem is that many people in their lives largely disregard the biological Laws of the Minimum and the Maximum due to a lack of adequate education regarding pharmaceutical interests, which are reflected through lobbying in legislation, medical curricula, and guidelines for doctors.

In this context, it is interesting to note that Mahatma Gandhi, although not a doctor, was aware of the importance of the Law of the Minimum for human health. In his *A Guide to Health*, published in 1921, he writes his version in Hindu metaphors:

> *The world is compounded of the five elements,—earth, water, air, fire, and ether.*
> *So too is our body.*
> *It is a sort of miniature world.*
> *Hence the body stands in need of*
> *all the elements in due proportion, —*
> *pure earth, pure water, pure fire or sunlight, pure air, and open space.*
> *When any one of these falls short of its*
> *due proportion, illness is caused in the body.*[22]

Our mental health suffers more than any other system in our organism from a disregard for this understanding of how nature works, and therefore also from a chronic deficiency of essential lithium, as we will see in detail in the following chapters. We should know how our mental immune system, in particular, functions in order to understand how lithium deficiency impairs its function and what consequences this has for us, our children, and ultimately for all of humanity.

Immunological Learning Aptitude—Physical and Mental

Every human being is unique—a point on which the teachings of Hippocrates and traditional European medicine (TEM), as well as traditional Chinese medicine (TCM), Ayurveda, and Tibetan medicine all concur. Nevertheless, each of these medical systems also has its own unique characteristics based on its specific typology. This allows practitioners to determine which preventive and therapeutic measures are most likely to succeed depending on the individual's type. But beyond all individuality and type, there is something fundamental that we all have in common. This includes all the essential physiological functions of our organs, the basic biochemical functions of our cells, and thus a long list of essential needs based on our biology; building on this, the crucial functions of our brain, which I summarize under the term "mental immune system," which all humans also possess.

But what exactly characterizes the mental immune system? Like the body's immune system, it may have initially developed as a purely protective mechanism. The ability we gain to empathize with others and recognize their intentions also enables us to empathize with ourselves and to see our own future in various options for action, to plan accordingly, and to consciously implement them. This is a deeply human trait. From an evolutionary perspective, the fact that you can read this book and understand its contents is not a given, but something special: No species living today other than humans has the ability to use symbolic language, to think and communicate in abstract terms, to grasp complex relationships, and to make the knowledge they gain useful for themselves and others. We attribute these particular abilities to a unique trajectory in the development of the human intellect. But the fact that you, and not someone else, are holding this book in your hands is not something to take for granted. Because you owe your existence not only to your parents, but also to your grandparents and ultimately to millions and millions, even billions of generations of direct ancestors before you, all the way back to the first living cell on our planet. So, if you are holding this book in your hands and reading these lines, it is only because all your ancestors successfully followed the evolutionary imperative: Be fruitful and multiply.

We must reach a certain age to be fertile and reproduce successfully. In humans, this is a developmental process that takes many years and is associated with countless risks. Ultimately, chance rules the world we live in. God rolls dice with quanta in the micro world, and in the macro world, too, things often turn out differently than one would expect. Nevertheless, the "magic" of the human mind lies in recognizing structures in natural chaos, making plans and implementing them. When it comes to survival in line with the evolutionary imperative, two closely related physiological functions take on particular significance: the physical immune system and the mental immune system. Figure 5 schematically illustrates how these two immune systems work together in accordance with the evolutionary imperative.

All organisms have their own immune system. Even bacteria are able to defend themselves against viruses (so-called bacteriophages, or bacteria-"eaters").[23] Nevertheless, new viruses repeatedly succeed in overcoming the defense mechanisms or immune system of their host cells and multiplying in them by adapting (randomly mutating) their genes. This applies not only to bacterial infections, but to all cells, including our own, which is why the human body's physical immune system must also be able to constantly adapt to changing pathogenic microorganisms.[24] Similarly, our mental immune system must also recognize and combat dangers from pathogenic *macro*organisms. These often take human form. The threat is usually not a direct physical one, but an indirect, mental one through so-called memes. In my book *The Methusaleh Strategy*, I explain the concepts of memes and genes as follows: "Our genetic material consists of units of information, of genes, and our cultural heritage consists of words and ideas, i.e. also of units of information. We call these memes. Only through the action of both, genes and memes, do we become what we are."[25] Genes that act together as a unit and are passed on accordingly are referred to as genetic material. We consider memes, which together form a narrative and are also passed on, to be cultural assets. Our genes control our cells, while memes or narratives influence our brain—that is, our thinking and behavior. Memes can also have a viral nature, both positive and negative: Viral is derived from *virus*, the Latin word for "poison." This underscores that narratives can indeed be harmful to us, but also that they possess a virus-like capacity for rapid spread, given our nature as social learning machines. It is our mental immune system that has the crucial task of distinguishing between helpful and harmful memes.

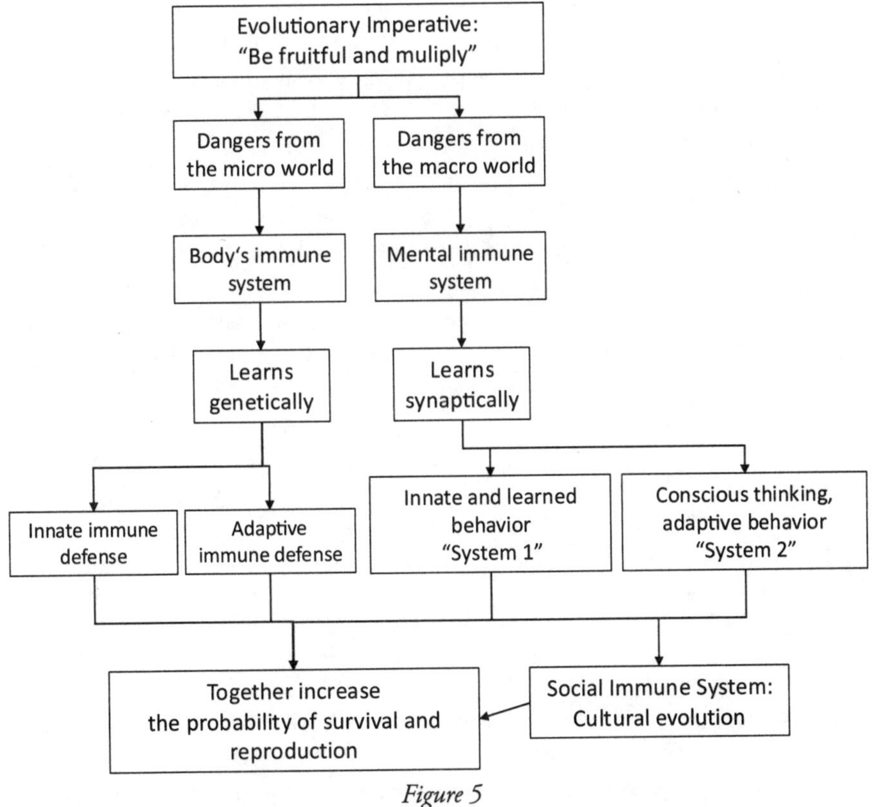

Figure 5

Viruses take control of foreign cells and force them to replicate their own viral genetic material so that these can in turn infect other cells. Viral memes or narratives, on the other hand—assuming a dysfunctional mental immune system—take control of our brains through propaganda so that we pass on the message and infect other brains. The main task of the mental immune system is to recognize, ward off, and neutralize life-threatening memes or dangerous narratives. Without a functioning mental immune system, we would be just as defenseless against life-threatening viral (harmful) memes as we would be against life-threatening viral (harmful) genes without a functioning physical immune system. Both the physical and mental immune systems are equally vital and, as we will discuss in detail, require essential amounts of lithium to function optimally. But if our mental immune system is vital—which I now would like to prove first—then any deficiency in a vital substance that impairs our mental immune system's function—as I will prove for lithium at a later date—also evidences its vital necessity!

Viral Memes versus Viral Genes. Memes can now spread faster through social media and mass media than viral genes can through human-to-human contagion. This is why, among other things, they are usually more dangerous, as can be clearly seen in their evolution since 2020: the meme or narrative of a deadly pandemic spread faster than SARS-CoV-2. Irrational fear ruled the world, even though a healthy physical immune system could easily have protected us from this threat. This meme, however, was not allowed to go viral because it posed a threat to the intentions of a minority of influential people and organizations—and so it was censored, as I document in detail in my book *The Indoctrinated Brain.*[26] Another toxic narrative was laboriously forced into circulation instead: The only way to survive the viral attack, we were told, was through an injection of a viral gene (mRNA) that programs every cell in our body to generate life-threatening viral spike proteins. It was due to suppressing on one hand and promoting on the other certain viral memes, that people died unnecessarily from a virus inherently harmless to those with healthy immune systems. Even more people died from fear and ultimately from genetic manipulation through poisoning with viral genetic material disguised as a vaccine. This shows that viral (toxic) memes can be more dangerous than viral genes—and only a healthy mental immune system can protect us from them.

Our two immune systems ensure our survival in a similar way, namely by learning from experience. The body's immune system genetically remembers experiences with pathogenic microorganisms (viruses, bacteria, etc.). Some of these memories originate from our ancestors, and our innate immune system possesses inherent genetic knowledge about them. For example, our immune cells recognize the spike protein of coronaviruses without ever having seen it before because our ancestors must have repeatedly encountered this type of virus millions of years ago; passing on the memory of these contacts obviously resulted in an evolutionary survival advantage. But the body's immune system could also learn from its own experiences. Cells of the so-called adaptive immune defense can, through recombination of their genetic material and subsequent selection, produce lifelong specific antibodies against invaders that neither they themselves nor our ancestors' immune cells have ever seen. The mental immune system, on the other hand, remembers synaptically, via altered neuronal networking in the brain. It has the useful experiences of our prehistoric ancestors. These exist in the form of innate instinctive behaviors as well as learned thought and behavior patterns from our direct ancestors (parents and grandparents) which we also mostly replay unconsciously. Since Daniel Kahneman's (1934–2024) discovery of two cooperating systems in human thinking, these innate and learned behavioral patterns have been summarized under the term System 1 thinking. However, the mental immune system also remembers new experiences from the macro world—at least from the time when our autobiographical memory center is fully developed toward the end of the second year of life. It allows us to consciously reflect and establish new patterns of thinking and behavior. This conscious thinking and acting is summarized under the term System 2. Kahneman received the Nobel Prize in 2002 for discovering these two systems of the mental immune system.[27] The two consciousness researchers Christof Koch and the Nobel laureate Francis Crick (1916–2004) wrote about this in an article published in *Nature Neuroscience* in 2003, in which they presented *A Framework for Consciousness* (a framework for researching the neuronal correlate of consciousness): It appears to be a major evolutionary advantage to have zombie modes [unconscious System 1] that react quickly and stereotypically, as well as a somewhat slower system [conscious System 2] that allows time for thinking and planning more complex behaviors.[28]

At the societal level, as social beings we also have a social immune system. It arises from the effective collaboration of many individuals' mental immune systems to shield society from harmful narratives by fostering a diversity of alternative thoughts, ideas, and explanations that then engage in healthy competition. Often a good thought is enough. Therefore, even with global challenges, humanity is better off seeking solutions as a system of many genuine individuals with unique mental immune systems. I tried to formulate this mathematically in my book *The Indoctrinated Brain*: "The sum of all individuals multiplied by their degree of individuality thus gives the degree of innovative capacity of a society."[29] The probability that at least one member of a society will find a solution to a problem that could threaten everyone increases with the number of members who differ maximally due to their individual life experiences and can make a unique contribution. Thus, the mental immune system of the individual, in the sum of all individuals, forms the social immune system of a society, which increases its chances of survival. But it is not only the brilliant ideas of an individual that give a particular group an advantage in the struggle for survival; much more often, it is the ability of the members of a community, accompanied by a functioning mental immune system, to cooperate peacefully, resolve conflicts, and survive together in order to protect one another.

The social immune system also learns. It is within the framework of this learning process that our artistic, scientific and technological achievements and, not least, our respective cultural characteristics emerge. This social immune system finds its closest counterpart in the family circle in the grandmother. Their existence has been proven to have far-reaching consequences for the survival of the associated clan. This will be examined in more detail below to clarify the social influence of System 2 thinking. The far-reaching role that an adequate supply of lithium plays will also become clear when we address this in the next chapter. The following generally applies: the transition from individual learning ability to the social level is the basis for the cultural evolution of humanity, which requires essential lithium for peaceful and thus healthy development.

Evolution of the Grandmother

The mental immune system includes both Systems 1 and 2. However, it is System 2 that makes our species unique. Reflected experiential knowledge, amplified by our social competence, became a primary driving force behind the human species' conquest of the Earth. We learn through our own experiences and learn from others' experiences.

Humans also developed, based on the individual mental immune system, the ability to positively influence reproductive success across generations through a social immune system. According to records in church books and birth registers, it was the presence of a grandmother that increased the probability of survival of her grandchildren until the end of the nineteenth century. For every decade that she lived within her family, an average of two more grandchildren reached adulthood than in families without a grandmother.[30]

As Fig. 6 illustrates, the grandmother accumulates more experiential knowledge by living longer. She can also support the family in general through her presence. Both her energy and her experience benefit the family. The more grandchildren who survive and grow up, the greater the probability that they will pass on the grandmother's (random) genetic makeup for longevity and mental fitness. This leads to the following assumption: If certain genes cause longevity and mental health, the likelihood of these genes producing a grandmother generation increases. This further strengthens the positive effect of genes, as the grandmother's experiential knowledge leads to a nongenetically determined higher probability of offspring survival; the resulting additional selective advantage of genes that cause longevity and mental health is known in science as the grandmother hypothesis. It is now common knowledge in schools. The "Evolution of

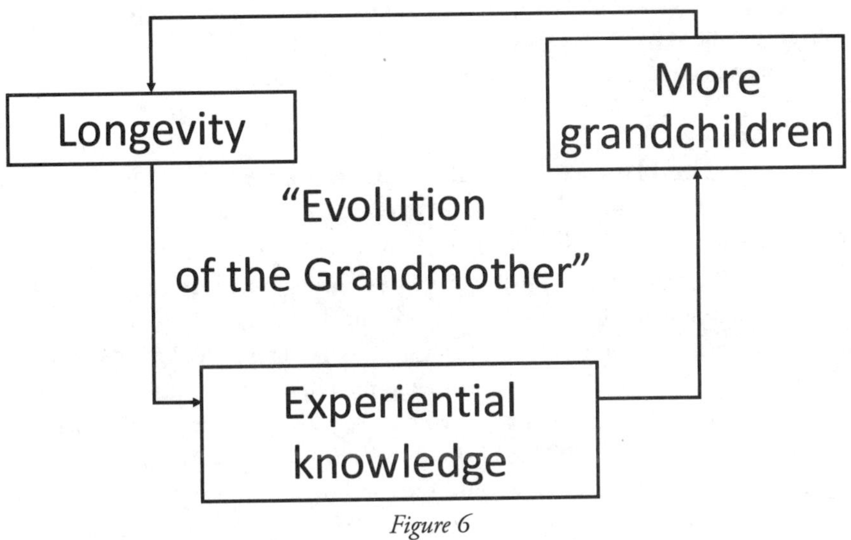

Figure 6

the Grandmother," as I like to call this phenomenon that shapes humans and thus all of humanity, explains the extraordinary longevity of humans with full mental fitness into old age, even when this age is far beyond female reproductive capacity.[31] For example, Frenchwoman Jeanne Calment (1875–1997) lived to be the oldest woman in the world to date, reaching the age of 122. She lived for about seven decades after her menopause and did not develop Alzheimer's disease.[32] On the other hand, chimpanzees, genetically our closest relatives in the animal kingdom, die just a few years after the females become infertile again. They have no grandmothers, but also no grandfathers, because after all, they inherit the same genes that determine their life expectancy. Life expectancy for almost all living beings is limited—in accordance with the evolutionary imperative—by the time of reproductive capacity. Humans are an exception, but so are orcas—the rulers of the seas. They have gone through a similar development to us. As a result of the social factor that a grandmother generation entails, longevity genes are more likely to prevail in their population.[33] In this species, as well, grandmothers can live more than twice as long as their own reproductive age would dictate and, through their experiential knowledge, increase the chance that their clan will survive in times of need, thus ensuring the survival of their genes for longevity and mental fitness into old age.

This also refutes earlier objections to the "Evolution of the Grandmother," according to which our extraordinary longevity is only due to our modern living environment. Human cultures also show evidence of this. The comparatively low *average* life expectancy of hunter-gatherer societies is due to high infant mortality—not to low *individual* life expectancy. Even in hunter-gatherer cultures largely isolated from modern life, according to the results of a corresponding study, around two-thirds of individuals who survived childhood lived to the age of seventy—and the authors noted that even encountering eighty-year-olds was not uncommon.[34]

The Human Mental Immune System Grows and Matures Throughout Life

The autobiographical memory center, which is responsible for the lifelong accumulation of essential experiential knowledge, is located in the paired structures of the cerebrum's temporal lobes. It is called the hippocampus due to its anatomical structure, which resembles a seahorse (see Fig. 7). The two hippocampi, which are about the size of a thumb, and referred to simply as the hippocampus, store what we consciously experienced, imagined, or were told, as well as the when and where, and which emotions were associated with it. This allows us to always know whether the remembered experience or thought was more advantageous or disadvantageous. Serving as our personal diary, the hippocampus can gather new experiences well into old age because it's the only structure in our brain able to form new nerve cells throughout life.

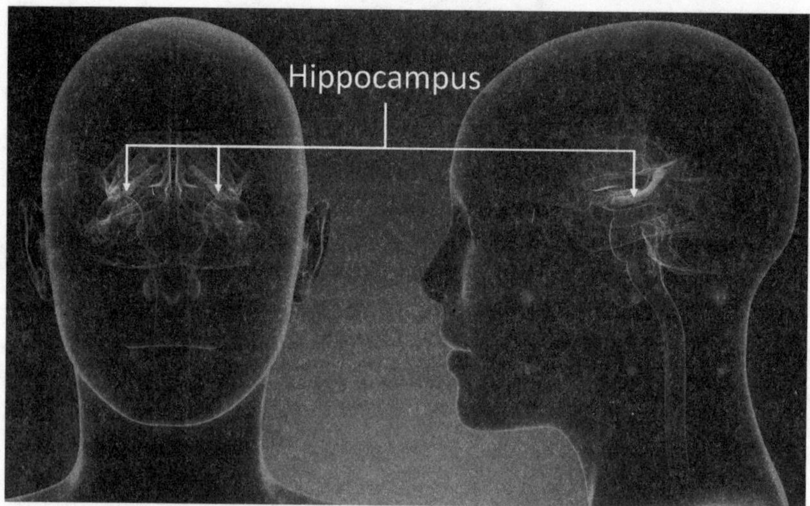

Figure 7

Only in this way is it possible, despite the hippocampus' limited storage capacity, to learn, experience, or think new things daily without overwriting and thus forgetting previous memories. Through so-called adult hippocampal neurogenesis, our "Diary of Life" adds a new page each day, preventing the need to overwrite previous entries and thus erase their content or our past experiences. In keeping with the "Evolution of the Grandmother," this unique process allows us to expand our wealth of experience throughout our lives. Even though science (as of spring 2025) is not yet entirely in agreement on how many new nerve cells are or can be formed in the hippocampus every day—depending on the methodology and the health of the person whose hippocampus is being studied, it could be several thousand or perhaps only a few hundred—they play a crucial, if not the central role in our mental immune system's functioning. "In contrast to the rest of the hippocampal network," wrote a group of neurogenesis specialists in a 2018 review article, "synaptic plasticity [the ability to learn and adapt] in the dentate gyrus [or the hippocampal input area, where neurogenesis occurs throughout life] is thus concentrated in a defined, functionally naïve [fresh and unused] subset of (new) neurons."[35] This unique mechanism of focusing plasticity distinguishes this neural network from all those already studied, from which they conclude that "the number of new cells required for a functional benefit is actually very low."

A study conducted using state-of-the-art methods, published in October 2024, came to a clear conclusion: "Our results suggest very low neurogenesis in the hippocampus throughout life and the existence of a local reserve of plasticity in the adult human hippocampus."[36] An expert report published at the end of February 2024 titled "Impact of Adult Neurogenesis on Emotional Functions in Mice and Humans" also concludes: "In sum, mounting anatomical, biochemical, and genomic evidence supports the presence of immature neurons in the hippocampus of healthy subjects throughout life."[37] The key word here is "healthy." In fact, a Spanish study identified thousands of immature and maturing neurons in the hippocampus of healthy subjects of different ages (even up to ninety years old!): "In contrast, the number and maturation of these neurons decreased with the progression of Alzheimer's disease. These results demonstrate that AHN [adult hippocampal neurogenesis] persists during both physiological and pathological aging in humans and provide evidence for impaired neurogenesis as a potentially relevant mechanism underlying memory deficits in Alzheimer's disease and potentially amenable to new therapeutic strategies."[38] But this study also left no questions unanswered methodologically, as the authors demonstrate in a subsequent review, another basic prerequisite for the clear detection of AHN: "We describe here in detail the methods used in our laboratory to *unambiguously* [emphasis added] demonstrate a population of immature neurons in the human hippocampus up to tenth decade of life. The criteria used to refine and develop the current protocol include obtaining postmortem human samples of remarkable quality and under strictly controlled conditions for immunohistochemical (IHC) studies, optimizing tissue processing and histological procedures, establishing criteria for reliable validation of antibody signal, and performing unbiased stereological cell counts."[39] The fact that tissue material from healthy deceased people is crucial for the good detection of AHN is now recognized by many AHN experts.[40] And the fact that the number of maturing nerve cells decreases drastically in the course of Alzheimer's disease, which I refer to as hippocampal dementia after its site of onset, as expected, was confirmed in another independent study conducted in Chicago on deceased people aged forty-three to eighty-seven: Thousands of new nerve cells have been identified in mentally healthy individuals; however, in deceased Alzheimer's patients, the more advanced the dementia, the fewer were found.[41] According to the authors of the aforementioned Spanish study, their findings "provide evidence for impaired neurogenesis as a potentially relevant mechanism underlying memory deficits in Alzheimer's disease that may be amenable to new therapeutic strategies."[42] This also aligns with my comprehensive systems theory of Alzheimer's development, published in 2016, in which I was among the first scientists worldwide to identify AHN disrupted by an unhealthy lifestyle as the core issue in Alzheimer's disease.[43] Furthermore, this line of thinking also offers a practical approach to prevention. My view that a disturbed AHN could be the cause and not the consequence of Alzheimer's disease is now shared by more and more researchers. In a commentary published in 2019 on these latest findings, gerontologists at Harvard University ask themselves—and in my opinion, rightly so: *Is Alzheimer's Disease a Neurogenesis Disorder?*[44]

These newly developing hippocampal nerve cells play an essential role in mapping and anchoring all autobiographical memory content as spatiotemporal information within the four-dimensional continuum of our existence. By creating a specific index for each experience, thought, and one of our plans, dreams, and hopes, they enable us to retrieve memories at any time. Consequently, they provide the groundwork for the development of self-awareness, the ability to reflect, and the creative planning and visualization leading to the implementation of future tasks. These place and time neurons, which I would also like to call index neurons because of their functional combinatorics, enable us to distinguish between past, present experiences and our future plans, which are also stored in the hippocampus.[45] This is the only way we know exactly what we have actually experienced ourselves, what we have only imagined, or what we have learned from others. Thus, they are crucial for telling the difference between others' experiences and our own, allowing us to develop into autonomous individuals.[46] If these functional properties of the index neurons were absent, what neurologists (and now AI researchers) describe as catastrophic interference would occur: Daydreams and real experiences collide and are perceived as indistinguishable. In people with schizophrenia, which is causally based on a developmental or maturational disorder of the hippocampus, this could be a major problem in their illness (more on this in Chapter 4 on psychiatric disorders).[47]

The main task of adult hippocampal neurogenesis is to strengthen our mental immune system through self-reflective life experience, in the spirit of the Evolution of the Grandmother. This includes the vital willingness to learn new things. This characteristic, known as natural curiosity, is closely tied to the function of these new hippocampal neurons, a link clearly demonstrated in animal studies.[48] These neurons are "hungry" for new experiences, as if their survival depended on it.[49] And in fact, it does:

If they aren't used due to a lack of new experiences or learning, they "starve" or undergo genetically preprogrammed self-destruction (apoptosis).[50] To encourage us to embrace our natural curiosity and not shy away from new experiences due to unwarranted fear, these new hippocampal neurons have the ability to regulate our mental resilience. Neuroscientists Antoine Besnard and Amar Sahay from Harvard University explain how this works in their 2015 article in *Neuropsychopharmacology*, "Adult Hippocampal Neurogenesis, Fear Generalization, and Stress": "In order to express fear only when it is appropriate, we have to constantly perform comparisons between previously encoded associations and what actually happens. Such a mechanism is adaptive in that it allows an individual to anticipate a potential threat by detecting relevant cues present in the environment."[51]

The new index neurons are ideally positioned to assess the initial stress impulse (and accompanying anxiety) triggered by each new situation. After all, in nature: Prevention is better than cure, i.e. it is better to anticipate danger than to blindly run into danger. Perceiving a new situation and simultaneously comparing it with index neurons in their hippocampal environment that have already stored comparable experiences (whether their own or learned from others), lowers the stress and anxiety levels following the all-clear. Thus, adult hippocampal neurogenesis significantly influences our ability and willingness to truly learn new things, venture into the unknown, and thereby gain new experiences. This unique functional interplay forms a crucial foundation for individual maturation. As social learning machines, a consequence of our mental immune system, beneficial innovations created by individuals or groups spread rapidly, fostering the cultural evolution of our social immune system at a societal level.

Adult hippocampal neurogenesis allows us to learn throughout our lives from others' experiences—our hippocampus loves stories—as well as from our own, even from our own thoughts and ideas. However, our autobiographical memory center is far more than a passive memory storage, as should be clear by now. The hippocampus is the active center of our creativity, individuality, self-image, and self-confidence, based on the new, eager-to-learn, stress-regulating nerve cells that are created there every day. Moreover, it grants us the awareness of how our plans and potential actions influence the lives of others, regardless of whether they are within our immediate temporal or spatial vicinity. I term this capacity "rational compassion," contrasting it with purely local and effective emotional empathy, which, despite its daily significance, doesn't always guide us to sound decisions because it overlooks potential negative impacts on those outside our immediate awareness. Rational compassion allows us to gauge the impact of our plans on those not in our immediate vicinity by consciously putting ourselves in their shoes and adopting their perspective. Consequently, this hippocampal capacity is a vital prerequisite for peaceful conflict resolution and social competence.

Essential Functions of the Mental Immune System

These diverse functions of the index neurons, newly formed through adult hippocampal neurogenesis, constitute the core of our mental immune system. In my view, this system encompasses all the fundamental characteristics that we imagine in an ideal human being, as depicted in Fig. 8.

A well-functioning mental immune system, fueled by productive adult hippocampal neurogenesis, enables us to be socially competent members of our respective communities. The uniqueness and creativity of each individual benefits the entire group. At the same time, autobiographical memory, as the central repository of our interpersonal experiences and thus of our relationships, is also the center of our social conscience.

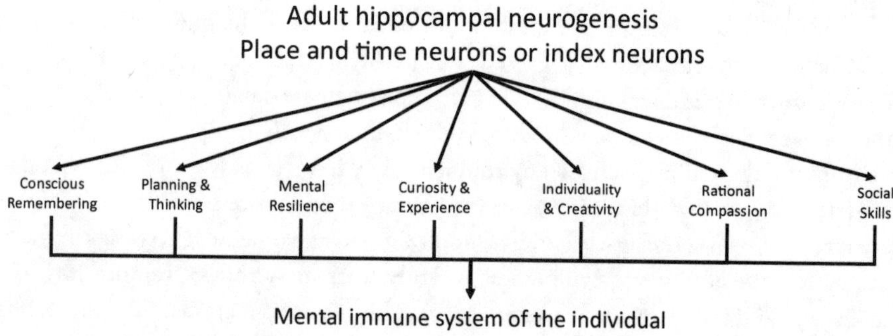

Figure 8

Paradoxically, the mental immune system, designed to protect us from life's dangers, is also our brain's most vulnerable and sensitive structure because its function depends on the lifelong production of new nerve cells. There is no brain function, and probably no bodily function, that suffers more when we ignore the Laws of the Minimum and the Maximum. Essential lithium is therefore not the only vital substance that we or our mental immune system need. But if this is lacking, then we have a (funda)mental problem that begins even before our birth, when our unique human brain develops.

CHAPTER 2

ESSENTIAL OR NONESSENTIAL?—THAT IS THE ESSENTIAL QUESTION

We share the same biology, regardless of ideology.
—Sting ("Russians," 1985)

How to Prove Essentiality?

Essential trace elements exert an influence on a multitude of physiological functions. Their effect is so vital that deviations from the requirements inevitably lead to deficiency diseases. This underscores their crucial nature and warrants their classification as essential trace elements, which ought to be obtained in minute quantities via nutrition. In a 343-page book titled *Trace Elements in Human Nutrition and Health,* published by the WHO in 2016, the six elements iodine, zinc, selenium, copper, molybdenum, and chromium are listed under the heading "essential."[1]

Two elements, cobalt and iron, had previously been identified as essential. The trace elements manganese, silicon, nickel, boron, and vanadium are listed under the heading "potentially essential." Fluorine, lead, cadmium, mercury, arsenic, aluminum, tin, and lithium are classified as "potentially toxic elements, some with possibly essential functions."

Lithium therefore appears here in the same category as lead, cadmium, mercury, aluminum, and arsenic, which, even in small quantities, are highly toxic. Nevertheless, the detailed chapter on lithium states that the trace element has been classified as essential for animals based on animal studies. Regarding toxicity, it was further recognized that this pertains solely to quantities rarely obtained through diet—quantities for which there has been no documentation, "as intake from food sources approximates 100 µg/day or less and typically constitutes less than 0.1 percent of therapeutic or potentially toxic doses." Moreover, one can even read there that WHO research on this supposedly toxic trace element found that "epidemiological studies in the USA have shown an inverse correlation between lithium in drinking water and mortality, particularly from heart disease[2], as well as the rate of admission to psychiatric hospitals."[3, 4] In a nutshell: These findings already suggest that lithium also plays an essential role in humans. Yet the chapter concerning the "potentially toxic element" culminates in the remarkably inconsistent conclusion: "Lithium may be vital for laboratory animals, but its necessity for humans remains unknown." This assessment is incomprehensible: Why is it classified as potentially toxic (a term according to Paracelsus, which ultimately applies to all substances if dosed high enough, even water) when animal studies suggest it might be essential, and beneficial health effects have even been observed in humans at natural intake levels? Considering the data available at that time, wouldn't it have been far more logical to classify lithium as at least "potentially essential"? This would not have been a mistake at all, as there was already sufficient evidence that a deficiency in this trace element has negative effects on humans and animals. This would have paved the way for this knowledge to be used directly and for substitution to be initiated for the general population—with immediate positive effects for everyone who would suddenly have been able to have an adequate supply of lithium.

The fateful path chosen continues to mislead researchers to this day. Thus, as of spring 2025, the currently most cited article on the essentiality of trace elements in the human diet takes another step in the wrong direction. This is a paper published in January 2020 by Aliasgharpour Mehri, a faculty member at the Reference Laboratory Research Center of the Iranian Ministry of Health and Medical Education in Tehran.[5] Mehri doesn't address the scientific data in the 1996 WHO report, which at least hints at lithium's essentiality, nor the vast body of studies accumulated over the past two decades on this trace element's crucial role in mental development, cognitive performance, and psychological stability with higher natural intake. There is also no longer any reference to the entire animal experimental evidence regarding the essentiality of lithium. Yet, during the evaluation of lithium for human use, warnings focus solely on its toxicity, despite this being undeniably relevant only at high doses, such as those used in the treatment of bipolar disorder or manic-depressive illness. However, this pharmaceutical use differs significantly in its objective and dosage from what we aim to achieve with trace element supplementation. Trace elements must simply be taken in a natural way or modelled on natural conditions, and in quantities that are also needed by the body.

Australian psychiatrist John Cade (1912–1980), who established and is considered the father of high-dose lithium treatment (though not actually the first), pointed in the right direction when, in his influential clinical study published on

September 3, 1949, he expressed amazement at lithium's extraordinary effectiveness: "The effect on patients with manic excitement is so striking and specific that it suggests that some factor in the body associated with the disease may be deficient in lithium ions."[6] But if, as Cade speculated, a lithium deficiency can cause or at least increase the risk of developing mental illness, then he rightly concluded: "Lithium may well be an essential trace element." Yet, Cade's speculative conclusion wasn't simply ignored. Suspiciously soon after Cade's groundbreaking publication, the U.S. Food and Drug Administration (FDA) banned even small amounts of lithium in sodas—where it had been marketed as a mood enhancer—citing toxicity. (We will discuss the perfidious way in which this ban was achieved in the last chapter). Yet, following this "success" of certain interest groups, things became quiet for a while around essential lithium, while the high-dose treatment method—despite severe dose-related side effects—became increasingly popular. Sixty-five years after Cade first suspected lithium's potential essentiality, Anna Fels, an influential psychiatrist and faculty member at Weill Cornell Medical College, published her personal opinion on the matter in *The New York Times* on September 13, 2014.[7] In response to her title question, "Should We All Take a Bit of Lithium?" Fels stated: "Lithium has been known for its curative powers for centuries, if not millenniums. Lithia Springs, Ga., for example, with its natural lithium-enriched water, appears to have been an ancient Native American sacred site. By the late nineteenth century Lithia Springs was a famous health destination visited by Mark Twain and Presidents Grover Cleveland, William Howard Taft, William McKinley and Theodore Roosevelt." Although Fels finds it astonishing "that the microscopic amounts of lithium found in groundwater could have any substantial medical impact, the more scientists look for such effects, the more they seem to discover." Indeed, "evidence is slowly accumulating that relatively tiny doses of lithium can have beneficial effects. They appear to decrease suicide rates significantly and may even promote brain health and improve mood," she continues. Fels also points out that scientists have proposed recognizing lithium as an essential trace element nutrient. And the psychiatrist asked herself then, as I do today: "Who knows what the impact on our society would be if micro-dose lithium were again part of our standard nutritional fare? What if it were added back to soft drinks or popular vitamin brands or even put into the water supply? The research to date strongly suggests that suicide levels would be reduced, and even perhaps other violent acts. And maybe the dementia rate would decline. We don't know because the research hasn't been done."

Although research on the subject has continued since then, it will of course never be completed. Every result will raise new questions; I know this only too well from my own time as an analytical research scientist. At some point, however, it becomes difficult to bear when people suffer unnecessarily from lithium deficiency or a compromised mental immune system, merely because certain individuals tirelessly insist on further experimentation before building practically on already existing results. "For the public health issue of suicide prevention alone, it seems imperative that such studies be conducted," Fels states in the last paragraph of her article, pointing out the urgent need for action, since in 2011 suicide was the tenth leading cause of death in the USA alone: "Research on a simple element like lithium that has been around as a medication for over half a century and as a drink for millenniums may not seem like a high priority, but it should be." More than ten years have passed since then, and once again we can hope that the time is ripe for a renaissance regarding knowledge of lithium. It is more important than ever to definitively prove lithium's essentiality, as I aim to do with this book. But first we need to ask: What criteria must trace elements typically meet to be recognized as essential?

Nutrition experts in the Department of Animal Nutrition Chemistry at Karl Marx University Leipzig have formulated an answer to this question. Under the leadership of the renowned trace element expert and agricultural scientist Manfred Anke (1931–2010), they published an article in 1984 titled "Importance of New Essential Trace Elements (such as Si, Ni, As, Li, V) for Humans and Animal Nutrition."[8] Based on a list of scientific studies and discussions that had been steadily growing since the 1970s, they established a catalog of criteria for proving the essentiality of new trace elements. According to this, essentiality is fulfilled if the following conditions are equally fulfilled:

1. A deficiency leads to growth disorders, reduced reproductive performance, and shortened life expectancy.
2. The trace element is a component or regulator of an essential body substance (i.e., a target).
3. A deficiency leads to pathological changes. Corresponding results must be reproduced in at least three different experiments on different species.
4. The deficiency symptoms disappear after supplementation of the element (intervention to correct the deficiency).
5. The symptoms reappear as soon as the animals no longer absorb the trace element (withdrawal test).

The arguments that I present in this book for lithium as an essential trace element show that all these criteria are met—but the spectrum of good arguments goes far beyond these minimal criteria. I will also introduce a disease syndrome resulting from lithium deficiency and, in addition to the criteria, will provide evolutionary evidence of the high degree of conservation of the interaction between lithium and its lithium targets. On the one hand, this evidence aims to appropriately locate lithium in the origin of life itself, and, on the other hand, it will allow us to demonstrate its essentiality not only in humans but—as suggested by the Leipzig nutrition experts—also in other species based on animal models.

Based on the understanding gained so far of the Fundamental Laws of Life and the biological processes necessary for life and healthy homeostasis, as well as on the guidelines of the Leipzig nutrition experts, the following core arguments for the essentiality of lithium as a trace element arise:

Argument 1: Neuropsychiatric developmental disorders and diseases caused by lithium deficiency can be described as mental immunodeficiency syndrome.

Argument 2: Dysregulation of key cell biological regulators due to lithium deficiency explains the clinical symptoms of mental immunodeficiency syndrome.

Argument 3: Lithium distribution is selectively regulated to maintain relatively stable concentrations in key organs, particularly the neurobiological center of the mental immune system, despite natural dietary fluctuations.

Argument 4: The lithium targets and their regulation by lithium are detectable in all organism kingdoms from bacteria, fungi, and plants, to humans and are thus highly conserved evolutionarily.

Argument 5: A lack of lithium shortens life in animals and humans, and a severe deficiency is incompatible with life. Conversely, correcting lithium deficiency has been shown to extend lifespan in several species, including humans.

Argument 6: Lithium was present in physiologically relevant quantities in food during human evolution's crucial phase.

The subsequent discussion will elaborate on these six arguments with maximum clarity, facilitating comprehension for specialists and ensuring understanding for interested nonexperts regarding the indispensability of lithium—without requiring blind faith. Even though each individual argument should be sufficient to demonstrate lithium's importance, as we know, the whole is greater than the sum of its parts. Furthermore, my aim was to provide you with a more profound understanding of the molecular biology of life, the origins of which lie with lithium. But let's begin with lithium's importance as an essential trace element for the mental immune system—the very core of our humanity and the system most severely impacted by lithium deficiency.

ARGUMENT 1: Developmental Disorders and Diseases: Mental Immunodeficiency Syndrome Due to Lithium Deficiency

Epidemiology of "Mental Immunodeficiency Syndrome" (MIDS)

As explained in the first chapter, this mental immune system, which aims to make us, both as individuals and as a group, less susceptible to external disturbances and dangers, is also perhaps our brain's most vulnerable and endangered function due to its reliance on the highly sensitive production of new nerve cells. This becomes easy to understand if we ask ourselves the rhetorical question: Which plant suffers more from a lack of water, the young seedling or the fully grown tree? Like hardly any other brain function, adult hippocampal neurogenesis is subject to the law of the maximum and the minimum: An excess of toxins, distress, and overstimulation inhibit new nerve cell production, as does a lack of essential nutrients or of other aspects of a species-appropriate lifestyle; I discuss this in detail in my book *The Exhausted Brain.*[9] Specifically, essential lithium deficiency, as we will see, results in a broad spectrum of hippocampal developmental and functional disorders that, in their various forms—from autism to Alzheimer's disease—impair the development and function of the mental immune system.

A large number of epidemiological studies have provided evidence of the essential importance of the trace element lithium for our mental and psychological fitness. Epidemiology, composed of the ancient Greek words *epidēmios* for "what comes upon the people" and lógos for "word" or "teaching," concerns the study of the causes of common diseases. What can befall a population when it leaves its ancestral homeland or alters its dietary habits is lithium deficiency, the effects of which epidemiological research can determine Lithium, like iodine or selenium, was washed out of the soil and into the sea over billions of years. It therefore only occurs in higher soil concentrations in mostly younger regions, which are usually of volcanic origin (more on this in Chapter 3). Owing to the frequently limited dietary consumption of lithium, lithium-bearing potable and tap water, especially originating from deep aquifers in volcanic areas, constitutes the principal source of lithium for most individuals. The often highly variable regional lithium concentrations in tap and local drinking water thus facilitate epidemiological studies on the impact of even small amounts—in the microgram per liter (µg/L, one millionth of a gram per liter) range—of naturally occurring lithium on the population's physical and, especially, psychological and mental health.

Neuropsychiatric Developmental Disorders

In 1970, the first epidemiological study was conducted in the USA, aiming to test John Cade's hypothesis that lithium deficiency leads to psychiatric developmental disorders. The epidemiologists found that "The incidence of patients' first admission and prevalence of readmission as well as the diagnosis of psychosis, neurosis and personality disorders was inversely proportional to the lithium content of their residential drinking water."[10] While the researchers stated this study neither confirmed nor refuted Cade's theory that lithium may be a an essential element, they noted it "does provide compelling evidence that naturally occurring lithium has a measurable and statistically significant impact on the frequency of psychiatric hospital admissions in this state, as well as on certain diagnostic categories of mental illness." Another study from 1972 confirmed the significant association between lithium deficiency in drinking water and the major forms of mental illness diagnosed, recorded, and reported by the Texas Department of Mental Health and Mental Disability at the time, such as schizophrenia, psychosis, neurosis, and personality disorders.[11] The results were similar to a study from North Carolina published in the same year.[12] Based on further such studies, the authors of a Japanese study published in *Schizophrenia Research* in 2017 suggested that due to the observed inverse relationship between the lithium content in tap water and psychotic experiences in adolescents, the administration of lithium could possibly have a preventive effect against such disorders.[13]

A systematic meta-analysis published in 2020, which included a total of approximately 113 million people, found a clear association between lithium in drinking water and neuropsychiatric disorders: Higher lithium concentrations in drinking water, in particular, have been associated with significantly fewer psychiatric hospitalizations—and, as with the Texas study just mentioned, this positive effect on mental immune system development also correlated, as expected, with lower suicide rates.[14] The mental immune system, which is supposed to protect us from life's dangers, appears not only less effective when experiencing lithium deficiency; psychological resilience and self-protection, ensured by a healthy mental immune system, may entirely cease to function. As a result, those affected become a danger to their own lives.

Suicide Rate

Although suicide is a complex and multifactorial phenomenon, it is not a mental illness. Nevertheless, a mental developmental and functional disorder with abnormal mood is a major risk factor for suicide, increasing the risk approximately twenty to forty times compared to individuals with a more balanced mood.[15] Lithium deficiency can both contribute to such disorders and independently cause a life-threatening mood imbalance. A 2011 Austrian tap water study, according to its authors, offered "conclusive evidence" that "an increase in drinking water lithium concentration of just 0.01 mg/L [i.e., 10 µg/L] correlates with a 7.2 percent reduction in the SMR [standardized mortality ratio; the ratio of observed to expected deaths in the general population] for suicide."[16] This incredibly small amount (more on lithium requirements in Chapter 3) is nonetheless astonishing, considering the prominent advertising of suicide hotlines and disclaimers with mental health support contacts accompanying nearly every article on psychological issues—yet the potential for a trace element to easily reduce suicide rates is not mentioned. Particularly spicy: There have long been numerous studies that have produced similar results. For instance, a 2020 systematic review and meta-analysis of prior suicide studies confirmed a dose-dependent association between lithium in drinking water and a reduced likelihood of suicide.[17] Another 2021 meta-study, "Consistent Findings Worldwide: More Lithium in Drinking Water, Fewer Suicides," analyzed fifteen previous epidemiological studies, concluding that the association between higher natural lithium concentrations in drinking water and lower regional suicide mortality can be considered established.[18] It's important to note that: "Considered established" means that there is no longer any doubt that inadequate lithium intake contributes to the development of psychological problems or difficulties in problem-solving. While epidemiological studies have statistically shown a protective effect of microdose lithium in drinking water against suicide, the link between individual body lithium levels and suicide was unknown until recently. A Japanese study measured cerebrospinal fluid (CSF) lithium levels in twelve suicide victims and sixteen non-suicide victims autopsied at the Tokyo Forensic Institute between March 2018 and June 2021. The lithium concentration in suicide victims was approximately two times lower, averaging 0.50 μg/l, than in non-suicide victims, who had an average value of approximately 0.92 μg/l "In summary," the authors state, "lithium, even in microdoses, plays an important role in suicidality."[19]

Consequently, those fortunate enough to inhabit areas where the tap water naturally exhibits elevated lithium levels benefit from a demonstrably reduced suicide risk—and typically enjoy a stronger foundation for mental resilience. What, then, prevents health-care authorities from utilizing lithium in essential quantities for the mental and psychological well-being of society, particularly for suicide prevention? A fundamental question that I try to answer in the last chapter—and we might not like the answer. However, I would like to point out here that time and again, studies with little conclusive information are used to avoid this important question. For example, reference is made to a study that found no correlation between lithium levels in tap water and a reduced risk of suicide.[20] Notably, the paper bears a title phrased as a leading question: *Is It Too Early to Add Lithium to Drinking Water?* However, as the authors of this "exception to the rule" study pointed out, the average tap water lithium level was extremely low at just 3.72 μg/L, and the various lithium concentrations exhibited minimal regional variation. The two limitations do not allow for a meaningful epidemiological calculation as compared to other international studies in which clear anti-suicidal effects were measured. Why, then, does this inconclusive result lead to questioning the widely established link between significantly reduced suicide rates and correspondingly higher lithium concentrations in drinking water, even in the title? This is at least questionable, if not incomprehensible, particularly since this study neither answers nor can answer this question. Rather, this study—within the now undeniable overall context—offers, in my view, the crucial insight that a lower threshold exists for lithium's health effects, which must be surpassed to observe a clinical impact (more on this in Chapter 3, where we'll attempt to calculate the natural lithium requirement). This partial result does not speak against, but rather for, adding sufficient lithium to tap water.

Devil's Advocate—The Cause of Fewer Suicides or Consequence of Suicide Prevention? Scientists from the University of Heidelberg analyzed the data of the Austrian study. They confirmed that lithium in drinking water was "a significant factor in the geographical pattern of suicide mortality."[21] However, they explored the hypothesis that a different causal explanation for the lower suicide rate might exist if "the distribution of lithium in drinking water is attributable not only to natural sources but also to prescribed lithium." The reason for this is that lithium has been used in medicine for over sixty years as a mood stabilizer and—albeit in very high doses—for suicide prevention. The implication, it is posited, could be that "the geographical distribution of suicide rates is not predicated on natural [low-dose] lithium effects, but rather on the geographical distribution of [high-dose] prescription patterns." Thus, the thesis posits that our protection from suicide doesn't stem from the serendipity of higher lithium levels in tap water, but merely from the happenstance of residing where physicians frequently prescribe lithium, resulting in elevated lithium concentrations in treated tap water due to patient excretion. The causality would therefore be the high prescription rate and not the natural lithium content in drinking water. Consequently, the relationship between natural lithium in drinking water, prescribed lithium medications, and suicide rates in Austria was reexamined in detail; the result was clear: "The [statistical] model calculations show no association between lithium from prescription drugs, presumed to accumulate in drinking water [. . .], and the suicide rate."[22] Thus, natural lithium levels in drinking water remain an explanation for the geographical distribution of suicide mortality in Austria, with higher levels correlating with lower risk.

But how exactly can lithium prevent something as multifactorial as suicide? The answer lies in lithium's capacity to directly and indirectly activate adult hippocampal neurogenesis through its targets (see Chapter 4, Additional Information: *How Lithium Deficiency Leads to Suicide*). In this way, it strengthens and stabilizes an essential function of the mental immune system: our psychological resilience. This is a fundamental prerequisite for processing negative experiences, even severe misfortunes, without developing clinical depression. Indirectly, it inhibits neuroinflammation—a process that can damage the brain and be triggered by extreme psychological stress. Such inflammatory processes in the brain are the primary cause of the neuropathological vicious cycle in depressive illness (these connections will be detailed in Chapter 4). Unsurprisingly, two Japanese studies conducted independently of each other provide further evidence for such a connection: Higher lithium concentrations in drinking water significantly reduce the occurrence of depressive mood values, according to the unanimous result.[23] Conversely, however, this also explains the increased likelihood of suicide in the case of lithium deficiency, because this fatal decision is usually preceded by severe depression or bipolar or manic-depressive disorder. The prevalence of both disorders (the total number of disease cases in a population at a given time) was significantly lower in US areas with lithium-rich water supplies, as shown by a large study using crude prevalence rates.[24]

Antisocial Behavior

The higher the lithium content in tap water, the less interpersonal violence occurred among adolescents, according to a large Japanese study published in 2017. The authors concluded that lithium might have not only antidepressant but also "anti-aggressive effects."[25] However, this finding wasn't entirely novel; in the 1990s, the prominent German lithium researcher Gerhard N. Schrauzer (1932–2014), then director of the Biological Trace Element Research Institute in San Diego, California, and his colleagues demonstrated an inverse correlation between lithium in drinking water and the incidence of crimes like robbery, burglary, theft, and illegal drug possession. They analyzed data from twenty-seven Texas water supply areas between 1978 and 1987.[26] The researchers also showed that the incidence of murder, suicide, rape, and overall crime was statistically significantly higher in water supply areas with low or no lithium in drinking water (mean 5 μg/L) compared to areas with higher levels (mean 123 μg/L). In this context, it is interesting that Gerhard Schrauzer and colleagues found that "the lithium content in the hair [which acts as a lithium memory, more on this in Chapter 3, Additional info: *Lithium Long-Term Memory—(Unfortunately) a Hairy Business.*] is low in certain pathological conditions, such as heart disease, learning disabilities, and incarcerated violent offenders."[27] From these extensive findings, Schrauzer and colleagues concluded that "lithification [adding lithium to] municipal drinking water supplies would be a simple, safe, and cost-effective means of reducing violent crime, suicide, and drug use."[28] To our continued astonishment, no such measure was seriously considered—despite numerous subsequent studies consistently indicating the same correlation. Similarly, scientists in Greece demonstrated an inverse correlation between lithium concentration in drinking water and homicide rates.[29] "Considering the findings of our prior study[[30]], which also indicated an inverse relationship between lithium content in drinking water and the incidence of suicide, murder, rape, and drug abuse," the researchers hypothesized "that natural lithium intake might influence impulsivity, a factor implicated in both suicidal behavior and aggression." *Japanese* researchers reached a comparable conclusion in 2020 when they correlated local drinking water lithium levels with crime rates across 274 communities on the island of Kyushu. They, too, were amazed at how even very low concentrations of lithium in drinking water led to a significant reduction in the crime rate among the population.[31]

Thus, lithium deficiency is not only a causal impediment to the individual development of the mental immune system but also evolves into a societal issue for the social immune system. This is the conclusion reached by Timothy M. Marshall, neurologist and professor of chemistry and pharmacology in Tucson, Arizona, in his 2015 overview *Lithium as a Nutrient*: "It appears that when people have deficient lithium intakes they experience poorer moods and are more easily agitated and reactive, as seen with increased rates of suicide, homicides, and violent crimes in areas with low lithium in their water supply."[32] Marshall goes one step further after evaluating the studies available at the time: "An optimal, nutritional intake of lithium may prevent or ameliorate many neurologic and psychiatric conditions through effects on nervous system metabolism and generalized anti-inflammatory and antioxidant effects. Pharmaceutical agents often mask symptoms without correcting underlying problems. Nutritional supplementation is safe—with wide-ranging neurological benefits—and should benefit overall health." In my opinion, general health also includes feeling safe in one's social environment. On the contrary, people prone to aggressiveness and violence (due to a lack of lithium) certainly do not contribute to this.

Depression and Alzheimer's Disease

As we have seen, lithium deficiency leads to many neuropsychiatric developmental disorders, which in turn are associated with increased rates of depression and even suicidal tendencies. Depression is an indication of impaired stress regulation due to deficient adult hippocampal neurogenesis and thus reduced psychological resilience. The resulting high release of neurotoxic stress hormones (see Chapter 4) further reduces the hippocampus' capacity along with that of the mental immune system. (A vicious circle that we will discuss in more detail in Chapter 4). I have already identified these connections in a scientific paper published in 2016 titled *"Unified Theory of Alzheimer's Disease (UTAD): Implications for Prevention and Curative Therapy."*[33] Also in my book *Alzheimer's Can Be Cured: Back to a Healthy Life in Time,*[34] which had already been published a year earlier, I pointed out the importance of microdosed (essential) lithium in the prevention and therapy of Alzheimer's dementia, in addition to many causal lifestyle factors. Two years later, in 2017, a Danish study of 73,731 dementia patients and 733,653 controls confirmed that the diagnosis of dementia was associated with lower lithium intake.[35] A link has been established not only for dementia in general, but also specifically for Alzheimer's disease, which can also be referred to as hippocampal dementia due to its neuronal origin in the brain. and vascular dementia, which originates from the blood vessels (from Latin *vas sanguineum*, the blood vessel). A large prospective US study on dementia development indicated that late-life depressive symptoms precede memory loss, but not vice versa, as UTAD accurately predicted.[36] Furthermore, higher depression scores predicted greater cognitive decline, even in individuals with normal baseline cognition—all independent of age, gender, education, and disease burden, including vascular disease. This finding corroborates the outcomes of earlier investigations demonstrating that comorbidity with depression is associated with a more pronounced degree and advancement of Alzheimer's pathology.[37] Consequently, depression, even when occurring within the framework of bipolar disorder, augments the risk of developing Alzheimer's disease. An increase in the risk of developing depression due to a lithium deficiency accordingly also causes an increased risk of developing Alzheimer's disease.

Mental Immunodeficiency Syndrome (MIDS)—A Lithium Deficiency Syndrome

Lithium would not be the first essential trace element for which a deficiency-related disease syndrome has been described and recognized. For example, selenium deficiency is considered the cause of Keshan disease, a form of heart disease endemic in China.[38] Iron deficiency leads to hypochromic (characterized by a lack of iron-dependent blood pigment), microcytic (small red blood cells resulting from an increased production rate to compensate for oxygen deficiency) anemia.[39] Iodine deficiency syndrome encompasses a broad spectrum of mental disabilities, ranging up to cretinism, which is an irreversible brain damage or a severe disorder of brain development.[40]

As we've observed, even minor variations in lithium concentration in drinking water, within a range of just a few μg/l, significantly affect the performance of the mental immune system, suggesting a substantial deficit. (With an adequate baseline supply, these small fluctuations in intake via drinking water would be negligible; more on this in Chapter 3). Individuals with such an insufficient lithium supply tend to exhibit increased restlessness, mood swings, elevated drug use, and depression. They also show pathological aggression accompanied by reduced impulse control, directed both inward (as evidenced by higher suicide rates) and outward, correlating with a measurably higher incidence of violent crime, rape, and homicide. Moreover, a lack of sufficient lithium in early childhood is strongly linked to the development of schizophrenia, psychoses, neuroses, and personality disorders, a connection that is reflected in increased rates of psychiatric hospitalization. A deficiency in lithium results in poorer (life) decision-making, as our mental immune system operates suboptimally. Bipolar disorder, likewise a disorder of the mental immune system, typically involves impaired decision-making, though this improves with lithium supplementation.[41] Long-term lithium deficiency also contributes to the development of neurodegenerative diseases like Alzheimer's, Parkinson's, and amyotrophic lateral sclerosis (more on this in Chapter

4), conditions that clearly contradict the "Evolution of the Grandmother" and thus reveal themselves to be anything but natural consequences of aging.

Given that micro amounts of lithium are clearly necessary for the development and maintenance of a functioning mental immune system throughout life, the logical question arises as to why a lithium deficiency syndrome describing this issue has not yet been defined. This also irritates Gerald Martone of the Alaska Psychiatric Institute. In his 2017 scientifically published mini-review on lithium supplementation, he writes: "There are numerous minerals that are critical to the human body. Are there vital elements that are also necessary for the optimal functioning of the human brain? Is lithium one of the essential minerals in the central nervous system?"[42] James M. Greenblatt, chief medical officer of Walden Behavioral Care in Waltham, Massachusetts, and nutrition expert Kayla Grossmann provide a clear answer to this question in the 2016 book *Nutritional Lithium: A Cinderella Story*. According to them, "the available scientific literature demonstrates the paramount importance of lithium for brain health and mental illness," and consequently, "lithium is an essential nutrient."[43] However, they also qualify this by noting that there is currently "no officially designated deficiency syndrome." This serious omission must now be remedied: Brain health is so intrinsically connected to lithium as a trace element that defining a corresponding deficiency syndrome would be straightforward. Therefore, considering the highly specific list of neuropsychiatric developmental disorders and diseases, all of which impair the functionality of our mental immune system, I suggest "mental immunodeficiency syndrome" as an umbrella term for the mental, psychological, and emotional consequences of chronic lithium deficiency (see Fig. 9).

Besides the deficiency symptoms and illnesses that directly affect individuals, this syndrome's effects also significantly extend into the social realm, making it particularly relevant for society as a whole. Ultimately, the social immune system emerges from the interplay of all individual mental immune systems. But it is more than the sum of these individual immune systems, because it creates new properties and abilities that its individual members lack. If a society, owing to a significant lithium deficiency, predominantly comprises individuals with dysfunctional mental immune systems, that is, with deficient System 2 thinking, a lack of curiosity and psychological resilience or tolerance, and rational compassion, democratic coexistence becomes untenable due to the fundamental absence of its prerequisites. Those with a weakened mental immune system contribute less to creative solutions for social challenges. They also risk falling into unjustified fear and panic, thereby exacerbating problems. Furthermore, those affected exhibit increased susceptibility to external manipulation or engage in stereotypical, maladaptive behavior—in extreme cases, even antisocial behavior—which makes life unpleasant or even difficult for many of their fellow human beings. Consequently, the mental immunodeficiency of many translates into a social immunodeficiency affecting all. This was clearly evident, for instance, when the staging of a pandemic in 2020 led to a division in society: In my opinion, the dysfunction of the mental immune system in a large part of the population—which unreflectively followed a life-threatening narrative (the harmful memes)—was more serious, threatening, and dangerous than the specter of an impending general dysfunction of the physical immune system (and thus of harmful viruses or their genes) that was constructed with immense media power. After all, practical solutions did exist, but few were even willing to consider them.

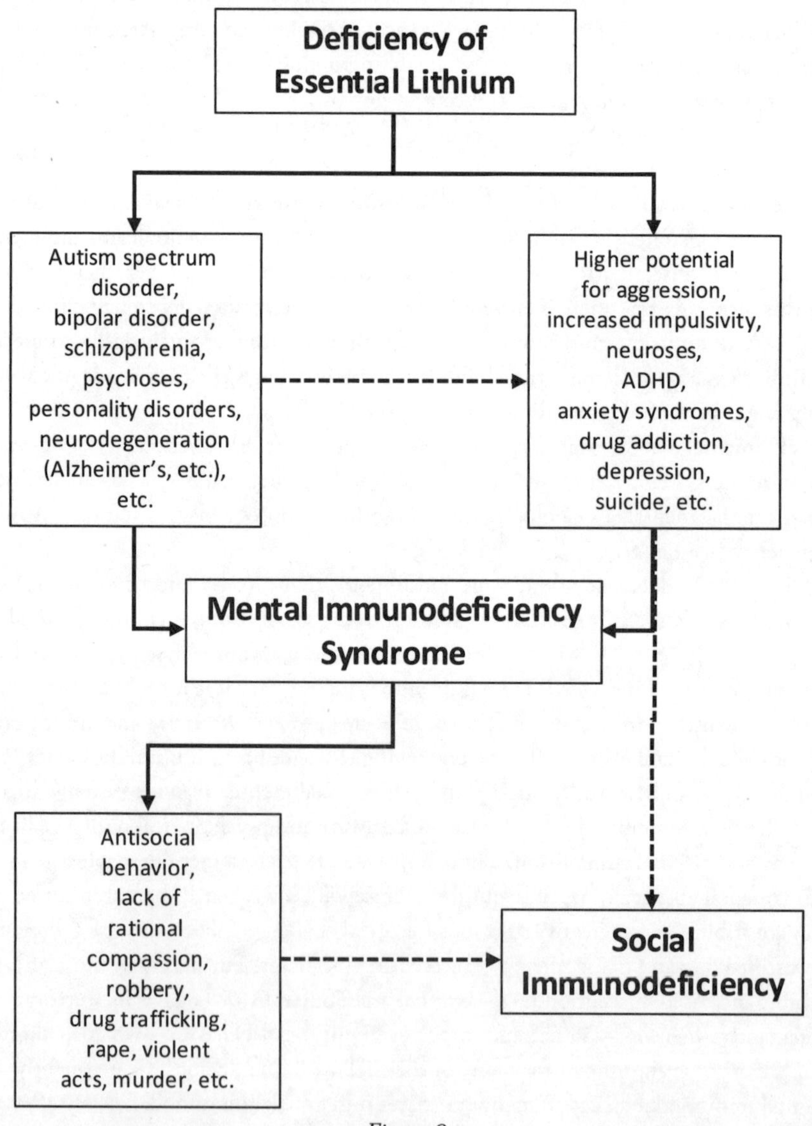

Figure 9

We will go into this in more detail, but it should already be clear what damage can be caused to individuals and society by ignoring the essential nature of a trace element. We will look at the underlying molecular mechanisms of mental immunodeficiency syndrome in the next subchapter. But it can already be stated that Argument 1 for the essentiality of lithium is well substantiated by the above statements. One can state:

The greater the lithium deficiency, the more likely psychoneural developmental disorders and diseases are.

ARGUMENT 2: Dysregulation of the Lithiome due to Lithium Deficiency Explains the Clinical Symptoms of Mental Immunodeficiency Syndrome

How Lithium Was Discovered the Wrong Way

In 1817, lithium was discovered as a new element. In the same century, lithium carbonate was first used to treat gout, based on the discovery that this lithium salt can dissolve the uric acid crystals responsible for gout symptoms. However, it was disappointing to discover that this only worked in the test tube, not in the patient. Nevertheless, lithium carbonate has also been tested for other, supposedly "gout-like" diseases. In 1859, Sir Alfred Baring Garrod recommended lithium salts for the first time to treat mania, considered back then as "brain gout."[44] But even this, as we know today, was a misconception. Extreme euphoria characterizing manic phases, along with an often-harmful drive for action in those affected, bears no relation to uric acid or gout. Nevertheless, John Cade, who rediscovered lithium treatment for mania following Garrod's initial attempts and published ten case reports on September 3, 1949, found that they could be treated surprisingly well with lithium carbonate.[45] The fact that something useful was discovered via a completely misguided approach would be scarcely believable if one didn't consider the extensive history of unsuccessful therapeutic trials—ultimately, lithium simply proved to be a lucky find. In any case, lithium in the form of lithium carbonate—less frequently also in the form of lithium acetate, citrate, chloride or sulfate—is the drug of choice for the treatment of bipolar disorder, in which manic and depressive phases alternate. It is now also occasionally used in the treatment of (unipolar) depression.

Due to the very high dosages that Cade considered necessary at the time, and which are still largely considered appropriate today, most patients complain of adverse effects due to the narrow therapeutic range: the narrow range between the necessary dose for typically purely medicinal intervention, and thus high dosage, and toxic overdose means lithium therapy can even be fatal. Cade wrote: "It is therefore extremely important that a patient receiving the maximum dose is examined daily and that nursing staff are instructed to look out for early signs of overdose."[46] However, the considerable side effects of lithium medication also lead to completely unjustified fears of poisoning when taking the very small amounts of essential lithium, especially as symptomatic treatment success is only achieved in around a third of patients despite the high lithium concentrations (the option of systemic causal therapy with much lower doses and potentially higher chances of success is discussed in Chapter 4). However, even when successful and the manic phase subsides with lithium monotherapy, the illness is by no means cured; lifelong maintenance medication is required to control symptoms.[47] Consequently, one typically doesn't refer to a therapy for bipolar disorder, but only as a treatment. In unipolar depression (major depression), high-dose lithium yields an even lower success rate, yet the side effects remain immense.[48] The poor ratio of side effect *costs* to treatment *benefits* stems, in my opinion, from the lack of systemic or causal therapy. Yet, this clear problem hasn't prompted a fundamental rethink; instead, it has spurred a lively scientific interest in lithium's mechanisms of action. This is because the hope of discovering lithium's target molecules or molecular targets is invariably linked to the hope of developing new, patentable pharmaceutical miracle cures— "magic bullets"—and of replacing non-patentable and therefore "much too cheap" lithium with more profitable preparations for lifelong treatment.

On the Trail of Lithiome

The first to be discovered, and therefore one of the most well-researched, lithium targets is glycogen synthase kinase 3 (GSK-3).[49] It was proposed as a direct target for lithium after the observation that this trace element, similar to insulin, stimulates glycogen synthesis.[50] Glycogen itself is a starch-like glucose storage molecule found primarily in liver and muscle cells, and to a lesser degree in fat cells. It was already known that insulin lowers blood sugar levels by inactivating GSK-3, so lithium must also inactivate GSK-3. Active GSK-3, on the other hand, inhibits the enzyme responsible for converting glucose into glycogen by phosphorylating it. Phosphorylation is the scientific term for the biochemical coupling of a phosphate group to a protein. Phosphorylation activates or inactivates proteins, depending on the attached phosphate group's impact on the enzymatic center's structure. When the GSK-3 kinase is active, it transmits a signal through phosphorylation, that says, "Stop the glycogen production!" In stressful situations, this is useful to raise blood sugar levels; after all, we need the energy to react (fight or flight).[51]

However, chronic stress and chronically overactive GSK-3 can lead to insulin resistance due to persistently elevated blood sugar levels, thereby increasing the risk of type 2 diabetes. Chronically elevated blood sugar, in turn, damages the brain and promotes neurodegenerative diseases like Alzheimer's—a first clear indication of how chronically hyperactive GSK-3 (under lithium deficiency) contributes to multiple lifestyle diseases simultaneously.[52] However, GSK-3 doesn't only affect blood sugar; it also phosphorylates many other proteins with diverse cellular functions. "Glycogen synthase kinase-3 (GSK3) may be the busiest kinase in most cells, with over 100 known substrates to deal with," write scientists from the Department of Psychiatry and Behavioral Sciences at the University of Miami's Miller School of Medicine.[53] Some even assume that there are well over two hundred GSK-3 targets, i.e. indirect lithium targets, as illustrated in Fig. 10.[54]

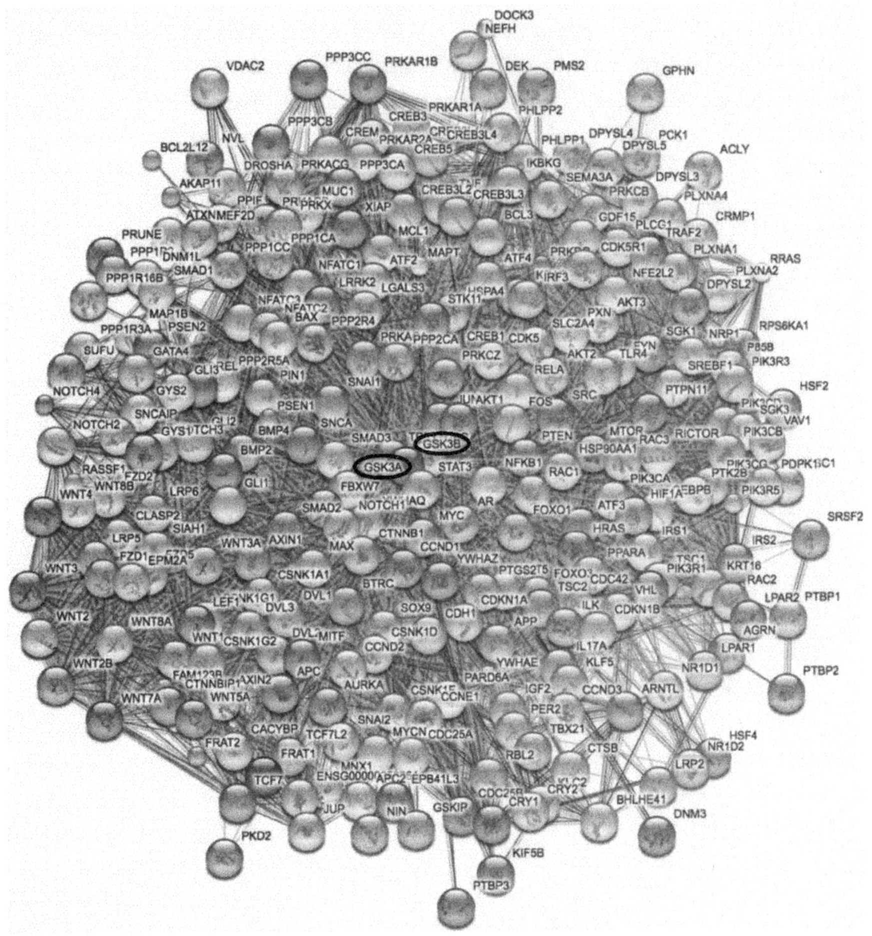

Figure 10

"Lithium direct targets are relevant for numerous cellular pathways involving a large variety of molecules," confirm systems biologist Magali Roux and geochemist Anthony Dosseto in their 2017 review. In it, they show that lithium is crucial for a variety of processes at all levels. This encompasses neuroprotection, protection against oxidative stress (through the activation of antioxidant mechanisms),[55] and mitochondrial energy production, which explains why, as stated, "it is rarely possible to establish a [simple] causal relationship between lithium targets at the molecular level and the resulting lithium effects at the system level."[56] Naturally, these complexities present poor prerequisites for those aiming to reduce lithium's effects to a single target and replace this trace element with a pharmaceutical "magic bullet."

It is already becoming clear why a lithium deficiency must have diverse effects: First, lithium acts on more than just GSK-3—a network of direct targets I'll refer to as the "Lithiome" from now on. Second, it also has a greater number of indirect targets, which are regulated by these primary targets. Third, these secondary targets then influence numerous other targets. Consequently, even highly sophisticated analytical and molecular biology methods cannot definitively link specific functions and effects of lithium to a single target, given that its action unfolds according to a multitarget principle. However, research is being continued to use lithium targets as pharmaceutical targets. The insights gained in this way continually support our argument regarding the essentiality of lithium in that we learn a lot about lithium's function and even more about the targets' essentiality. This leads to a very interesting thesis: if lithium-regulated targets have essential functions, then logically lithium must also be essential.

Somatic Cells Are Biological Information Processors. You can imagine each of our body cells as a kind of biological computer that constantly processes information. This occurs due to the individual body cell's inherent will to survive and to facilitate its coordinated interaction with all other body cells, for the benefit of the entire organism—always with the evolutionary imperative in mind. Each cell receives signals (messenger substances, nutrients, etc.) both internally and from its surroundings, processes them, and then relays its interpretation back to its environment via signaling molecules—essentially the data packets of cell biology. In this complex network of information analysis and transmission, there are several nodes where many signals converge, are biochemically evaluated, and the results are relayed. These node or key molecules are of utmost biological importance, and many of them are regulated by lithium. These functionally important nodes include many of the known lithium targets, not least the "intracellular information processor," GSK-3.[57]

All members of the Lithiome have in common the most important aspect: Lithium deficiency leads to their dysregulation. As a result, there are always negative effects on all physiological systems, which inevitably lead to developmental disorders and diseases. The entire Lithiome serves, as we shall see, the evolutionary imperative and the means to an end optimized for humans in the sense of the "Evolution of the Grandmother," i.e. longevity and the development and maintenance of the mental immune system's functionality into old age.

So, we have learned: The Lithiome comprises a large network of lithium targets, all of which seem to serve an overarching goal: to build up our brain and maintain its function and performance (and thus also that of the rest of the organism, see Chapter 6) into old age. These include cellular receptors, structural proteins, signaling molecules, gene regulators, and many more. Consequently, this subchapter could easily span thousands of pages if the aim were to present the entire current understanding of the Lithiome, target by target, including the infinitely complex interactions among its components. But that would be neither sensible nor effective. Since lithium affects all of its targets simultaneously, a systemic approach is ultimately necessary anyway. The significant advantage of its evolutionarily optimized effect on multiple targets simultaneously is that even small amounts suffice to achieve the naturally "desired" systemic health benefits. This stands in opposition to a pharmaceutical "magic bullet" which, being designed to target usually just one component regardless of the system's complexity, necessitates high dosages and consequently carries a high risk of side effects.

Thus, the magic of lithium as an essential multi-talent is its capacity to bring the interaction of all these targets into harmony, mirroring the species-specific optimization or perfection that has evolved—including for us humans. This is especially true for protein structures and their production; that is, our genes and their regulation, which have undergone optimization or fine-tuning for the presence and regulation of lithium through small evolutionary adjustments (random mutations) over countless billions of years for all species alive today. This interaction results in an overall function that no single target can account for: The whole is clearly more than the sum of its parts—much more! Therefore, there will never be a drug that can take over essential lithium's role. Even if it were possible to develop a lithium-like chemical that regulates a particular target similarly to lithium, this chemical won't be able to regulate all other lithium targets like lithium does; only lithium itself is this all-rounder. Lithium was there first. The targets have therefore adapted to lithium, not the other way around! Essential lithium lacks any undesirable effects that would justify trying to find a chemical with isolated similar effects. Ultimately, the justification for such attempts is once again merely the disdainful pursuit of profit by large pharmaceutical companies, which cannot take natural order into account because their business model opposes it.

The Complexity of the Lithiome

Despite the Lithiome's complexity and countless interactions always demanding a systemic perspective, I would nevertheless like to introduce some specific lithium targets or target classes below. Don't let their complicated names deter you; they aren't important for the following discussion and a deeper understanding. I mention them here solely to offer a starting point for readers who wish to delve deeper into the Lithiome from a scientific perspective. However, it is relevant to use a few illustrative examples to develop a basic understanding of how lithium, as an essential trace element, has exerted its health-constituting effect—from the very beginning of life on our planet. Thus, the Lithiome currently (as of spring 2025) encompasses six distinct molecular biological categories, each containing a few to numerous lithium targets (a detailed description can be found, for instance, in the article "From Direct to Indirect Lithium Targets: A Comprehensive Review of Omics Data"[58]):

1. Magnesium-dependent, lithium-sensitive protein kinases (e.g., GSK-3)
2. Magnesium-dependent, lithium-sensitive protein phosphatases (e.g., inositol monophosphatase, IMPase)
3. Magnesium-dependent, lithium-sensitive G proteins and adenylate cyclase
4. Magnesium-dependent, lithium-sensitive phosphoglucomutase (PGM)
5. Allosterically regulated enzymes and ion channels (e.g., AMPA and NMDA receptors)
6. Lithium-sensitive gene regulation: via DNA binding site modification (e.g., BCL-2) and transcription factor binding (e.g., CRE and AP1)

In what follows, GSK-3 and other lithium targets, such as IMPase, will continue to be relevant as we examine the physiological functions of lithium and the effects of lithium deficiency. Lithium-dependent and lithium-regulating ion channels (see point 5 in the list above), which govern the lithium concentration within our body cells, will be discussed in detail in the following chapter. However, this should not give the impression that these targets are more important than the others. The primary reason we have substantial material on these targets, unlike others, is that they were identified significantly earlier than lithium targets, allowing for the accumulation of considerably more scientific knowledge. Thus, a combinatorial search query (lithium plus the name of each lithium target as of late 2024) in PubMed, the database covering the entire range of biomedical publications indexed by the U.S. National Medical Library, yielded the following counts of publications:

- Lithium & Glycogen Syntase Kinase 3 (GSK-3): 1,383
- Lithium & Inositol Monophosphatase (IMPase): 270
- Lithium & Adenylate Cyclase: 268
- Lithium & AMPA: 86
- Lithium & Phosphoglucomutase: 36

The comparatively high number of publications on GSK-3 and lithium could stem from the fact that this lithium target, besides its early discovery and multifunctionality, is also considered "drugable." This implies the expectation that its enzymatic function and molecular structure make it amenable to being influenced or, in this case, inhibited by specifically developed chemicals (i.e., drugs). It is the hope that they could be at least as effective as lithium, the natural inhibitor of this target, that has enormously propelled GSK-3 research. This could be one reason why lithium is not recognized as essential and is even banned in many countries as a dietary supplement—otherwise, why would one go to the trouble of trying to replace it? Unfortunately, there is no end in sight, although the attempt is fundamentally doomed to failure. As no one seems ready to concede this, the absurd search for a substitute for the already optimal and adequately available solution that is essential lithium, continues.[59]

Lithium Reduces the Activity of Some of Its Key Targets

Lithium affects the Lithiome's different targets in a multitude of ways. GSK-3 and IMPase are two of the most well-studied lithium targets, a fact evident from the figures mentioned earlier. These two have become prime examples of how lithium manifests its activity within four of the six previously noted target classes of the Lithiome. They all have the following in common (see Figure 11):

1. They function as signal regulators when a magnesium ion is bound to their active site.[60] This highlights why magnesium is an essential trace element.
2. A lithium ion can displace the magnesium ion, not only due to a similar ionic radius and thus comparable size, but also because it fits perfectly into the same binding site.[61] This displacement naturally downregulates the otherwise consistently high activity of the lithium target, further establishing lithium as an essential trace element.

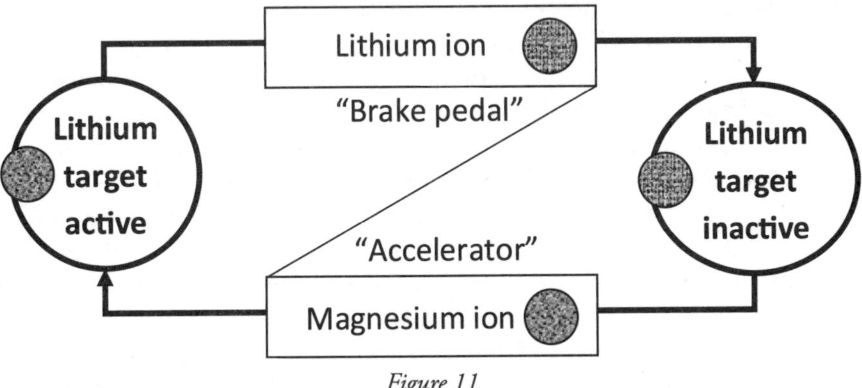

Figure 11

To understand this specific biochemical and essential function of the trace element lithium, I'd like to illustrate it with an "impractical" example. It is well known that the cultural and technological development of humanity parallels the biological evolution of life in its adherence to the same basic mechanisms. Random human ideas (culture) or genetic mutations (nature) are tested for their usefulness and either selected or discarded. But since we don't live in a world where cars only drive uphill, inventors have equipped cars not only with an accelerator pedal but also with a brake pedal. Cars that can only accelerate but not brake are so absurd that no reasonable person would even consider them. The same applies to the evolution of key proteins of life, as illustrated in Figure 11: If they had only an accelerator pedal (magnesium ions) and no brake (lithium ions), they, like cars without brakes, would only function well under conditions requiring "acceleration." But life is complex—living beings are constantly exposed to changing influences from different directions. To maintain balance (homeostasis), it is therefore necessary to have options for regulation in both directions. A medium level of activity is ideal. The reason this is achievable is that our body cells possess more than a single active copy of each respective lithium target. Were this not the case, one lithium ion displacing one magnesium ion would result in the entire system shutting down. Rather, thousands of copies of each lithium target exist in cells. The overall system does not have an absolute active or inactive state, but rather a variable mixed state consisting of an active part (bound magnesium ions) and an inactive part (bound lithium ions). In this sense, to extend the automobile metaphor, I'd like to describe lithium's action as regulating activity to a slightly active level. For comparison: a typical car engine is set to idle at around 800 revolutions per minute to ensure it remains drivable at all times, even when at rest, while minimizing fuel consumption and wear.

For lithium targets, the system would be hyperactive if the intracellular environment contained only magnesium ions and virtually no lithium ions. In this case, the system would operate at full power. But just as an engine overheats and sustains long-term damage from constant overrevving, so too does the organism suffer long-term damage if healthy homeostasis cannot be achieved due to a lithium deficiency. Consequently, a chronic, serious lithium deficiency leads to developmental disorders and life-threatening diseases or—continuing the car-without-brakes analogy—to unnecessarily increased wear and tear (e.g., neurodegeneration, Chapter 4) and even life-threatening accidents (e.g., life-threatening cytokine storms, Chapter 5). Given that most people have a relatively stable magnesium intake by default (as the mineral is sufficiently present in many foods), the system's overall activity depends almost entirely on daily lithium intake. But since we left the "Garden of Eden," where the mental immune system was able to develop to its full strength, the basic supply of lithium is usually no longer guaranteed (more on this under Argument 6).

Low Dose, Tremendous Effect? A previously unmentioned but potential objection to recognizing lithium as an essential trace element is that biochemical studies indicate lithium inhibits its targets (like GSK-3 or IMPase) with very low efficiency; this is also cited as an explanation for why treating bipolar disorder necessitates such high lithium blood levels. However, if this objection held true, it wouldn't explain why such minor differences in lithium concentration, as found in drinking water, can demonstrably have such a large impact on our mental health. This contradiction occupied me for a long time—until I penetrated deep enough into the matter.

The efficiency with which an active ingredient (such as lithium in this case) inhibits an enzyme or biochemical catalyst (which most lithium targets are) is expressed by the so-called IC50 value. The *inhibitory concentration*, i.e., the concentration of an inhibitory active ingredient that reduces an enzyme's activity to 50 percent of its original level, is investigated and measured in vitro using biochemical methods. Determined biochemically, the IC50 of lithium for GSK-3 is around 1 to 2 mmol/l (millimoles per liter), while for IMPase it's about 0.8 mmol/l.[62] For treating bipolar disorder, recommended blood lithium levels are approximately 0.6 to 1.2 mmol/l, with recent clinical studies indicating that 0.4 to 0.8 mmol/l may be adequate.[63] A blood lithium level of 0.8 mmol/l thus represents about 40 to 80 percent of the GSK-3 IC50 and roughly 100 percent of the IMPase IC50. This results in a pharmacological inhibition of the two targets by approximately 30 and 50 percent, respectively. However, these values only apply if the results from the test tube can be transferred to the situation in a living cell. (An explanation of the unit of measurement "mol" can be found as additional information in Chapter 3: *Lithium Levels in Molar Magnitude*).

To achieve a blood concentration of about 0.8 mmol/l when treating bipolar disorder with lithium, the usual daily dose is 1½ to 3 extended-release lithium carbonate tablets, containing approximately 18.3 to 36.6 mmol or 128 to 256 mg of pure lithium. In contrast, the lithium concentrations observed in the previously discussed epidemiological drinking water studies regarding the impact of lithium deficiency on the mental immune system and overall mortality were merely in the range of approximately ten to a few hundred µg/l. This equates to a lithium intake that is only about one-thousandth of those clinical doses, assuming a daily drinking volume of approximately one liter. This comparison highlights even more why we need to explain how such a relatively small amount can have a measurable effect at all. However, several individual explanations exist, which in my opinionhelp us understand lithium's biochemical effect even at comparatively low concentrations. The answer is regrettably complex, like the Lithiome itself:

1. The relatively high IC50 values for GSK-3 and IMPase measured in vitro don't reflect the natural cellular environment because the magnesium ion concentrations used in determining the IC50 were higher than those found in nerve cells, for example.[64] Visualizing this as a brake (lithium ions) and an accelerator pedal (magnesium ions), as in Fig. 11, it's like testing braking performance at full throttle. This doesn't reflect the natural situation; consequently, the braking performance is underestimated, and lithium's IC50 value is overestimated. Lithium is much more potent in cellular reality.
2. As we will see under Argument 3, lithium ions are enriched by approximately tenfold in tissues that produce nerve cells. This mechanism would significantly shift the ratio of magnesium ions to lithium ions in these cells in favor of lithium. This selective enrichment ensures that an otherwise serious deficiency is compensated for as effectively as possible in tissues with high demand.
3. As depicted in Fig. 12, all lithium targets are part of a complex network involving direct and indirect interactions (dashed arrows), where some targets are inactivated by lithium, while others, such as Akt (also a kinase), are directly activated (number 5).[65] Some of these lithium targets, like active Akt (target D), in turn inhibit other lithium targets such as GSK-3 (number 6), thereby further enhancing the overall trace element effect—akin to a brake booster, to revisit the car metaphor.[66]
4. Beyond the competitive inhibition observed in the test tube, lithium also blocks the biochemical self-activation of GSK-3 within living cells (see Fig. 12, Number 2).[67] This also means that small amounts of lithium have biological effects in living cells.
5. Some GSK-3 reactivating enzymes, such as certain phosphatases, are also directly or indirectly inhibited by lithium targets, which makes intracellular self-inactivation more permanent and thereby further enhances the lithium effect (Fig. 12, numbers 3 and 4).[68]

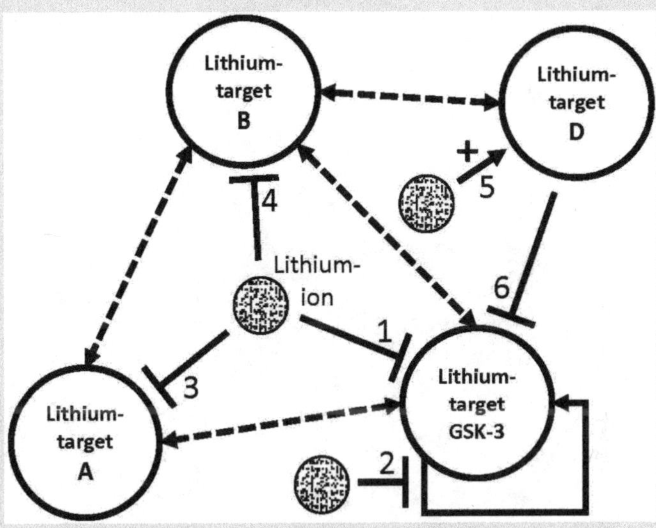

Figure 12

The influence of lithium on a lithium target is therefore not a single effect, but a network effect in the entire lithiome, which leads to an enormous increase in effect.[69] Therefore, the intracellular inhibition at therapeutic doses is probably at least one, if not two, orders of magnitude more pronounced than the IC50 values determined in the test tube would suggest. This interplay might also explain the constant threat of toxicity associated with lithium treatment for bipolar disorder. In any case, the molecular genetic elimination (a genetic knockout, as opposed to a pharmacological knockout with high-dose lithium, for example) of lithium targets such as GSK-3 or IMPase in animal experiments is not compatible with life (more on this later). The concerted action of the entire Lithiome could explain why the human organism is so sensitive, on the one hand, to lithium deficiency as well as to the small concentration differences observed in drinking water studies.

The Lithiome at the Center of Life

Lithium regulates all basic biological functions via the interactive Lithiome in a network-like manner, and Fig. 13 illustrates some of these examples. All of them are closely coordinated through an evolutionary optimization process. All of them play a central role in the regulation of adult hippocampal neurogenesis in diverse and synergistic ways. These biological functions, and thus lithium itself, are essential for the mental immune system's development and lifelong maintenance.

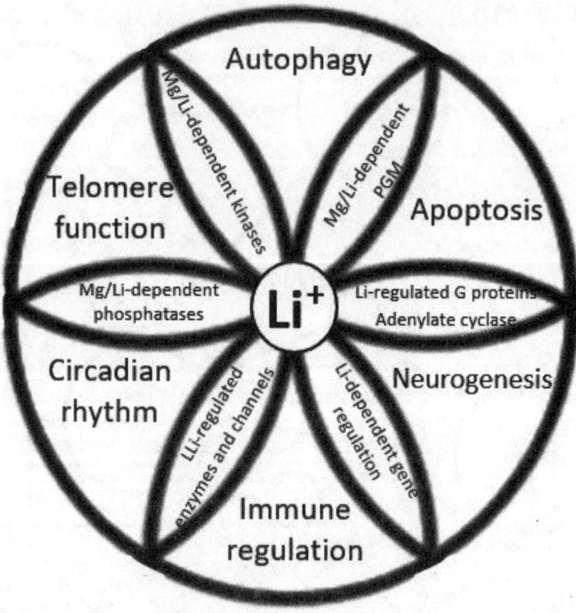

Figure 13

Autophagy: The word autophagy comes from the ancient Greek *autóphagos* and literally means "self-consuming." What sounds horrible is vital here. Specifically, this involves the breakdown of misfolded and potentially toxic proteins, as well as aged cell organelles like mitochondria (the powerhouses of our cells), when they no longer function optimally. The controlled destruction of damaged mitochondria is closely linked to the production of new mitochondria, a process known as mitochondriogenesis.[70] Mitophagy and preventing the cell-damaging accumulation of dysfunctional proteins, in conjunction with mitochondriogenesis, is crucial for maintaining neuronal energy metabolism and cell function; consequently, they are also the foundation of psychological and mental performance. These processes are important for all body cells, but especially for our nerve cells because they are irreplaceable. This is due to their often having several thousand to hundreds of thousands of highly functional connections (synapses) with each other and therefore preventing their replacement with new nerve cells without serious functional impairment. This complex synaptic network represents our individuality, which is why our nerve cells must be capable of lasting as long as we live. This is only possible through the process of intracellular autophagy. As we see in Chapter 4 in detail, autophagy that does not function optimally plays a decisive role in most neurodegenerative diseases, such as Alzheimer's or Parkinson's. Ultimately, this can be traced back to a hyperactivity of the lithium targets GSK-3[71] and IMPase[72] under today's "normal" conditions, a hyperactivity that cannot be logically explained without lithium deficiency. Only a sufficient supply of lithium ensures that activity is regulated to a healthy (homeostatic) level. An unnatural and unfortunately "normal" high activity of GSK-3 and IMPase due to "normal" lithium deficiency thus leads to the blockage of two mutually complementary molecular signaling pathways, inhibiting the vital process of autophagy.

Essential lithium would naturally unlock these synergistic signaling cascades that stimulate these vital cell-, tissue-, and brain-rejuvenating processes.[73] Lithium thus promotes cellular self-cleansing, which, as explained above, is crucial for our mental immune system, as it keeps our non-replaceable nerve cells functional into old age. Together with the influence on adult hippocampal neurogenesis, this explains why the brain is particularly sensitive to lithium deficiency and why the drinking water studies (Argument 1) reveal the full disease spectrum of mental immunodeficiency.

Apoptosis: Our genome contains a programmed cell suicide mechanism that triggers apoptosis, a term originating from the ancient Greek *apóptosis*, meaning "to fall off." The process is actually as unspectacular as leaves falling in the autumn wind. The suicidal somatic cell digests itself from within and quietly disappears. Like the autophagy of intracellular organelles, such as dysfunctional mitochondria, apoptosis ensures that the organism is freed from dysfunctional cells, and that their components can be returned to (recycled within) the metabolism. Mitochondrial dysfunction (due to deficient mitochondriophagy or impaired autophagy, among other things) is one of the triggers for the cellular suicide program.[74] This demonstrates that these two systems (autophagy and apoptosis) form a functional unit via the Lithiome.[75] Even unneeded or no longer needed cells activate the apoptosis program, reflecting the "use it or lose it" principle because our organism is optimized for economy.

In my book *The Alzheimer's Lie*, I specifically address the importance of apoptosis for our brain development and self-realization: "Overall, approximately twice the number of neurons and synapses are formed in the first years of life as we will need later in life. This ensures that sufficient neurons are available for all vital adaptations (all possible manifestations of our future self). This initial surplus guarantees our theoretical capacity to become anything we desire (or require), enabling us, for instance, to acquire any language, along with any faith. Whether we grew up in the Stone Age or today, whether in the Middle East, Africa, or modern North America: The human brain is prepared for anything; it simply needs to maintain the neurons and synapses it's using, as selected through an evolutionary adaptation process, while all others are eliminated by apoptosis. It is our individual experience that makes synapses survive or disappear, become stronger or weaker. Which these are is determined by what our brain experiences. Ultimately, it's this resulting unique synaptic combinatorics that makes us who we are and determine how we feel, think, and act. An essential aspect of our self-development therefore consists of removing all superfluous neurons unable to establish functional connections with others—or their self-removal."[76] In particular, the highly vulnerable neurons newly formed through adult hippocampal neurogenesis must be protected from apoptosis until they are protected by new experiences that they store through synaptic connections with other neurons and thus validate their existence.[77] If not needed, they disappear again.[78] Lithium inhibits apoptosis via its target molecules such as GSK-3 and thereby increases the survival chances of these young neurons until they are integrated into the hippocampal network.[79]

Immunoregulation: Infections inhibit adult hippocampal neurogenesis. The mediators responsible for this are pro-inflammatory messenger substances, such as Interleukin-1β (IL-1β),[80] Interleukin-6 (IL-6),[81] or tumor necrosis factor α (TNF-α)[82] which is also highly upregulated in *acute* inflammation.[83] The evolutionary biological advantages of this mental immune system inhibition in acute illnesses are described in Chapter 4, as are the resulting disadvantages in chronic inflammation when these mechanisms permanently block adult hippocampal neurogenesis. Then there is a risk of neuropsychiatric developmental disorders, depression, and even fatal courses such as Alzheimer's disease.[84]

Telomere Function: Bacteria-like single-celled organisms were most likely the first living organisms on Earth (see Fig. 17, Argument 4). They lack a nucleus, and their genetic material exists as one or more circular chromosomes. In contrast, eukaryotes (from ancient Greek *eu* for "right" and *karyon* for "nucleus") have a cell nucleus that contains their genetic material. These include the other four of the five major kingdoms: Protists (unicellular animals), plants, fungi, and all animals. The genetic material of all eukaryotes, including humans, consists of linear chromosomes with ends known as telomeres. Telomere is derived from the ancient Greek *télos* for "end" and *méros* for "part." The latter refers to their repetitive character: They consist of a large number of many identical, repeating sections or sequences of DNA. These serve a specific function, the necessity of which arises from the mechanism by which eukaryotic genetic material replicates prior to cell division. Since the necessary enzyme (a DNA polymerase) can't start copying directly at the end, chromosome ends shorten with each cell division (bacteria with their circular chromosome don't have this problem). Due to the high number of repeating sequences, telomeres protect the genetic program between them from degradation during cell division. Therefore, in germ cells, from which new generations have continually emerged since the beginning of life, a special enzyme complex called telomerase ensures that telomeres always retain their original length, allowing every new living being to start life with the longest possible telomeres. Without telomerase, each cell division would result in a successive telomere shortening until this ultimately destabilizes the entire genome and leads to a halt in cell growth, known as senescence (from the Latin *senescere*, "to grow old, to age"), and ultimately even to cell death (apoptosis).[85] After about fifty to fifty-two cell divisions, the "Hayflick limit," named after the US gerontologist Leonard Hayflick (1928–2024), is reached.[86] This makes it clear: if germ cells were unable to maintain their telomere length through telomerase activity, the Hayflick limit would also apply to them, and the evolution of eukaryotic life would have ended after only fifty to fifty-two generations.

Not only germ cells, but also the somatic cells that develop from them and form our organism require telomerase. The theory states that the longer the healthy telomeres are maintained, the higher the life expectancy. The telomere length in our somatic cells is therefore a measure of our biological age, influenced by our lifestyle through telomerase activation.[87] Telomere shortening isn't necessarily an indicator of unavoidable chronological aging but primarily of biological aging, which we can influence to a certain extent—after all, no one lives forever.[88] Once telomeres reach a critical length, our ability to replace old or damaged cells diminishes, increasing the risk of many diseases we've previously attributed solely to aging itself, rather than the way in which we age.[89] A supposedly inevitable shortening of telomeres in our somatic cells with each cell division, and thus rapid biological aging, also contradicts the concept of the Evolution of the Grandmother, which suggests we can still contribute meaningfully to family survival in old age—a contribution largely hindered by the many supposedly "age-related" diseases that are now an unnatural norm. The vital substance lithium also regulates telomerase via its specific targets and thus ensures telomere lengthening. In this way, it creates a basic prerequisite for longevity vis-à-vis the "Evolution of the Grandmother."[90]

Somatic cell senescence results in a senescence-associated secretory phenotype (SASP), marked by an elevated release of pro-inflammatory messengers and numerous other bioactive compounds. Besides its role in cancer, this phenotype also encourages neurodegenerative processes. In fact, it has been shown in human brain cells that microdosed lithium directly suppresses both senescence (by activating telomerase) and the SASP.[91] This is crucial for maintaining the functions of the mental immune system, which is centered in the hippocampus and relies on productive adult hippocampal neurogenesis. Therefore, in order to remain cognitively competent into old age, the hippocampus must be able to form new neurons throughout life. This is why, unsurprisingly, it has the highest telomerase activity in the brain.[92] Longer telomeres also correlate with hippocampal volume. For example, one study shows that telomere lengthening increases the ability of human hippocampal progenitor cells to divide, which has a positive effect on human cognitive functions and helps to keep the mental immune system intact into old age—also in the spirit of "the Evolution of the Grandmother."[93] In an animal model of depression, it was shown that dysregulation of telomeres in the hippocampus increases the likelihood of depressive behavior.[94] In humans, it has been confirmed that a loss of function of the mental immune system is the result of a loss of hippocampal telomerase activity due to an inappropriate lifestyle.[95] Current longitudinal data also support this fatal correlation and show that a greater shortening of telomeres in older people is associated with a greater loss of hippocampal volume.[96] In this context, it should also be mentioned that bipolar disorder, for example, is associated with shortened telomeres and that lithium has a positive influence on this disease by activating telomere lengthening.[97] Shorter telomere lengths correlate with the development of anomalies in brain structure and are functionally partly responsible for mood disorders.[98] So that adult hippocampal neurogenesis and thus the mental immune system at least do not suffer from a short-term diet-related lithium deficiency—and this underlines the evolutionary significance of these functions—this nerve cell–forming region has the ability to selectively enrich this trace element (See Argument 3).[99]

Circadian Rhythm: Lithium regulates the circadian rhythm through so-called "clock proteins," which refers to our organism's ability to synchronize all physiological processes to a period of about twenty-four hours. The center of the circadian system is the suprachiasmatic nucleus: This refers to a cluster of nerve cells situated directly above (Latin: *supra*) the crossing point (ancient Greek: *chiasma*) of the optic nerves. Through them, the central circadian regulator receives information about daylight in order to synchronize our internal clock with the day-night rhythm. Specifically, the lithium target GSK-3 seems to play a pivotal role, since it regulates numerous other clock proteins and is therefore fundamentally important for the periodicity of the internal clock mechanism.[100] Dysregulation of GSK-3 leads to disruptions in the circadian rhythm, causing, among other issues, sleep disturbances and, consequently, hippocampal disorders. These, for example, increase the risk of developing Alzheimer's disease. Furthermore, various clock proteins involved in the generation of the body's internal rhythm have also been causally associated linked to the development of bipolar disorder. Lithium also influences our sleep quality in this manner, which is, in turn, vital for normal adult hippocampal neurogenesis and consequently for a functioning mental immune system, offering protection against Alzheimer's disease.[101] Furthermore, lithium's regulatory effect on the circadian rhythm may be a critical component of its therapeutic action in patients with bipolar disorder; indeed, a dysfunctional sleep-wake cycle is considered a primary factor in this mental illness.[102]

Neurogenesis: Lithium controls the formation of new nerve cells in the brain.[103] This occurs both indirectly through inhibiting neuroinflammation (see above and, in detail, in Chapter 4), but also directly: Lithium activates both nerve growth factor (NGF)[104] and another key growth factor for brain nerve cells, brain-derived neurotrophic factor (BDNF).[105] Both are not only crucial hormonal activators of adult hippocampal neurogenesis but also protective factors against apoptosis.[106] In contrast, when GSK-3 is hyperactive or not naturally inhibited by lithium, it blocks the release of BDNF and, via this mechanism, also suppresses adult hippocampal neurogenesis, resulting in inadequate functioning of the mental immune system and a potential host of lifestyle diseases.[107] Thus, a Lithiome dysregulated by lithium deficiency inhibits adult hippocampal neurogenesis and promotes cellular suicide. This is particularly problematic for the new hippocampal neurons until their complete integration into the hippocampal network protects them from the activation of the self-destruct program.[108] Neurons are, in a sense, "social beings" that possess a drive to survive as long as they are synaptically connected to other neurons and receive a signal through them indicating: "we need you."

Lithium at the Core of the Mental Immune System

All biological functions of lithium establish the fundamental conditions for life itself, as well as for healthy brain development and the preservation of our mental abilities into old age. This is particularly true for lifelong productive adult hippocampal neurogenesis. Consequently, lithium ultimately exerts fundamental effects on all aspects of an effective and functioning mental immune system via the Lithiome and its functions, as schematically shown in Figure 14 using the lithium atomic nucleus (see Essential Opening Remarks).

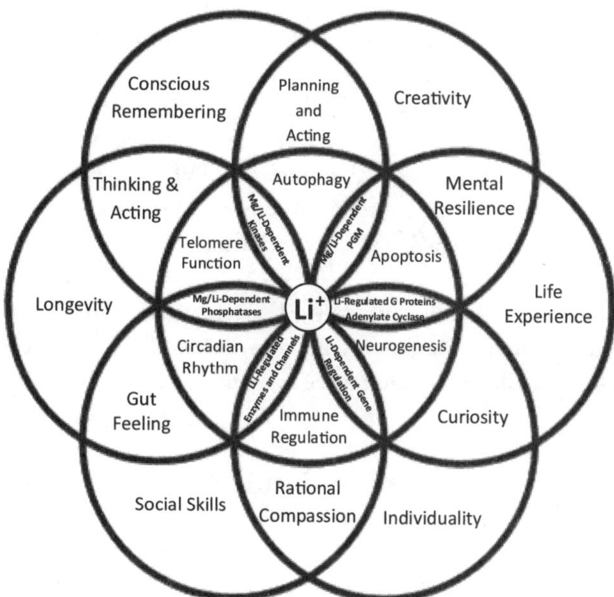

Figure 14

The Fundamental Laws of Life must of course also be observed here: Numerous deficiencies and various pollutants can impair the crucial functions mentioned here, as well as many others, and consequently affect the mental immune system. However, even when we meet all recognized natural needs and minimize known toxic influences, an existing lithium deficiency, often fatally undetected, can still lead to a dysregulation of the Lithiome. I would like to emphasize once again that essential lithium is not a panacea and should therefore always be seen in the context of a systemic understanding of human nature. Nevertheless, because its importance as an essential trace element is unrecognized, a lithium deficiency is a primary factor in the dysfunction of the mental immune system for many people today. The significant influence of lithium on its targets, in the context of an existing, often severe deficiency, explains the enormous impact that even small differences in lithium concentration in drinking water have on our mental health, as detailed in the previous subchapter based on epidemiological studies.

Thus, a complete causal chain exists, extending from essential lithium through the trace element-regulated Lithiome and its various functions, ultimately leading to the development and function of the mental immune system. This makes it easy to explain why a chronic lithium deficiency has such devastating effects on mental health and mental performance. This also holds true for the numerous pathophysiological effects that, depending on when the deficiency occurs, can lead either to a disruption in hippocampal development or to a deactivation of adult hippocampal neurogenesis—often both. This also highlights how the issues grouped under the term "mental immunodeficiency syndrome" have unpleasant consequences not only for the individuals affected but also for society as a form of "social immunodeficiency."

Conversely, however, this means that all these neuropsychiatric developmental disorders, behavioral abnormalities, neurological and neurodegenerative diseases could very likely be avoided through essential lithium as part of a species-appropriate lifestyle, as we will discuss in detail in Chapter 4. Argument 2 for the essential importance of lithium is also well substantiated. We can conclude:

Dysregulation of the Lithiome caused by a deficiency of essential lithium is causally linked to neuropsychiatric developmental disorders and mental and even neurodegenerative diseases, which we summarize as mental immunodeficiency syndrome.

ARGUMENT 3: Lithium-Ion Distribution Regulated According to Demand

Lithium Accumulation in Neurogenic Zones

Essential trace elements are characterized by their absolute necessity for vital biological functions. Despite daily and diet-related turnover fluctuations (due to increased or decreased intake or excretion), the organism must be able to maintain a relatively stable concentration in vital organs. Consequently, the concentrations of ions such as sodium, potassium, calcium, iron, chloride, or magnesium in the blood and extracellular fluid are necessarily strictly regulated by highly selective channel systems and active transporters in cell membranes. However, it can be shown that lithium distribution within organisms is also selectively regulated. This is a basic prerequisite for the essential function of this previously disavowed trace element. It was already known at the end of the nineteenth century that a natural accumulation of lithium occurs in human embryonic tissue. In a 2002 publication, Gerhard Schrauzer already suspected that this phenomenon should be understood as evidence of lithium's essential role, when he wrote: "The fact that embryonic lithium concentrations are highest during early fetal development suggests that it is specifically needed." But what could be the reason for this increased requirement? Since embryonic and especially fetal cells have a high cell division rate, they naturally require high telomerase activity.[109] Lithium activates telomerase via proteins of the Lithiome, e.g. by inhibiting GSK3, as discussed in detail in the previous chapter.[110] Animal studies have also shown that lithium plays a role in enlarging the stem cell pool (which is only feasible with high telomerase activity) from which all somatic cells develop.[111] In mice, lithium thus stimulates the proliferation of stem cells and progenitor cells of the hematopoietic system.[112] Without functioning telomerase and sufficient lithium, and the associated immediate repair of chromosome ends after each cell division, fewer and fewer new cells would be formed that could contribute to organ growth, and no fetus could develop healthily. This is why an artificially induced lithium deficiency had fatal consequences in animal experiments: In pregnant rats and goats experiencing lithium deficiency (though not yet absolute!), significantly fewer offspring were born, and as expected, these had an enormously reduced birth weight, as we will discuss in more detail later.[113]

In these animal experiments, organ-specific enrichment was investigated, among other things, in the few surviving offspring that continued to be kept on the low-lithium diet. Yet, the first generation of lithium-deficient animals showed only comparatively moderate decreases in concentration in certain organs. In the brain tissue (and thyroid) of these animals, there was even an overcompensation, i.e., the lithium concentrations were even slightly higher than in animals that had not grown up with lithium deficiency; finally, due to the extreme deficiency situation, there was also a decrease in concentrations there from the second generation onward. This is a strong indication that brain function is lithium-dependent and that nature employs a very efficient active transport mechanism to ensure that lithium's regulatory influence in these tissues is not immediately lost, at least until the deficiency is resolved, even with a dietary lack of lithium. However, it also shows that less of the essential trace element is excreted under lithium deficiency, otherwise the organ-specific decreases in concentration would have been much more pronounced. There must therefore be an active return transport of lithium in the kidneys.

The lithium concentration in rocks, soils, and fresh water varies greatly from location to location.[114] If lithium is therefore essential, it would have to be assumed that organisms have developed the ability to keep the lithium concentration in the organs or cells in which it performs its functions reasonably stable even with fluctuating supply; goats and rats have clearly already achieved this under extreme experimental conditions. The decisive factor for lithium's effect, given its intracellular site of action, is not the blood concentration (which drops relatively quickly when there is a deficiency), but the cell concentration, which is evidently maintained at a higher level through regulation—at least until the lithium-ion transport system is overwhelmed by the persistent deficiency in the blood. The blood concentration, which we will discuss in more detail in the next chapter, is therefore only a surrogate parameter (from the Latin *surrogatum* for substitute). This means it's an easily measurable proxy for the difficult-to-determine cell concentration, which is what truly matters—but due to the regulatory mechanisms described, it only provides limited information about the intracellular lithium concentration.

Therefore, compared to most other essential trace elements, such as selenium or zinc, lithium has a very wide "essential breadth" due to this regulation of cellular concentration (we will discuss this in more detail later). In fact, depending on food supply, lithium intake levels can fluctuate between a few µg and a few mg, representing a roughly thousandfold difference (that is, three powers of ten or orders of magnitude). Within this relatively wide range, regulatory systems ensure, at least short-term, that intracellular concentrations don't dramatically change, thus preventing significant alterations in the control of key bodily functions that shouldn't depend on daily dietary lithium fluctuations. However, intake quantities in the medicinal range of several hundred mg are completely unnatural and overwhelm these systems, which can even result in severe poisoning.

The ability to accumulate lithium in tissue did not first appear with evolution in goats and rats; it already occurred with marine life, as mussels and fish are also able to do so. This already points to an answer to the question, which we will explore in detail later, of where and when in the evolution of humans and their mental immune system a sufficient supply of lithium was still naturally guaranteed. Here is just this much: While the average lithium concentration in seawater is around 0.175 mg/l, levels as high as 2.8 mg/kg have been found in the head and spinal column (or brain and spinal cord) of species like sardines and anchovies. This represents an approximately sixteen-fold selective enrichment compared to seawater and up to a twenty-fold enrichment of lithium compared to muscle tissue—where the lithium concentration is even slightly lower compared to seawater.[115] Therefore, nervous tissue appears to have already developed an increased lithium requirement in marine organisms, suggesting that those primitive fish with random mutations leading to a correspondingly efficient lithium transport system may have gained a selective advantage. It could even be assumed that the development of more powerful brains was contingent on the evolution of efficient lithium transport systems. A 2017 study on mice very impressively shows that highly selective brain region-specific lithium accumulation occurs in mice. The accumulation was shown particularly for those nerve cell areas in which active neurogenesis still takes place in adult animals.[116] In mice, as in humans, these regions include the dentate *gyrus dentatus* in the hippocampus. In mice, the subventricular zone (SVZ) is also included. New neurons are formed there even in adult animals.[117] From there, both new, not-yet-integrated neurons and their precursor cells, which continue to divide or multiply, migrate via the so-called rostral (forward) migration stream (RMS) into the olfactory brain, where they contribute to olfactory memory. Following the administration of lithium at dosages typical for the treatment of bipolar disorder, all four neurogenic brain regions in mice show an approximately tenfold higher—and thus selective—accumulation of the trace element compared to other brain regions, such as the cerebellum, where no adult neurogenesis occurs (see Fig. 15).

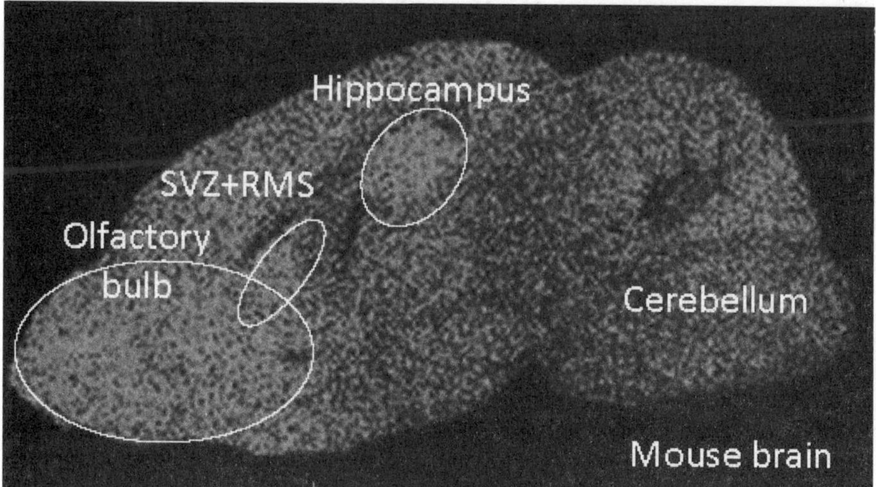

Figure 15

The brain regions where adult hippocampal neurogenesis occurs are also the central areas for the development and maintenance of the mental immune system's function. A corresponding accumulation of lithium in the area relevant to humans was demonstrated in 2020.[118] Here, a relatively higher accumulation was even shown in the left hippocampus as compared to the right. We will see why this is so shortly, but for now we can conclude from this fact: The accumulation of lithium in certain areas of the brain is not random. This finding therefore supports the "hypothesis of active lithium transport in the brain," according to the study's authors.[119] The key word for understanding the significance of this finding is "active," because selective accumulation through ion transport requires energy. A fundamental rule of evolutionary biology states that a mechanism for which diverse organisms at various evolutionary stages have continuously expended energy for hundreds of millions of years can only persist over such an extensive period if no mutation can eliminate it without impairing the organisms' ability to survive or jeopardizing their existence. As this active accumulation has been proven not only in humans, but also in numerous other animal species, it can be assumed that this is accompanied by a selection advantage.

The active accumulation of lithium, especially in the left hippocampus, obviously serves an evolutionary biological purpose. A clue as to what this might be good for can stem from taking a closer look at the functions of the left hippocampus as compared to those of the right, as shown below.

Directional Lateralization of a More Efficient Brain

Directional lateralization occurs when organs, functions, or activities (e.g., linguistic processes in the brain) tend to be located, either predominantly or entirely, on one side of the body. While humans, like most animals, present an external symmetry, their internal organs, including the liver, spleen, and intestines, are "lateralized."[120] In her article "Brain Lateralization and Cognitive Capacity," Lesley J. Rogers, professor emeritus of brain and behavioral research, explains that the brain's symmetrical appearance caused a long-held assumption that the left and right hemispheres were the same and functioned identically. But this was a misconception, writes Rogers: "From the small brains of insects to the different sized brains of vertebrates, including humans, the left and right sides of the brain process information differently and control different patterns of behavior."[121] According to Rogers, this lateralization avoids the duplication of computational processes and thus increases overall cognitive capacity.

It is irrelevant which brain hemisphere performs which function, and yet a "directional" tendency seems to have developed in humans. Even if every person's brain is different and the supposed rules only reflect statistical tendencies, it can be roughly assumed that in most people the left hemisphere of the brain tends to consciously process rational, linguistic, analytical, temporally linear, and logical processes. The right hemisphere, on the other hand, processes stimuli holistically, pictorially, intuitively, timelessly, but in a spatial, emotional and more physically oriented way.[122] This tendency is also reflected statistically in the functions of the two hippocampi. It is therefore assumed that the human hippocampus' lateralization is primarily due to the cerebral hemispheres' functional lateralization. Thus, while the left hemisphere is responsible for language processing, the right hemisphere makes a greater contribution to visual attention. When semantic information becomes relevant, the left hippocampus becomes active. If spatial information is important, the right hippocampus supports its processing.[123] The right hippocampus appears to be primarily involved in spatial memory while the left hippocampus is more involved in situation-dependent memory, whether episodic or autobiographical. Thus, the right hippocampus is considered a repository for spatial relationships, whereas the left hippocampus' role is storing relationships between different linguistic content in the form of narratives.[124] This is also confirmed by studies on patients with either unilateral or bilateral hippocampal lesions.[125] Accordingly, the authors write in the article "Lateralization of the Hippocampus: A Review of Molecular,

Functional, and Physiological Properties in Health and Disease," published in 2023, that both sides of the hippocampus function differently and are affected differently by neuronal diseases.[126]

Left Hippocampus: Center of the Mental Immune System?

Through its contribution to conscious, rational, and linguistic thinking, the left hippocampus (due to directional lateralization) appears to be more crucial for the mental immune system's function than the right. Although the study data showed, on average, an increase in lithium accumulation in the left hippocampus, in rare cases—and I would like to emphasize this again—the right hippocampus can take over this function. However, this is rather rare. This is consistent with the clinical observation that, in neuropsychiatric disorders, the left hippocampus exhibits abnormalities statistically more frequently than the right. This suggests that the left hippocampus—which, as we have seen, accumulates the most lithium most consistently—also plays a more crucial role in a functioning mental immune system and thus a significant part of people's cognitive capacities. In contrast, if the body actively accumulates lithium in this region, this provides an explanation for why a lithium deficiency causally contributes to mental immunodeficiency syndrome.

Regarding the question of a causal relationship between lithium deficiency and the disorders that I summarize under the term mental immunodeficiency syndrome, it is interesting that the authors of the human study point out that their functional imaging of the left hippocampus identified precisely the subcortical (beneath the cerebral cortex) structure that plays a central role in current neural models of emotional dysregulation in bipolar disorder.[127] The authors also state: "The identification of a consistently high normalized [calculated using statistical methods] lithium concentration specifically in one region [left hippocampus] is therefore an interesting finding and supports the hypothesis that lithium promotes neurogenesis." And the authors of the corresponding study in mice also write: "The reason why lithium preferentially distributes to limited brain regions is unknown, but [. . .] our findings support the hypothesis that lithium regulates the homeostasis of neurogenic [nerve cell-producing] regions, thereby contributing to a better understanding of the elusive mechanisms by which lithium acts in the brain."[128]

As previously discussed, one of the many lithium functions that could play a role here is the lithium-dependent activation of telomerase.[129] In their 2019 article, "The Polygenic Nature of Telomere Length and the Anti-Ageing Properties of Lithium," the authors argue "that lithium may be catalysing the activity of endogenous mechanisms that promote telomere lengthening."[130] A 2020 review article provides a mechanistic explanation for this link: "Collectively, our results suggest that telomere shortening could represent a mechanism that moderates the proliferative capacity of human hippocampal progenitors, which may subsequently impact on human cognitive function and psychiatric disorder pathophysiology."[131]

This further indicates why lithium accumulation in this specific brain region is functionally important for a healthy mental immune system. In particular, the volume of the left hippocampus is usually significantly smaller on average compared to healthy controls.[132] These findings are supported by the previously mentioned studies, which demonstrated that patients with bipolar disorder treated with lithium had a larger left hippocampal volume than those who received no or alternative treatment.[133] This is of great importance because, as a German multicenter research team found in 2022, it is very likely that not only bipolar disorder stems from a hippocampal maturation and dysfunction disorder potentially linked to lithium deficiency; major depressive disorder, schizophrenia, and schizoaffective disorder also share overlapping symptoms, risk factors, and biological features, despite initially appearing as distinct clinical entities.[134] According to their research, the common neuropathological denominator is a selective volume loss of the left hippocampus compared to healthy controls. In order to identify the common cause of neuropsychiatric disorders—such as bipolar disorder, depression, and schizophrenia—and to better understand these disorders, it is now important to focus on the similarities between them rather than the diagnostic distinctions. In addition to the selective developmental disorder and/or shrinkage of the left hippocampus, the selective lithium accumulation that occurs there is another common denominator. If these two phenomena are indeed causally related—which appears to be the case based on the Lithiome's function—this would provide a further explanation of how and why lithium deficiency causes mental immunodeficiency syndrome and also point to the most obvious therapeutic approach. The list of neuropsychiatric developmental disorders and mental illnesses within the mental immunodeficiency syndrome spectrum, in which the left hippocampus experiences a greater volume loss compared to the right, is considerably longer. These include:

- Autism Spectrum Disorders (ASD)[135]
- Bipolar Disorder
- Schizophrenia
- Conduct Disorders[136]
- Severe depression (Major Depressive Disorder)
- Post-Traumatic Stress Disorder[137]

We will discuss some of these hippocampal dysfunctions in more detail in Chapter 4 in the context of systemic or causal prevention and therapy. In my opinion, the highly interesting question that still needs answering at this point is: Are lithium-specific transporters known?

Lithium Transporters and Lithium Channels

The authors of the review article, "Towards a Unified Understanding of Lithium Action in Basic Biology and Its Significance for Applied Biology," write, "It appears that lithium has had a significant effect on first molecular and then biological evolution, and that major elements of lithium transport and modulation of biochemical function are shared across all classes of living systems."[138] Indeed, lithium may have played an essential role in the origin of life. The prerequisite for this is that corresponding lithium transport mechanisms had already developed at the beginning of evolution. Their development could even have been a necessary prerequisite for life's emergence. Can this thesis be presented credibly based on our knowledge about life's origin? The theory that the precursor to life consisted of networks of self-replicating RNA molecules is now gaining ground.[139] This idea is plausible from a molecular biology perspective because RNA is an all-rounder. Although DNA, the material of our genome, only stores information and proteins, the blueprints for which are encoded in genes, typically have only structural or enzymatic functions without encoding information, RNA can do both: store information and perform enzymatic functions. In such a primordial world, initially only biochemically active, self-replicating, and to a certain extent "living" RNA would occur—it would be the precursor of cellular life as we know it. According to the theory, the self-production and self-replication of sufficiently long RNA molecules would be crucial for the emergence of such a prebiotic RNA world. Simulations of such self-catalyzed RNA production have shown that lithium is better suited as an essential co-factor than sodium.[140] Lithium could indeed have played a decisive role at the very beginning of life-forming processes. This may then have continued with the emergence of cellular life with the first primordial cell (see Argument 4). However, lithium had to be able to penetrate cell walls or cell membranes to participate in intracellular processes. And indeed, there is now good evidence that bacteria can absorb lithium from an aqueous medium and thus possess active transport systems for this trace element.[141]

For some of these bacterial lithium transporters, corresponding orthologous family members have also been identified in humans.[142] Orthologous proteins or their orthologous genes are derived from two species' common ancestors. They generally possess highly conserved sequences and typically retain the same or a similar function, as we'll see in more detail with further examples concerning the Lithiome in Argument 4. The existence of orthologous lithium transporters throughout the animal kingdom should no longer surprise us, considering that lithium can be actively concentrated in the neuronal tissue of fish, mice, and humans, for example.

As discussed at the beginning, this ability to regulate lithium concentrations in a tissue-specific manner is crucial from an evolutionary biology perspective, especially in acute but temporary deficiency situations. This condition alone allows vital lithium functions to be sustained, even in the event of diet-related fluctuations in intake. The extent to which lithium transport disorders lead to disease isn't yet definitively clear, but it remains a possibility. Indeed, evidence for this exists: Well over half of bipolar disorder patients, for example, show no or only a partial response to lithium therapy, despite their serum lithium concentration being within the therapeutic range. According to a French group of neuropsychopharmacologists, a change in lithium distribution, e.g. due to "regional and variable transport mechanisms of lithium through the blood-brain barrier," could be the cause.[143] The blood-brain barrier consists of an inner lining of the blood vessels, formed by so-called endothelial cells, that is impenetrable to many substances. It ensures that only substances essential for maintaining brain functions enter the brain, and that those which become toxic upon accumulation leave it. In this context, scientists point out that lithium itself modulates the properties of the blood-brain barrier, for example, by influencing the production and activity of various ion transporters, possibly to induce its own transport. A number of different ion transporters were, in fact, identified in 2018, seemingly playing an important role in transporting lithium ions across the plasma membrane of (brain) endothelial cells, as depicted schematically in Figure 16.[144]

Even if it's still unknown how nature achieves selective lithium accumulation (as of spring 2025), there are at least eight distinct types of ion channels and transport systems in cell membranes (a-h in Fig. 16) with many family members that could facilitate regulated lithium ion transport.[145] As shown here, all of them are demonstrably capable of actively influencing the intracellular lithium concentration. These channels are found in varying densities within the membranes of different cell types, including those in the brain, but also in renal tissue or the thyroid gland. This accounts for renal or thyroid toxicity when lithium is administered in unnaturally high pharmacological doses, far outside the essential range (more on this in the next chapter).

Yet, the names of these transport systems offer no definitive insight into their full range of capabilities. While named after sodium, potassium, calcium, and hydrogen ions, they can also maneuver lithium ions across a cell membrane. The terminology sometimes stems from purely scientific-historical reasons: Ion channels and ion transporters were first.

First identified for their importance in transporting other ions, ion channels and transporters were thus accordingly named. Even though their ability to channel or transport lithium ions is relatively new from a scientific perspective and hasn't yet led to them being named accordingly, this in no way indicates the evolutionary priority for which ions the channel or transport system originally developed. Lithium transport is usually attributed to an ionic radius that happens to be similarly sized, rather than a deeper biological purpose—partly because it's not recognized as an essential element. The only genuinely comparable case is the magnesium ion: 134 versus 130 picometers. In my opinion, it's not only possible but, given the many other indications of lithium's essentiality, highly plausible that some of these developed from the start primarily as lithium-ion transporters or lithium-ion channels. Yet, the fact that they were not named lithium-ion transporters or lithium-ion channels, for the reasons cited, is more than just a semantic quibble—it carries very practical consequences: as long as no transport system is named after lithium, it will be much more difficult for the scientific community to accept the realization that the active transport of this trace element not only occurs but is an essential part of the evolutionarily orchestrated function of the entire Lithiome.

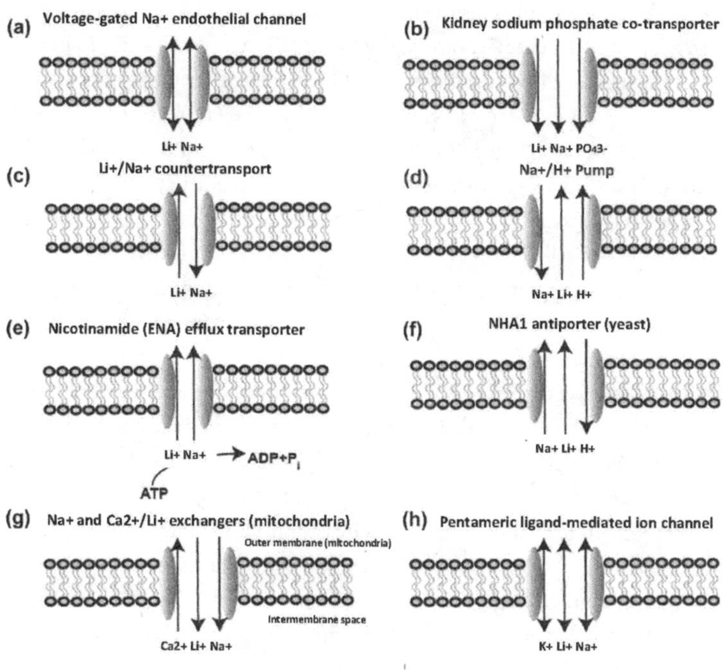

Figure 16

Lithium very likely has such diverse and fundamental effects (perhaps more than any other trace element) because it has been present since the dawn of life (Argument 4). Evolution simply couldn't ignore this trace element—always present in the diet of our prehistoric ancestors (see Argument 6)—and therefore had to factor it into all bodily processes, actively utilizing it! As a trace element, lithium's effect on humans isn't coincidental; rather, this effect is a biological necessity that has developed since the beginning of life and persisted through to us. Argument 3 for the essentiality of lithium is, therefore, also well supported.

The active regulation of lithium distribution and tissue-specific lithium concentration suggests that the body strives to maintain stable lithium levels, especially in lithium-dependent organs, even amid fluctuating dietary intake. Lithium notably accumulates most intensely in the neurological core of the mental immune system.

ARGUMENT 4: Highly Conserved Evolutionary Regulation of Essential Lithium Targets

Evolution of Life's Diversity

This subchapter holds immense importance, both for a deeper understanding of lithium's biological function and for further evidence of this trace element's essential role for humans. But it also provides a unique insight into the evolution of life that is noteworthy. My hope is that Argument 4 will contribute significantly to preventing the continued suppression of essential lithium. The evidence supporting the essentiality of lithium targets is simply too clear, especially within the context of billions of years of evolution that has never achieved one thing because it's incompatible with life: producing mutations that make organisms independent of essential lithium. However, the following explanations are, by necessity, also very scientific and could also be placed in a grayed-out "Additional Information" section for experts. However, as I consider them very important, I hope that laypeople will also be interested in delving a little deeper here. To ensure that even readers who prefer not to navigate through the following scientific explanations can still grasp their results in a simplified way, I've summarized the three key findings and conclusions at the end (see *Summary* at the end of Argument 4).

Some of the basic biological mechanisms essential for life, and thus found in all existing life-forms today, must have developed more than several billion years ago in the common ancestor of all current life-forms. An unimaginably long time, considering that 10 to 15 million (some even estimate 100 million) different species arose from the genetic material of this primordial being through countless independent copying processes—generation after generation. In this process, nature selected random copying errors (mutations), duplications of individual genes or entire chromosome sections from the original genetic material in the respective germ line, which gave the newly formed generation a small advantage, or mercilessly eliminated whatever was disadvantageous. This mechanism led to genetic adaptations and, inevitably, to changes that resulted in all the diverse living creatures emerging from that original cell (see Fig. 17). This includes not only those still existing today, but also all those that have since become extinct—always in the spirit of "survival of the fittest."

When observing nature, our attention usually focuses on the changes that have occurred and are still occurring during life's evolution: for example, the emergence of new species, the development of new characteristics, and so forth. In my opinion, however, the functions that all life-forms share are just as interesting. These are represented by the parts (sequences) of their genetic material that haven't changed over life's long evolutionary history precisely because they are not allowed to change! Why are these cases so interesting? Any random change to genetic code at these unique, unaltered genomic sites would likely be incompatible with life. These genetic regions, which are thus unchanged across species and either code for specific protein sequences or are

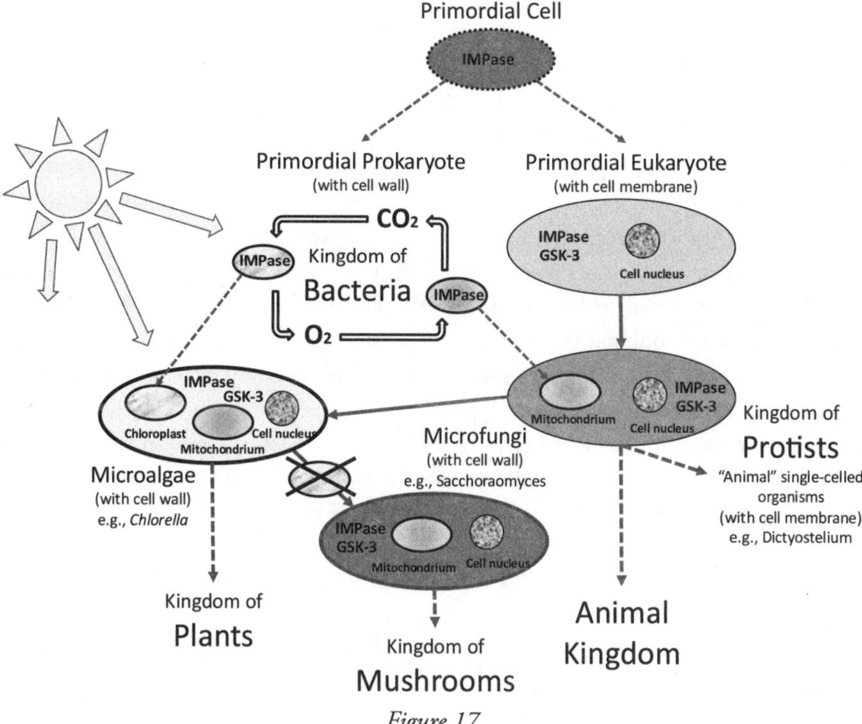

Figure 17

themselves active in gene regulation, are called conserved, from the Latin *conservare* meaning "to preserve" or "to keep." "Highly conserved" refers to gene sequences that have persisted through hundreds of millions of years of evolution, possibly since the beginning of life itself. The presence of genetic conservation in all life-forms, therefore, serves as an indication of something indispensably fundamental. Because if these biological functions, encoded in their respective genes, have been firmly engraved in the book of life, so to speak —that is, if they've endured countless copying processes to this day across an incredible diversity of species unscathed—then these genes truly must be essential for all beings. This applies to you, to me, and to all people, regardless of individuality or type affiliation. The "key genes of life" encode "key proteins of life" that, given their ancient evolutionary origin, have usually undertaken many different functions in the development of increasingly complex life-forms: No body system can function without them, whether energy, metabolism, or physical or mental immune system; ultimately, every organ function is decisively regulated by them. In fact, in addition to their durability, their versatility is another important indication of their indispensable importance for life itself. Such conserved gene sequences and key proteins, indispensable for life, are discovered by comparing the genomes of different species. If the same gene sequences are found in several kingdoms, e.g. in the plant kingdom and in the animal kingdom, the gene must be very old. Some of these highly conserved protein sequences or key proteins are even found in all five kingdoms of life (bacteria, protists, plants, fungi and animals; see Fig. 17). Their origin therefore goes back to the origin of life itself. We do not know what the primordial cell (in the upper part of the image) looked like and what properties it possessed. It's even uncertain whether it—and thus life itself—actually originated on Earth.[146] Life's germ, for instance, might have arrived on our planet via a meteor from another region of space, given that all building blocks of life have been found in meteors.[147] Nevertheless, we can assume that the initial primordial eukaryote and prokaryote (cells with and without a true nucleus) shared a common ancestor, dubbed the primordial cell, since they employ an identical genetic code. It is very unlikely that they developed several times completely independently of each other. The presence of identical genes across all highly diverse life-forms further supports a common origin in the primordial cell. IMPase, for example, is found in all five kingdoms of life and must therefore have developed in the primordial cell, as illustrated in Fig. 17. GSK-3, however, is found only in the four kingdoms of life that evolved from the primordial eukaryote, indicating a later genetic origin.

As also seen in Fig. 17, the large kingdom of bacteria developed from the primordial prokaryotes. Characterized by a stable cell wall and the absence of a clearly defined cell nucleus, they're also referred to as precaryotes (literally: pre-nuclei). They include the cyanobacteria, which break down carbon dioxide through photosynthesis—that is, with the help of sunlight—and convert the carbon into energy-rich carbon compounds to build their cells. The oxygen released in the process is released into the environment. Other bacteria incapable of photosynthesis, in turn, can utilize this oxygen for energy production by "burning" energy-rich carbon compounds like sugar or cellulose with it, thereby releasing carbon dioxide. This completes the energy cycle, which is kept going by sunlight, as shown in Fig. 17.

The symbiotic incorporation of such an oxygen-utilizing primordial bacterium into a eukaryotic primordial cell (which possessed a flexible cell membrane instead of a rigid cell wall) initially led to the emergence of the kingdom Protista. The vast animal kingdom very probably evolved from a protist, possibly an ancestor of the amoeboid unicellular organism Dictyostelium, given their identical cell structure. We call the primordial bacterium living symbiotically in these eukaryotic cells the mitochondrion. Our somatic cells' energy-generating powerhouse retains its own bacterial genetic material, enabling it to reproduce autonomously. We refer to this process as mitochondriogenesis. Mitochondria are responsible for enabling all our cells to use oxygen to release energy, which was originally stored by sunlight in complex carbon compounds like sugar or fatty acids. This is the most effective form of energy production, indispensable not only for our muscles but, above all, for our energy-hungry brain. Without this symbiosis billions of years ago, complex life, requiring a great deal of energy, could not have developed.

As indicated in the illustration, the first primordial plant cell possibly originated from a protist with a stabilizing cell wall and a symbiotic integration of a primitive cyanobacterium that could harness sunlight for energy. Perhaps the microalgae of today (e.g., Chlorella) are very similar to this primitive plant cell. In the course of further evolution, the entire plant kingdom eventually developed. The integrated primordial cyanobacterium, present in all plant cells, is known as the chloroplast. This green solar powerhouse, like mitochondria, still retains parts of its original bacterial genetic material and can continue to reproduce independently within plant cells. Chloroplasts are responsible for leaves being green and plants' ability to convert carbon dioxide into complex, energy-rich molecules such as sugar or cellulose with the help of sunlight. However, plants also have mitochondria, allowing them, like animal cells, to release the energy stored in sugar during daylight hours, even at night or without light, with the help of oxygen.

As Fig. 17 illustrates, the first unicellular microfungus likely evolved from a primitive microalga after its chloroplasts had been eliminated, as indicated. In the course of evolution, the complex kingdom of fungi developed from such a microfungus. A typical representative that is still alive today is baker's yeast (*Saccharomyces cerevisiae*). Like all plant cells, it has a cell wall, but at the same time—like all protists or animal cells—it has only mitochondria for energy production. Baker's yeast, therefore, digests sugar and uses the carbon dioxide released in the process to make the bread dough rise.

IMPase, A Lithium-Regulated Primordial Gene

Through autophagy, dysfunctional proteins and cell organelles (such as mitochondria or chloroplasts) can be digested and their components recycled. This function serves to rejuvenate the cell. It is particularly important for our nerve cells, which cannot be replaced. This is why the brain reacts particularly sensitively to lithium deficiency, as seen in Argument 1; the mental immunodeficiency syndrome is a potential consequence. The recycling function, a vital part of cell metabolism, was long only suspected in eukaryotes, but has now also been identified in bacteria and prokaryotes and termed "protophagy. "From a bird's eye view," write the discoverers of these cell biological parallels, "the process of protophagy follows the same rules as eukaryotic autophagy, exhibits a number of similarities with it [. . .] and also achieves the same ultimate goal (survival of a biosystem under stress conditions and maintenance of homeostasis)."[148] However, the extent to which IMPase, also present in bacteria (see below), is involved in protophagy has not yet been clarified.[149]

The IMPase plays an essential role in sugar and thus energy metabolism, as well as in many other vital functions of the cell.[150] It's therefore not surprising that at least one gene coding for an IMPase can be found in representatives of all five kingdoms of life. This places the origin of the IMPase gene at the beginning of all life or in the primordial cell, as shown schematically in Fig. 17. GSK-3 is found in four kingdoms of life, excluding the bacterial kingdom (my gene bank search for a bacterial ortholog was, at least, unsuccessful) GSK-3, therefore, likely did not originate in the primordial cell of all life, unless it was eliminated from the first prokaryote's genome during its evolution—a possibility, albeit unlikely. It's far more probable that GSK-3 evolved within the first primordial eukaryote. From there, the GSK-3 gene was passed on to all four eukaryotic kingdoms of life, as shown in Fig. 17. The common origin can be recognized by the fact that its protein sequence is conserved. The most highly conserved sequence region is IMPase or GSK-3, each of these enzymes' active center. FYI: Enzymes are biochemical catalysts. By lowering the activation energy, they ensure that biochemical processes run efficiently even at body temperature. This also gives these vital reactions a direction, which early in evolution predestined these two lithium targets to also function as information processors and signal mediators. They, therefore, not only regulate energy metabolism, which is essential for all life, but are also directly or indirectly involved in all biological processes, as discussed in Argument 2.

However, the multifunctionality of GSK-3 or IMPase also has its price: Any changes in the gene sequence that would lead to functional disorders are ruled out. This limits evolution when it comes to assigning new tasks to these information processors. This ensures such a high degree of functional conservation that, for example, the genes of all proteins involved in glucose degradation in baker's yeast (*Saccharomyces cerevisiae*) could be exchanged for the corresponding human genes without any loss of function, even though both species have undergone over a billion years of independent evolutionary development.[151] This raises the question of how it's nevertheless possible that lithium targets like IMPase or GSK-3 could acquire so many new functions in the evolution from unicellular organisms to complex life-forms, as we saw in Argument 2. How can they regulate autophagy and glucose metabolism, yet also control apoptosis or neurogenesis, to name just two additional functions? Last but not least, the entire mental immune system is completely under the control of the lithiome. The evolutionary expansion of functions for master regulators like GSK-3 and IMPase, purely through genetic optimizations, is limited by these gene-specific and many system-immanent aspects. Beyond a certain optimized outcome in the evolutionary process, any further expansion of functions toward even more complex beings (like humans with a mental immune system) could only detrimentally affect the overall system. To achieve further complexity, the functional expansion of master regulators like GSK-3 and IMPase was only possible through another trick of nature: gene duplication. In this type of genetic modification, the gene for the key protein isn't altered by selective mutations. Instead, it's randomly duplicated due to a much larger copying error during the production of new germ cells. From that point on, two copies are passed down: the original gene (Gene #1) and the duplicate (Gene #2). Gene duplications—or the doubling of even larger chromosomal regions containing many genes—are a powerful driver of genetic innovation.[152] This newly created Gene #2 is initially superfluous. Unlike Gene #1, which has many essential functions and must not be altered further, Gene #2 can therefore be changed by mutations. In this way, Gene #2 can take on new functions in the organism, allowing it to evolve further to greater complexity and performance.[153]

Despite Gene #2 continuously changing and evolving independently of Gene #1 through evolution, their kinship typically remains evident. This is due to the preservation of their fundamental structure, particularly in the active center, like lithium binding in lithium targets and the regulation it influences (though this preservation isn't strictly mandatory). GSK-3, for example, has doubled in the course of evolution. The nematode *Caenorhabditis elegans* and the fruit fly *Drosophila melanogaster* are controlled by only a single GSK-3. With few exceptions, the same holds true for fish, amphibians, and reptiles. In all mammals, including humans, GSK-3 exists in two forms, GSK-3α and GSK-3β, encoded by two genes on different chromosomes (IMPase has also duplicated into IMPase-1 and IMPase-2).[154] Their overall protein sequence remains about 85 percent identical, with the lithium-binding sequence region showing an even higher 97 percent identity. This holds true despite well over 200 million years of evolution having passed since the emergence of mammals and the duplication. This further indicates that the fundamental function of regulation by magnesium and lithium ions had to be retained even for the duplicated protein in its new roles.[155] One can also speculate that such a gene duplication might have first made possible the evolution of mammals with their more complex brain, the neocortex. In fact, GSK-3α and GSK-3β have partially overlapping but also very different functions in brain development, particularly in the cerebral cortex and hippocampus.[156]

The 97 percent high conservation in the lithium-binding sequence region between GSK-3α and GSK-3β is, on the one hand, a strong indicator of this functional region's importance. On the other hand, it doesn't allow us to clearly discern which amino acids are truly functionally significant in the active site. It might be all or merely a few that are not yet crystallized, as insufficient time has elapsed since GSK-3's duplication, on the evolutionary clock, for mutations to affect all the less critical ones. To get a clearer picture, we would therefore have to go further back in time. Since IMPase, unlike GSK-3, must have already existed in the primordial cell, I decided to take a closer genetic look at the enzyme's active center, which, like GSK-3, is activated by a magnesium ion and inactivated by a lithium ion.

Fig. 18 shows a comparison of the protein sequences of the IMPase-1 active site. The twenty different amino acids, or protein building blocks, are presented using their international letter code. Of the respective proteins in their total length of approximately 280 amino acids, a centrally located section of the active center of approximately 40 amino acids in length and a section of 21 amino acids in length located further back can be seen. Due to the high sequence similarity (bolded letters), it's evident that only one amino acid, the sixth from the back, differs between humans and chimpanzees. The biochemical and functional difference between isoleucine (I) and valine (V) is not substantial. However, their common ancestor lived perhaps 10 million years ago, which isn't that long a time ago for evolutionary biology.

Inositol Monophosphatase 1 (IMPase-1)
Magnesium and Lithium-ion binding site in the active site

```
                    #64                          #84#86#87                                    #212
Human          // GEESVAAGEKSILTDN..PTWIIDPIDGTTNFVHRFPFVAVSIG // YEMGIHCWDVAGAGIIVTEAG //
Chimpanzee     // GEESVAAGEKSILTDN..PTWIIDPIDGTTNFVHRFPFVAVSIG // YEMGIHCWDVAGAGIVVTEAG //
Mouse          // GEESVAAGEKTVFTES..PTWFIDPIDGTTNFVHRFPFVAVSIG // YEMGIHCWDMAGAGIIVTEAG //
Goat           // AEEAAAAGAKCVLTPS..PTWIVDPIDGTCNFVHRFPTVAVSIG // YQFGLHCWDLAAATVIIREAG //
Roundworm      // GEESVAGGAKIEWTDA..PTWIIDPIDGTTNFVHRIPMIAICVG // VEYGIHAWDVAAPSIIVTEAG //
Drosophila     // GEEETAKNNNVSGELTNAPTWIIDPIDGTSNFIKQIPHVCVSIG // YIEDMYPWDCAAGSLLVKEAG //
Baker's yeast  // GEESYVKGETVI.TDD..PTFIIDPIDGTTNFVHDFPFSCTSLG // WDGGCYSWDVCAGWCILKEVG //
Dictyostelium  // GEESTKDGIYNWGNE...PTWVIDPIDGTTNFVHRFPLFCVSIA // YEWGIHPWDIAAASLLITEAG //
Tomato         // GEETSAATGDFDLTDE..PTWIVDPVDGTTNFVHGFPSVCVSIG // LIGYGGPWDVAGGAVIVKEAG //
Rice           // GEETSAALGATADLTDD.PTWIVDPLDGTTNFVHGFPFVCVSIG // EIGFGGPWDVAAGALILREAG //
Bacterium      // GEEGGGPADVTATPSD..VTWVLDPIDGTVNFVYGIPAYAVSIG // YEHGVQVWDCAAGALIAAEAG //
```

Figure 18

To identify which amino acids are crucial for the basic function of IMPase, we must travel further back in time, i.e. gain information about the IMPase sequence of older ancestors. Fig. 18 illustrates that, compared to the human sequence, the IMPases from genetically more distant species feature highly conserved amino acids (not only shown in bold, but also framed for improved visibility). As observed in plants, fungi, and unicellular protists such as *Dictyostelium*, highly conserved amino acids are found in IMPase's active center. As shown, even the IMPase of *Mycobacterium tuberculosis* possesses highly conserved amino acids. These were present in the common ancestor of humans and this bacterium and have thus been preserved over several billion years, making them crucial for the fundamental function of IMPase-1. This illustrates what has already been described above (see also Fig. 17): that IMPase must have originated in the primordial cell, the common ancestor of all five kingdoms of life, and has been partly completely preserved over trillions of generations. An analysis of the three-dimensional functional structure of bacterial IMPase revealed that five of these highly conserved amino acids are particularly crucial for binding the magnesium ion in the enzymatic center of both proteins.[157] These include the glutamic acid (E) at position #64, highlighted in gray, two aspartic acids (D) at positions #84 and #87, the intermediate amino acid isoleucine (I) at position #86, and a third aspartic acid (D) at position #212.

In fact, the enzymatic center of IMPase forms as the protein chain folds, creating a pocket from the two sequence segments shown in Fig. 18. This can also be imagined as a hollow hand, with each of these five highly conserved amino acids acting as a finger, jointly holding a magnesium ion. This ion, in turn, fixes their spatial structure, as shown in Fig. 19. However, the hand, or IMPase-1, shown here doesn't hold a magnesium ion, but a lithium ion. This is because the scientists who performed this structural analysis investigated not only the binding of the magnesium ion in the bacterial IMPase's active center, but also that of the lithium ion, comparing their results with existing knowledge of human IMPase-1:

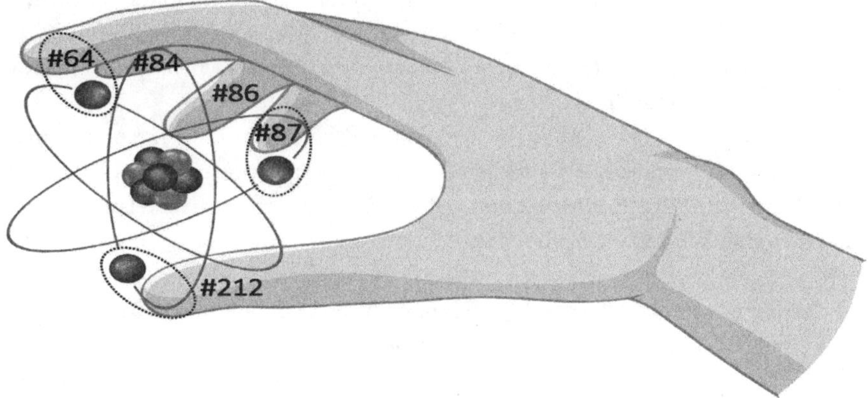

Figure 19

Like its human counterpart, "SuhB [the name of the bacterial IMPase] is activated by a magnesium ion and inhibited by a lithium ion."[158] IMPase-1 is therefore not constantly (constitutively) active; rather, its activity depends on whether a magnesium ion, instead of a lithium ion, is bound in its enzymatic center. A further study using the IMPase of the bacterium *Staphylococcus aureus* came to the same conclusion: "The structural investigation of the lithium ion binding site in the staphylococcal IMPase (SaIMPase I) using X-ray crystallography," the scientists explain, "indicates common or overlapping binding sites for magnesium ion and lithium ion in SaIMPase I's active center." This proves that both ions bind to the active center with different effects—magnesium activates, lithium inhibits—and that this mechanism has been preserved over billions of years of evolution.[159]

This extremely high conservation of the five fingers is found across all kingdoms of life, beyond humans and bacteria, as seen in Fig. 18. This includes, for example, tomatoes, rice, amoeboid slime molds such as Dictyostelium, worms, fruit flies, mice, goats, and chimpanzees. This indicates that any alteration to this structure, and thus its function, would be incompatible with life. It is, therefore, the consequence of a biological necessity, as a third experimental study was able to elucidate even more clearly. In this case, the scientists examined the IMPase, or SuhB, of the intestinal bacterium *Escherichia coli.* They write: "In this case, the scientists examined the IMPase, or SuhB, of the intestinal bacterium *Escherichia coli.* It exhibits significant sequence similarity to human IMPase-1 and possesses most of its major active residues [or the functionally crucial structures of the respective amino acids, above in bold font and underlined]. Here we show that [. . .] the bound form of the lithium ion in the active site [. . .] can be clearly detected, and based on our data and other biochemical data, the lithium ion [. . .] binds to aspartic acids #84, #87, and #212."[160] Thus, three of the five amino acids that grasp a functionally activating magnesium ion like five fingers of a hand are also capable of "taking hold of" a lithium ion in a functionally inhibitory manner, as schematically illustrated in Fig. 19. However, the illustration shows a lithium atom, not a lithium ion. Common electron clouds between lithium and each "finger" (or amino acid), as depicted, stabilize the bond or "grip."

Summary: As mentioned at the beginning of the explanations for Argument 4, these relationships aren't easy for the molecular biology layperson to grasp. I'd therefore like to summarize the fascinating findings we can derive from these relationships in simpler terms. We have learned:

1. The primordial cell (Fig. 17), the common ancestor of humans, bacteria, and all known life, must have already possessed an essential protein from the Lithiome, IMPase, around 4 billion years ago.[161] Due to the complete conservation of the lithium binding site across all life-forms, this protein was very likely already lithium-regulated even then!
2. The functionally relevant structures have been inherited unchanged over myriad of generations, across the most diverse directions of evolving species, since the existence of this primordial cell. This means billions upon billions of random mutations couldn't harm these protein sequences because their function is so essential that any structural change would no longer be compatible with life.
3. The targeted functional binding of a lithium ion in the IMPase's active center proves lithium's importance for this protein class, and thereby also establishes the essential nature of the competing mechanism between magnesium ion (activating) and lithium ion (inhibiting)—thus proving lithium's own essentiality!

The necessity of the last conclusion arises against the background that a functional binding of magnesium ions in the active center would also be possible without simultaneously allowing a functional lithium-ion binding site. In fact, our genome codes for thousands of different proteins that all functionally depend on being able to bind magnesium ions, yet very few of them are simultaneously regulated by lithium ions.[162] In other words: the *magnesiome*, i.e. the set of all proteins regulated by magnesium ions, and the *lithiome*, i.e. the set of all proteins regulated by lithium ions, overlap only very slightly. Thus, compared to the large magnesiome, there's only a relatively small subset of less than a dozen lithium targets (found in four of the six previously mentioned target classes of the lithiome; see Argument 2). In these, the competitive mechanism between magnesium and lithium ions for the same binding site is highly conserved, with opposite functional consequences (magnesium ion activates, lithium ion inhibits). In my opinion, this can only mean that the function of lithium targets such as IMPase actually depends essentially on the possibility of regulatory binding and influence by a lithium ion, which lithium's essential function. This also proves Argument 4 for lithium's essentiality. It could be shown that:

Lithium targets are evolutionarily highly conserved key proteins of all life, and their regulation by lithium is also completely conserved evolutionarily. This clearly indicates the essentiality of lithium regulation. This means that the trace element itself must also be of essential importance.

ARGUMENT 5: Life-Shortening and Offspring-Reducing Effects of Lithium Deficiency

Evidence of Essentiality from Animal Studies

Lithium is found, at least in very small traces, practically everywhere in nature and thus also in food and drinking water. However, this is generally insufficient to ensure an adequate supply even with an otherwise balanced modern diet. There are numerous epidemiological studies on severe deficiency states (see Argument 1), but no comparative or observational studies on the consequences of an absolute lithium deficiency in humans. For ethical reasons, experimentally inducing an absolute lithium deficiency in humans is categorically ruled out; although it seems downright ironic to write this, knowing about the serious deficiency that is tolerated in reality, which not only makes people's lives more difficult than necessary in many respects but even shortens them. In this sense, the non-recognition of lithium as an essential trace element—and, at least as grave—the ban on offering it as a dietary supplement in large parts of the world (more on this in the last chapter) must also be regarded as unethical. While I also consider attempts to withhold or even deprive animals of lithium through a special diet ethically problematic, we no longer need new experiments to answer this question. The animal experiments conducted several decades ago have long since yielded meaningful results.

The Essentiality of Lithium in Rats

From a scientific perspective, to obtain a clear animal experimental result regarding the essentiality or vital necessity of a trace element, it would be ideal, of course, to create a complete deficiency. Despite the aforementioned fact that small amounts of lithium can be found in almost all foods, scientists succeeded in the mid-1970s in developing a lithium-reduced food for laboratory rats. Instead of the usual approximately 350 μg/kg, its lithium content was merely about 15 μg/kg.[163] While this is significantly less than typical, it's still far from being lithium-free. This is mainly due to what we know from epidemiological studies on drinking water: that even the smallest amounts of lithium have a detectable influence on the human mental immune system, i.e. produce effects. Despite this limitation, however, some highly interesting observations have been made. For example, in rats that grew up with this nutritional lithium deficiency, the lithium content was found to be reduced by 20 to 50 percent in almost all tissues, as compared to control animals. This was surprising, as the lithium intake (assuming the same amount of food) was reduced by a factor of 22. Since animals cannot produce lithium, there must inevitably be a so far unknown but highly effective regulatory process leading to less lithium excretion in the event of a deficiency. Under these deficiency conditions, the pituitary gland and the adrenal glands were, surprisingly, able to keep the lithium concentration in their tissues almost constant. This clearly indicates that lithium is deliberately and actively accumulated in these hormone-producing organs, presumably to maintain their vital functions even during at least temporary deficiency conditions. However, despite these regulatory countermeasures, the lithium deficiency was by no means without consequences. It led to a drastic reduction in litter size of up to 30 percent fewer offspring, with a survival rate only half that of the control group litters. Chronic lithium deficiency therefore reduces reproductive success in rats and clearly leads to reduced population growth. Given that telomerase, for instance, is lithium-dependent and that this, in turn, limits the growth potential of stem and somatic cells—keyword: Hayflick limit (see Argument 2)—these results are not surprising.

In 1992, another study was published in which rats were fed an even further reduced concentration of lithium via their diet: One group of rats received just 2 μg/kg of lithium in their diet for five consecutive generations while three generations of another group received as little as 0.6 μg/kg.[164] Compared to the previous study, this intensified lithium deficiency resulted in an even more drastically reduced litter size in comparison to the control animals fed a 500 μg/kg lithium-rich diet.

In addition, an even greater reduction in birth weight and further limited probability of survival of the few newborns were documented. Milk production was also severely hindered. After weaning, from the first generation under lithium deficiency to the third, the rats' weight dropped from an average of 513 g to 167 g. The percentage that survived at all fell from 84 percent to 36 percent. The authors of the publication summarize their results succinctly, but nevertheless quite clearly: "The impairment of reproductive success and lactation performance indicates that lithium is an essential element."

In 2005, animal nutrition expert Prof. Manfred Anke (1931–2010) and his team reported that "abnormal behavior was observed in lithium deficient rats, and this disappeared after lithium supplementation."[165] And as early as 1989, Japanese researchers published that lithium-deficient rats could no longer perform learned behaviors under psychological stress, but regained this ability in a dose-dependent manner by administering microdoses of lithium.[166] The two research teams thus achieved two things: firstly, they provided another building block in the evidence for essentiality by successfully intervening to remedy the deficiency. Secondly, they also documented a concrete indication of lithium's special significance for "higher" functions, especially brain functions and, by extension, behavior.

The Essentiality of Lithium in Goats

In 2005, the same working group led by Manfred Anke also researched and published the health effects of a low-lithium diet in goats over fifteen generations.[167] The control goats received feed with a lithium content of 10 mg/kg, while those in the low-lithium diet group received feed containing 1.7 mg/kg. Although this represented a sixfold reduction, it was still a relatively high amount of lithium compared to the rat studies. Despite it not being an absolute lithium deficiency, this lithium reduction nonetheless severely impacted growth, reproductive performance, well-being, and mortality: "The kids of lithium-deficient goats had a 9 percent lower birth weight than those of control goats. Furthermore, a low-lithium diet led to reproductive disorders in both goats and rats."

However, the lithium reduction not only led to a significantly lower pregnancy rate, but also to a miscarriage rate fourteen times higher than that of the control animals (14 percent versus 1 percent). In addition, the animals' life expectancy was drastically reduced: 41 percent of the lithium-deficient goats died in the first year of life, most likely from infections, indicating a dysfunction of the body's immune system (the authors report conspicuous skin changes), compared to only 7 percent of the control animals. The mortality rate with lithium deficiency thus increased sixfold(!) despite the animals not even suffering from an absolute deficiency! This leads to the conclusion that absolute lithium deprivation would most likely have been fatal for all animals. "In summary," according to the authors, "the results of a lifelong low-lithium diet over more than 15 generations of goats show that lithium can be essential for fauna [or wildlife] and thus also for humans."

The Essentiality of Lithium in Humans

Gerhard Schrauzer wrote in 2002 in his masterpiece *Lithium: Occurrence, Dietary Intake, Essential Importance for Nutrition:* Since "lithium deficiency in humans will likely never reach the severity observed in animals experimentally deprived of lithium, the symptoms of lithium deficiency in humans—if they appear at all—will probably be mild and manifest more as behavioral disturbances than physiological abnormalities."[168] We have already discussed such behavioral disorders in humans at length, as resulting from the impairment of "higher" functions due to lithium deficiency (key term: mental immunodeficiency syndrome). They are much more serious than Schrauzer assumed, and to describe them as mild does not do them justice. Of course, being human means more than just the ability to survive and reproduce; the lack of lithium must be severe to have such grave and fatal consequences that even the slightest differences in drinking water concentration demonstrate such profound effects, as we discussed in Argument 1. If it were not for the ability described in Argument 3 to accumulate lithium in the brain regions relevant to "being human" (especially the hippocampus), the effects on the human psyche would certainly be even more dramatic than they already are.

It should be pointed out here that we can't yet definitively rule out that a lithium deficiency also leads to reduced fertility or a higher miscarriage rate in humans. After all, birth rates and fertility have been continuously falling for a long time, and a causal connection could be functionally explained via a dysregulated Lithiome. But this is pure speculation. What we do know for sure is that life expectancy is reduced because of a deficiency, as seen in animal studies. A sufficient supply of lithium increases the number of healthy years of life as well as the lifespan in comparison to people with a deficiency. Conversely, an insufficient supply reduces the number of healthy life years along with life expectancy. This finding is thus expected to intensify with worsening deficiency, potentially leading to a complete incompatibility with life due to a total Lithiome dysregulation. Fortunately, however, an absolute deficiency is rarely reached, given lithium's ubiquitous presence, even in micro-quantities. In fact, for me, the most dramatic indication of lithium's essentialness is that chronic deficiency is associated with an increased rate of violent acts, both against others and oneself, which can prematurely end life. In particular, the significantly increased suicide rate with lithium deficiency is clear evidence of the mental immune system's essential dependence on an adequate supply of this trace element. These fatal consequences are maddening when considering how easily the deficiency could be remedied. Every victim is therefore a silent cry for help.

As early as 1994, Gerhard Schrauzer and a colleague were able to demonstrate a causal relationship between mood and the intake of even a very low lithium dose in a controlled clinical intervention study.[169] All study participants were former drug users (mainly heroin and methamphetamine), violent offenders, or had a history of domestic violence, i.e. a history not exactly indicative of a balanced mood. For the study, they were randomly divided into two groups: one took 400 µg of lithium daily (a lithium-enriched yeast capsule), and the other received an identical-looking placebo (a yeast capsule without lithium). After four weeks, the groups were reversed. Weekly, the subjects completed self-assessment questionnaires with twenty-nine questions covering parameters like mental and physical activity, capacity for thought and work, mood, and emotionality. In contrast to the control group, the overall mood test scores in the lithium group increased continuously and significantly during treatment. Even more interesting than the statistics, however, are the comments from those who—and this bears repeating—were taking a mere 0.4 mg of lithium per day. For example, a former crystal meth and alcohol user with a history of domestic violence said: "I feel a big difference in that I don't fly off the handle so quickly anymore." Other subjects commented one to two weeks after compensating for their lithium deficiency: "My mood is stable and I can think more clearly now. I feel pretty good," or "I used to be irritable, but since taking these vitamins that has changed drastically." One person who was suicidal before taking lithium said after four weeks: "I am no longer depressed. I see light at the end of the tunnel." This is also roughly the feedback I now receive daily from people who have started to take essential amounts of lithium as a result of the information I have given them. Dr. Christian Schellenberg, a pediatrician practicing in Potsdam, Germany also reports similarly remarkable improvements in mood, concentration, and overall outlook on life among his young patients.[170] Also of interest is the comment of a person who took her lithium-containing yeast capsule irregularly, against the instructions—unaware, however, that it was not a placebo: "When I take these vitamins regularly, my mood seems to improve." This observation aligns with findings from discontinuation trials conducted in a pediatric practice in Potsdam—there, too, psychological disturbances reemerged immediately when the intake of significant amounts of lithium was halted on medical advice (Dr. Christian Schellenberg, personal communication). A former heroin addict, alcoholic, and violent offender stated during placebo treatment: "I am always extremely moody and often argue with my girlfriend." After switching to lithium, he stated after one week: "I have noticed changes in my behavior." After four weeks: "I don't get as upset about little things as I used to," and "I am usually in a much better mood." The study authors state, "Based on these results and the analysis of the participants' voluntary written comments, the two study authors conclude that lithium, at the selected dosage [0.4 mg/day—approximately 300 to 500 times lower than that used in the treatment of bipolar disorder], has a mood-enhancing and stabilizing effect."

The Anti-suicidal Effect of Lithium Is Causal. A first indication that lithium deficiency actually increases the risk of suicide—not just a coincidental correlation as shown by the drinking water studies—is the following clinical observation: In contrast to all other therapies, both for unipolar (major depression) and bipolar disorders (manic-depressive illness), lithium therapy reduces the risk of suicide.[171] Lithium is thus qualitatively very different from other frequently prescribed antidepressants (such as carbamazepine, olanzapine, or amitriptyline, etc). While these can reduce depressive symptoms, they do not safeguard patients from suicide; quite the opposite: In fact, the risk of suicide exists especially when individuals, through such drug therapy, are just regaining the capacity to act but haven't yet overcome their depressive mood. A large survival analysis, using anonymous patient data from the UK Biobank, even showed that therapeutic lithium is associated with an almost fourfold higher probability of survival compared to other antipsychotics.[172] The neuroinflammatory cause of suicidality and lithium's anti-inflammatory function offer a mechanistic explanation for the observed causality behind the suicide reduction by this trace element, which we'll discuss in detail in Chapter 4.[173]

However, the life expectancy of people with mental disorders, which can be the result of a chronic lithium deficiency, is not only shortened by suicide, but also to a far greater extent by other causes. Reasons for a much too early end of life include, for example, eating disorders often linked to mental health issues, from anorexia to obesity, which in turn shorten life by an average of more than sixteen years due to numerous secondary diseases, as a meta-study from 2023 showed, including a total of more than 12 million people with mental disorders.[174] Drug use, frequently associated with mental and social problems, reduces life expectancy by over twenty years on average. The results of an Australian study published in 2019 are summarized in the University of Queensland's press release as follows: "The risk of an early death was higher for people with mental disorders across all ages—apart from an increased risk of death due to suicide, we confirmed increased risk of death due to conditions such as cancer, respiratory diseases and diabetes."[175] Premature death from Alzheimer's disease was also directly linked to the lithium content of drinking water, as seen in a study from Texas.[176] Last but not least, we must note at this point that if the mental immune system does not develop healthily or becomes increasingly dysfunctional in the course of life due to lithium deficiency, the probability of not making healthy life decisions increases.

However, this relationship between lithium deficiency and shortened life expectancy can be reversed: If large parts of the population are chronically deficient, a higher lithium intake should increase life expectancy. In fact, "the regular intake of the trace element lithium [. . .] can significantly extend life expectancy," according to a 2011 press release metaphorically titled *Lithium—Fountain of Youth from the Tap*, which summarized the results of a study conducted by nutritionists at Friedrich Schiller University Jena in collaboration with Japanese epidemiologists.[177] Study leader Michael Ristow—then Chair of Human Nutrition at Jena—worked with Japanese colleagues to examine data from more than 1.2 million residents across eighteen neighboring communities in Japan's Oita Prefecture. They found that lithium, even at low concentrations in drinking water (59 µg/l versus 0.7 µg/l), reduced overall mortality and thereby contributed to increased longevity.[178] This protective effect remained clearly evident even when excluding deaths by suicide from the analysis.

"Indeed, the effects of trace amounts of lithium on life expectancy are impressive," wrote a team of Canadian researchers who conducted a similar follow-up study in Texas (see below), referring to the Japanese study by Ristow and colleagues. However, the researchers were particularly impressed by the extremely low dose, which was sufficient to achieve an effect, and apparently "lithium concentrations in water can be a factor of 1000 lower than the therapeutic doses for bipolar disorder."[179] At that time, however, this was the only study that had examined the life-prolonging effect of low lithium intake in humans. The Canadian team, therefore, deemed it essential to investigate the negative correlation between trace lithium in drinking water and overall mortality in other population groups as well. Only then could the results be generalized. Therefore, they investigated the association between trace lithium and all-cause mortality, expressed in years of life lost, using a much larger dataset. This data was obtained from readily available information from 234 counties in Texas. The dataset was not only larger than the one from Oita, Japan, and included significantly more data points, but it also showed a much wider distribution in terms of lithium concentrations across the individual counties: the average lithium levels in the Texas counties ranged from 3 µg/l to 539 µg/l. In Oita, on the other hand, they only ranged from 3 µg/l to 539 µg/l. The result was clear, as the authors write: "Thus, our current results extend and corroborate the life-prolonging effect of lithium in humans, as postulated [based on the Japanese study]." They also indicate that an even longer lifespan could be achieved in Japan if more lithium were consumed there, a point we'll delve into in the following chapter when estimating how high the optimal intake should be.

However, epidemiological studies only point to possible links, not proof. "It should be noted, however," wrote the scientists who did the Japanese study, "that the results in humans are observational in nature and therefore cannot establish a causal relationship between high lithium intake and reduced mortality. Theoretically, a lifelong intervention study in humans would be required to provide causal evidence for the mortality-reducing effects of low-dose lithium supplementation. Such a study cannot be conducted for obvious reasons."[180]

Animal Studies Applicable to Humans

The scientists used a different species to prove causality: the classic model organism for antiaging studies and now well-known nematode *Caenorhabditis elegans*. Given the Lithiome's widespread evolutionary conservation, even between worms and humans (see Chapter 1), numerous animal experimental findings can serve as strong indicators for addressing this same question in humans. For this reason, animal experiments that have attempted to demonstrate a life-shortening effect of lithium deficiency in humans are also included here.

An earlier study, using a very high lithium concentration, which, when converted to higher body weight, would be typical for treating bipolar disorder in humans, extended the lifespan of *C. elegans* by 46 percent.[181] What's more impressive, the nematodes lived longer even at a thousand-fold lower lithium concentration, such as that found in the tap water of Japan's Oita prefecture, which is associated with the highest life expectancy. According to the authors, this "reflects unexplained molecular effects on human mortality."[182] However, given our discussion above on lithium's effect on the entire Lithiome, these molecular effects are no longer unexplained; instead, there are very good indications of the molecular relationships that can account for this. These range from a reduced tendency toward inflammation and improved autophagy to increased telomerase activity[183], demonstrating just a few of the rejuvenating effects we've already discussed. These processes are regulated in many animals and humans using the same lithium targets, allowing us to draw conclusions from animal experiments applicable to humans. In older nematodes, the number of mitochondria decreases when lithium is administered, as the trace element stimulates autophagy and thereby mitochondriophagy, i.e. the breakdown of dysfunctional mitochondria.[184] However, this observation also coincides with improved energy production. Lithium contributes to extending life by stimulating mitochondriogenesis, which increases the number of well-functioning mitochondria, even if their overall total count is reduced. The age-related decrease in mitochondrial energy production can also be specifically counteracted in humans, e.g. through physical activity.[185] Among other things, this increases the activity of the mitochondrial master regulator PGC1α, also activated by lithium.[186] In the meantime, lithium's life-lengthening effect has been found in many other organisms, such as the fruit fly *Drosophila*.[187] This indicates already that the molecular processes that lead, with a sufficient supply of lithium, to an extended lifespan have remained highly conserved over very long evolutionary periods. Given their conservation across evolutionarily distinct species (mammals, worms, and insects), similar effects can be expected in humans as those observed in the studied animal species, as shown by the Japanese and Texan studies. In fruit flies, it was also shown that the extended lifespan is attributable to the lithium-dependent inhibition of GSK-3.[188] For instance, genetic upregulation of GSK-3 or a constitutively active GSK-3 mutation shortened the lifespan of fruit flies by 30 percent and 50 percent, respectively.

In this context, "constitutive" means permanent. The same situation holds true for human lithium deficiency. Since the natural "brake" is missing (see Argument 2), this constitutive activity of GSK-3 (even without artificially induced mutation) is a huge health problem, as we will further discuss in more detail in the last chapter. A brief preview: This unnatural state of a constitutively active GSK-3 is considered natural because lithium is not deemed essential. A lack of this trace element, responsible for the constant activity of GSK-3 (but also of IMPase, see above), therefore, however nonsensically, can also play no role. Consequently, from the pharmaceutical industry's point of view, all chronic diseases are also a "natural" consequence of being human (and not, among other things, of a lithium deficiency)—at least that's what they want us to believe—which is why, as a further consequence of this erroneous assumption (that lithium is not essential), a long and healthy life is only possible with their medications. But let's save this debate for the last chapter, when all the evidence is on the table, such as that from the fruit fly experiment: The mutation-induced GSK-3 activation and subsequent lifespan reduction were almost entirely counteracted by lithium treatment. In addition, low doses of lithium extend the lifespan of genetically unmodified fruit flies by up to 20 percent compared to a lithium-deficient state, whereas very high (toxic) doses severely shorten life. This is precisely what's observed in humans but, from the pharmaceutical industry's perspective, shouldn't be the case. Ultimately, even the authors' commentary doesn't emphasize lithium's importance for human longevity, but rather focuses solely on GSK-3 as a pharmacological target: "The discovery of GSK-3 as a therapeutic target in the aging process will likely lead to more effective treatments," the authors write, "that can modulate the aging process in mammals and further improve health in later life."

In mice, however, which were typically kept in cages in completely unnatural experimental conditions, no statistically significant, but at least a graphically visible, extension of lifespan could be determined. Furthermore, a statistically significant extension of the period during which the mice lived healthy lives was found.[189] This study can be used as an example to illustrate that lithium alone isn't sufficient to significantly extend an unnaturally maintained life. As already noted several times, lithium is not a miracle drug. However, it is effective insofar as a deficiency shortens life in many ways. Ristow and colleagues therefore assume "that these life-prolonging properties of low-dose lithium, based on observations [in animal models], can be translated into lower overall mortality in humans exposed to comparable amounts of lithium in a similar manner over a longer period of time."[190] Accordingly, an international team of researchers focusing on aging concluded the conserved function of lithium as part of an animal study, e.g. against Alzheimer's pathology, concluded: "The far-reaching life-prolonging effects of lithium in various model organisms suggest that it exerts evolutionarily conserved mechanisms that promote healthy aging."[191]

Evidence of Essentiality in Three Different Species

This also fulfills another requirement of the list of criteria for proving the essentiality of new trace elements, compiled by researchers at the Karl Marx University in Leipzig and presented at the beginning of the chapter: Evidence was provided across three different species, with humans as the third—even though, in this case, we had to rely on highly significant animal studies to prove the causality behind the correlative results of the epidemiological studies. Even if some animals with a relative lithium deficiency remain viable, experiments clearly show that this inadequate supply leads to a massive reduction in reproductive capacity and a high mortality rate during the first year of life. If it is so obvious that a lithium deficiency is incompatible with preserving the species and the evolutionary imperative "Be fruitful and multiply," then establishing lithium's essentiality should not really be an obstacle.

The reasons for lithium's lifespan-extending effects stem from the diverse functions of the trace element-regulated Lithiome, some of which we've already examined in connection with the mental immune system. Many of lithium's other functions, seen in Fig. 20, will be discussed in detail in later chapters. At this point, I simply want to state that these epidemiologically identified correlations stem from causal biochemical, physiological, and clinically verifiable causes. This shows that lithium is clearly an essential trace element. Given that it performs its evolutionarily conserved essential functions, evident in all model organisms, through the Lithiome, I find no plausible scientific basis to suggest that humans alone should be an exception to this evolutionary progression. As we have already seen from the mental immunodeficiency syndrome, this is precisely not the case.

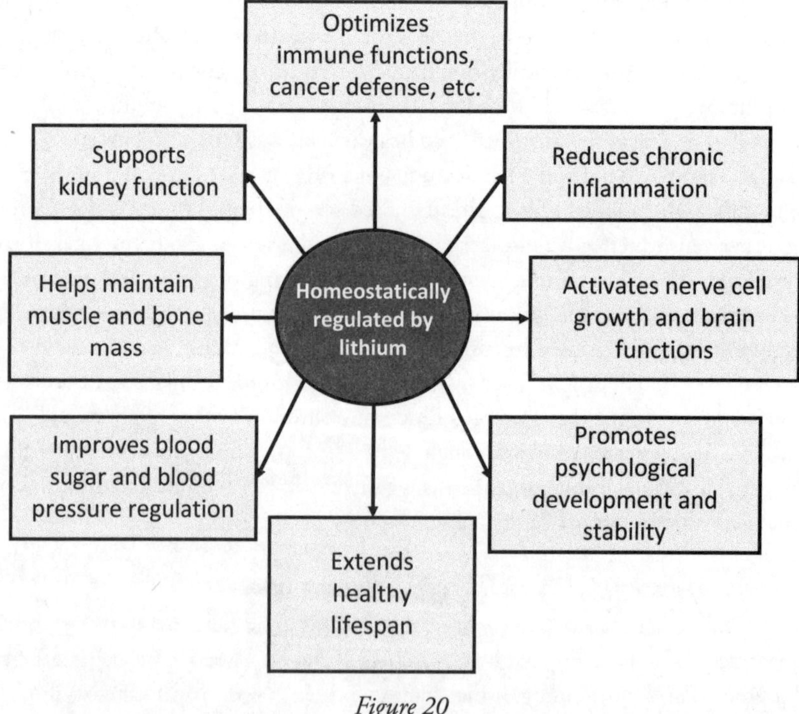

Figure 20

The development of species possessing an evolutionarily conserved Lithiome must have occurred under conditions of a stable and sufficient lithium supply over an extended developmental period. The question, therefore, is no longer whether humans have a natural need for lithium—all species do, as we learned in Argument 4. Instead, the focus is on when and where Homo sapiens was able to develop its unique mental immune system without a lack of lithium. We will deal with this in the next subchapter. It's the final link in the chain of evidence for lithium's essentiality. At this point, however, it should be concluded that Argument 5 for lithium's essentiality has also been proven:

Lithium deficiency shortens life in animals and humans, and compensating for the deficiency lengthens life.

ARGUMENT 6: Lithium Present in Physiologically Relevant Amounts During Crucial Human Evolution

The Evolution of the Human Mental Immune System

Vitamins and trace elements, for instance, are vital substances indispensable for human life. However, this has only become apparent due to humans' high cultural and mental adaptability to new habitats. This is because whenever humans drastically changed their eating habits or moved into new habitats where local conditions meant that a lack of one of these essential vital substances, developmental disorders and diseases increased. For example, long sea voyages without fruit and vegetables on board inevitably led to scurvy, which a sufficient supply of vitamin C or ascorbic acid would have prevented.[192] Peeled rice, which is almost devoid of vitamins, as the main food can lead to beriberi if no other sources of vitamin B1 are available. If the causative vitamin deficiency is not corrected, the disease—like scurvy—can be fatal and is accompanied by paralysis, cramps, and anxiety.[193] In the Alpine countries, iodine deficiency is endemic. There, in addition to goiter formation, in extreme cases cretinism (derived from the French *crétin*, meaning weak-minded) occurs, an extreme form of mental developmental disorder. Likewise, lithium intake is also subject to enormous local fluctuations—sometimes deficits arise for which the human organism cannot compensate. As a result, mental immunodeficiency syndrome develops, a condition I describe for the first time in this book. It is based on damage to the mental and psychological abilities that characterize us as human beings. Our ability to develop according to our natural potential relies on the availability of diverse species-appropriate environmental conditions and essential nutrients, the presence of which can no longer always be taken for granted today. Conversely, there must have been enough time in a particular place for our ancestors' genetic material to adapt to the conditions that offered a decisive selective advantage. Over a long period, a genetic makeup could develop there whose basic functions we share with all other humans and which still encodes our mental potential for higher intellectual and social performance today. As a result of this adaptation, however, those past conditions must still be met today if we wish to fully develop and maintain this potential. The question we must now ask ourselves is where to find the evolutionary biology equivalent of the biblical Garden of Eden, the place where all aspects of human natural needs were sufficiently available over a long, stable period. The answer would also lead us to a deeper understanding of the vital properties of the "Barrel of the Minimum" (Chapter 1), relevant to humans. In short, we would better understand why we need what in order to fully develop and exploit our innate potential.

Human longevity's lithium-dependent evolution and the mental immune system's evolution are two sides of the same coin, both culminating in the concept of the Evolution of the Grandmother, as a deficiency in essential lithium ultimately diminishes the likelihood of success in the evolutionary struggle for survival and reproduction. The central question therefore arises: When, and over what period of human evolution, was our mental immune system able to rely on a sufficient lithium supply to ensure its life-sustaining function? Since approximately 1,300 ethnic groups inhabit Earth today, all possessing the same genetic potential to develop a high-performance mental immune system despite their varied outward appearances—and it's unlikely this developed independently multiple times to the same level—there must have been such a formative phase that shaped humanity as a whole. It is therefore time for us to examine when and where in human evolution the development of our mental immune system, in particular, reached its peak and made us what we can all be today: a spiritually, psychologically, mentally stronger, more open-minded, and consequently more peaceful human race.

The archaeological search for this evolutionary "Garden of Eden" led to a discovery at the end of the twentieth century. It is located on Africa's south coast. The borders were formed by the sea on one side and an inhospitable desert on the other. Thanks to research, we also know quite precisely how long this habitat that shaped humanity must have existed. An ice age, beginning around 194,000 years ago and lasting a total of about 70,000 years, caused the icebergs at Earth's poles to grow skyward, while central African regions dried out and became hostile to life. Only the few of our ancestors who found their way to the coasts of Africa at that time, or who happened to already be living there, survived this drastic phase.[194] Genetic analyses show that the roughly 1,300 different ethnic groups comprising humanity today are descended from this group of people.[195] They not only survived this phase but also developed the genetic potential for the mental abilities that all people worldwide still possess today. The living conditions in the African coastal region were ideal for this; people lacked nothing. A significant peculiarity relevant to our inquiry is that this location offered virtually inexhaustible mussel beds, which ensured a basic maritime diet and also fostered brain development, as archaeologist Curtis W. Marean of the Institute of Human Origins at Arizona State University discovered.[196] Science journalist Ann Gibbons, who specializes in prehistory and early history, subsequently explained in her article, "Humans' Head Start: New Views of Brain Evolution," which appeared in 2002 in the prestigious US scientific journal *Science*: Our cultural self-image—still influential today—that portrays humans as "stocky hunters bringing home wild animals and carving their meat with stone tools" is flawed.[197] The derived and therefore equally flawed concept of the Paleo diet is nevertheless still frequently invoked to justify our modern consumption of meat products, including those from factory farming, as entirely natural.[198] However, this type of diet leads to deficiencies that negatively affect mental health—a conclusion supported by brain researchers such as Michael A. Crawford of Imperial College London, based on evolutionary biology. Crawford commented in a *Scientific American* article that explores the question of where our ancestors obtained the vital nutrients essential for brain development: "The animal brain evolved 600 million years ago in the ocean and was dependent on DHA and compounds essential to the brain such as iodine, which is also in short supply on land."[199] So you know: DHA (docosahexaenoic acid) stands for docosahexaenoic acid, the most complex of all omega-3 fatty acids. Our bodies can barely produce it, making it essential. According to Crawford, "To build a brain, you would need building blocks that were rich at sea and on rocky shores." The discoveries of Marean and colleagues confirm this, as Crawford further explains: "Accessing the marine food chain could have had huge impacts on fertility, survival and overall health, including brain health."[200]

Gibbons drew a much more relevant picture of the last decisive steps in human evolution in her *Science* article, namely "that of fishermen—and fisherwomen—scouring the shores for fish, seabird eggs, shellfish, and other seafood."[201] This is also confirmed by Stephen Cunnane from Sherbrooke University in Quebec, Canada: "Once we were able to access the coastal food chain in Africa—far more rich and reliable than inland sources of fish—brain and cultural evolution exploded."[202] After extensive research by nutritionists at Imperial College in London, the new finding that "There is now incontrovertible support of this hypothesis from fossil evidence of human evolution taking advantage of the marine food web."[203] In fact, in the following 40,000 to 45,000 years (i.e., about 150,000 years ago), the human brain reached a historical maximum with an average 10 percent larger volume compared to people living today.[204] Whether the comparatively larger volume also entailed a higher mental capacity can, of course, no longer be determined. But there are indications that this is the case: after all, our ancestors had to survive in a complex world without the technological aids we have today. For example, they were already capable of producing fire-hardened stone tools using complex techniques.[205] And, according to the latest findings, they were able to navigate the open sea in boats as early as 130,000 years ago.[206] "Widespread ethnographic and archaeological evidence from many areas of the world shows that modern humans living on coastlines often ratchet up the use of marine foods and develop social and technological characteristics unusual to hunter-gatherers and more consistent with small scale food producing societies," writes Marean.[207] What are these characteristics, and how did they come about?

Developing the human brain's creative abilities was the result of a genetic adaptation process not exclusively based on a rich supply of brain-healthy food. Other factors also contributed to the mental development of *Homo sapiens*, as life in coastal caves not only changed the diet, but also led to drastic changes in all areas of life. We know, for example, that communities consistently grew larger once they predominantly relied on aquatic rather than terrestrial food sources—that is, when they lived as fishers and gatherers instead of hunters and gatherers.[208] And the larger these closely knit social groups became, the more internal conflicts arose and needed resolution. Asserting oneself under these circumstances was no longer a matter of physical strength, but of social intelligence. As communities became sedentary, the need for successful self-assertion intensified within both social and reproductive competition, particularly within increasingly large and dense populations. Particular preference was given to an individual who excelled through friendliness and social competence: Foresight, peaceful conflict resolution, and emotional intelligence became selective advantages in these settings. The Dutch historian Rutger Bregman coined the term "Homo puppy" to describe humans exhibiting these characteristics.[209] In his book *Humankind: A Hopeful History*, Bregman discusses a historical animal experiment from the late 1950s that inspired this concept. The experiment took place on a Siberian animal farm and aimed to transform naturally highly aggressive silver foxes into friendly animals through selection.[210] By the fourth generation, the first animals were already wagging their tails, and with each subsequent selectively bred generation, even adult animals retained their puppy-like appearance. "The friendly foxes produced far fewer stress hormones, but more of the 'happy hormone' (serotonin) and more of the 'cuddle hormone' (oxytocin)."[211] In other words, exactly the combination of neurotransmitters that makes the hippocampus grow.[212] Decades later, American Brian Hare, now a professor of evolutionary anthropology, discovered that the friendly silver foxes were significantly more intelligent than their aggressive, wild ancestors. "If you want a clever fox," he concluded—a revolutionary insight—"you don't select for cleverness; you select for friendliness."[213] The process underlying this rule is comprehensible and, therefore, in all likelihood, applicable to other mammals: If you want to survive by being friendly, you have to be good at interpreting the behavior of others around you. According to Bregman, Dmitry Belyayev (1917–1985), the Russian geneticist who initiated this unusual project, was already certain at the time that these findings could also be applied to humans.[214] I believe we can assume prehistoric clans on the coasts of Africa also selected for friendliness. Antisocial, aggressive offspring were probably ostracized from the group or, at minimum, struggled with mating and reproduction. This selection brought with it increasingly more social intelligence, empathy and, not least, rational compassion—all characteristics of modern humans' mental immune system.

Evolution of the Lifelong Learning Mental Immune System—A Baldwin Effect? Humans are characterized by their highly efficient mental immune system and its capacity—in line with the Evolution of the Grandmother—to enable lifelong adult hippocampal neurogenesis (AHN) to make them become wiser and wiser (*sapiens*), i.e. *Homo sapiens* even into old age. Critics who dismiss any deeper mental significance of AHN claim it's merely a rudimentary remnant from the evolution of originally more primitive brains. They contend it only serves to repair comparatively simple and constant neural pathways in the adult organism—a capability our brain still masters, but from which it no longer benefits.[215] So, what evidence can be presented to show that AHN did indeed bring with it a newly acquired higher function?[216] A group of neuroscientists from Bordeaux, France, may have found the answer. The Baldwin effect, named after the US philosopher and psychologist James Mark Baldwin (1861-1934),[217] describes a mechanism of "ontogenetic adaptation": a behavior acquired during individual development (ontogenesis) is initially passed on to the next generation through imitation, but this profoundly changes the social environment for subsequent generations. Over time, the learning effects (or learning abilities) may be optimized by genetically anchored skills. New sociocultural developments, such as those made possible by life in the rich "Garden of Eden" on the African coast, may therefore have given a selection advantage to those people who had accidental genetic adaptations, which in turn led to better handling of these same sociocultural circumstances.

The Baldwin Effect, therefore, differs from the Lamarck Effect, named after Jean-Baptiste de Lamarck (1744–1829), in which behaviorally acquired traits are epigenetically inherited for a few generations.[218] However, it's also not pure Darwinism, whose discovery is attributed to Charles Robert Darwin (1809–1882), where random but advantageous mutations are selected through improved reproductive opportunities. And here comes the crucial point for our question: The Baldwin Effect enables individuals to learn or develop new behaviors that are essential for survival or very useful, but cannot be inherited as automatisms or instincts, with the help of lifelong active hippocampal neurogenesis. The French neuroscientists, therefore, assume that the initial function of adult neurogenesis was most likely actually tissue repair in simple brains, in which neurons with relatively few connections are easier to replace. However, the more complex a brain becomes, the more difficult it is to repair it with new neurons.[219] On this basis, they rightly suggest: "there may therefore have been a point in evolution where the advantages of complexity outweighed the advantages of reparability in certain areas of the brain." And the researchers from Bordeaux go on to explain: "At this point, [neurogenesis] must have lost its function of tissue repair. However, this does not mean that it has lost all its functions. In fact, it may have regressed in almost all parts of the brain, with the exception of certain privileged areas where its maintenance at a certain low level [of daily production rate of new neurons] has increased the individual's fitness despite the costs associated with maintaining neurogenesis. This is the case in the hippocampus' *Gyrus Dentatus*."[220]

According to Marean, the process of this coastal adaptation in Africa thus marks a turning point. It is "important to separate occasional uses of marine foods, present among several primate species, from systematic and committed coastal adaptations."[221] This is why—perhaps for the first time in human history—a sense of ownership emerged, as the vital food source needed to be secured for one's own clan. The need to defend shared territory against invading foreign groups both required and fostered the evolution of cooperative behavior within one's own group. This was another basic prerequisite for the "social conquest of the Earth" by *Homo sapiens*, as sociobiologist Edward Osborne Wilson (1929–2021) called it.[222]

The development roughly summarized here is based on archaeological studies and should not be romanticized. It is not a question of whether man developed certain moral standards at that time or whether he was intrinsically good or evil. However, something special happened regarding his mental constitution: The mental immune system that emerged during this epoch enabled humans to cooperate efficiently and increasingly tend toward peaceful solutions in social conflict situations. However, a consistently paradisiacal state of universal peace certainly didn't materialize in this regard, as violent conflicts have occurred in all epochs of human history, and that hasn't changed to this day. Our ability to consider the long-term consequences of our current thoughts and actions does not mean that we always make decisions beneficial to all parties involved. But the likelihood that a "*Homo puppy*"—characterized by high intelligence and kindness—will recognize a long-term advantage in peaceful compromise and cooperative behavior over affect-driven violent conflict is supported by the data: A lack of lithium impairs our mental immune system's functioning, increases aggression, and promotes antisocial behavior (see Argument 1).[223] Physician and psychiatrist James Greenblatt even describes the combination of irritability, anger, and rage as a lithium deficiency syndrome.[224] It could therefore be—and we will return to this topic in the last chapter—that the recorded history of human violence is also a history of lithium-deficient societies.

The "Expulsion" from the "Garden of Eden"

Back then, on the African coast, our direct ancestors were thus compelled by their environment to develop the mental abilities encompassed by the mental immune system: natural curiosity paired with high psychological resilience and social empathy (emotional intelligence), enormous memory capacity, creativity and inventiveness, and planning-oriented thinking. Gradually, however, the metaphorical expulsion from the "Garden of Eden" occurred, possibly impelled by steady population growth alone. Or perhaps their mental immune system simply wanted to know what lay beyond the horizon. Whatever the motivation, our ancestors spread out along the coasts and eventually inland, moving into areas less rich in vital nutrients, which meant that obtaining food entailed more and more work—and, keeping with the biblical reference, also increasingly involved a life of scarcity and therefore disease. The consequences of this "expulsion" from earthly paradise can thus be understood as a symbol for the health problems that arose from the altered living situation, particularly nutritional deficiencies. There were basically four options in the struggle to survive:

1. Genetic adaptation to deficiency situations
   I would like to provide two examples of this: The dark skin that protected our ancestors in the African cradle of humanity from intense sunlight impedes the body's own vitamin D production in more northern climes. This problem was overcome on the journey to the far north because, in the long term, only those whose skin became lighter due to random mutation of pigment genes survived. This process was merciless, for ultimately, vitamin D deficiency leads to life-threatening birth difficulties when a pelvis is rachitic and pathologically narrowed. So, the further north you look, the lighter and blonder the indigenous population is. However, the body's own vitamin D production only functions optimally in the summer months. In prehistoric times, when people spent almost all their time outdoors, this was likely still largely enough to ensure a winter supply from the body's own stores of the fat-soluble vitamin. In our modern world, however, where cultural practices lead us to less sun exposure or even to deliberately avoid it, these stores aren't sufficiently replenished in summer, and the 25-hydroxyvitamin D level then drops further in the winter months. This risks seasonal colds, which should actually be referred to as "darkening diseases" due to their cause.[225]

   Oxygen, as is well known, is also essential for life, but it's scarce in high mountains, especially the Himalayas. We now understand how people in the high altitudes of Nepal or Tibet solved this problem through genetic adaptation, and how selection continues to progress today. For example, Tibetan women living in regions above 3,500 m have genetic modifications that allow them largely normal reproductive success despite the lack of oxygen. This, in turn, means these high-altitude-optimized genes are more likely to be passed on.[226] However, the genetic optimization of vitamin D self-synthesis or genetic adaptation to low-oxygen air have their limits. Our genetic adaptability for these systems ceases at the Earth's poles or in extreme altitudes.

2. Maintaining the original, essential diet
   This is what our ancestors tried to do. Initially, they spread out mainly along watercourses to continue feeding predominantly on aquatic sources: along the African coasts, along the Nile, or along the waters of the Great Rift Valley, which stretches for about 6,000 km from present-day Mozambique in the south to Syria in the north.[227] Bone analyses confirm that the European descendants of South African fishers and gatherers continued to subsist primarily on aquatic foods, namely mussels and fish, until about 12,000 years ago.[228] This contrasts, incidentally, with *Homo neanderthalensis*, a prototypical hunter-gatherer who fed mainly terrestrially on meat and plants. As is well known, it did not survive the competition from *Homo sapiens* and died out.[229] Whether a connection exists here is unclear, but it is evident that fish is also a good source of vitamin D3. In fact, a sufficient supply of vitamin D can also be ensured through food, as the Inuit in Greenland must do all year round. The UV-B component is filtered out of the solar radiation there, which always hits at a very oblique angle, due to its longer path through the atmosphere. Since the natural synthesis of vitamin D is fundamentally impossible at these latitudes, the essential vitamin, which the body urgently needs, must be ingested with food. Although the Inuit are among the northernmost peoples, they have retained the dark skin and dark hair of our common African ancestors. This is because, throughout their entire migration to the far north, this ethnic group simply solved this problem by consistently consuming aquatic food sources.
3. Acclimating to a state of suboptimal health
   The biblical account of the expulsion from paradise meant that humans became mortal. The expulsion from the coastal "Garden of Eden," which is actually documented in the history of mankind, did indeed lead to deficiency diseases and early death as a result of suboptimal living conditions. However, living under inadequate conditions is only conceivable where there is no absolute shortage of a vital substance. This is easily understandable in the case of vital oxygen: above an altitude of 8,000 meters, the so-called death zone, no human being can survive without technical devices. But even in the gray zone, which begins at an altitude of around 5,000 meters, life becomes so difficult that only a few people reside there permanently. Humans, therefore, only penetrate those habitats where reproduction is still possible for a sufficient number of societal members. This may not be ideal, but in these marginal areas, a relative lack of vital substances isn't necessarily linked with the corresponding health consequences. After all, the unhealthy deficiency didn't arise overnight; it was the result of humans' slow expansion into these areas over many generations and was, therefore, presumably regarded as normal, fated, or not even questioned in the first place.

A good example of the unquestioned acceptance of suboptimal health conditions over long periods is chronic iodine deficiency, which shows many neuropathological and medico-historical parallels to the widely accepted lithium deficiency. Iodine is, of course, an essential component of thyroid hormones, whose function is, in turn, indispensable for our brain's development.[230] In Switzerland, for example, there are virtually no available sources of iodine. Elisabeth Pacher Wiedmer and Dr. Ulrich Joseph Woermann from the Swiss Institute of Medical Education, in their *Reminiscence of a Forgotten Disease*, write: "Until the beginning of the 20th century, the disease known as cretinism was widespread in Switzerland. Due to an iodine deficiency, the affected people suffered from a developmental disorder that led to short stature and mental retardation." Many of those affected developed a goiter or an enlarged thyroid gland. In their almost futile attempt to produce thyroid hormone despite the deficiency, this took on "sometimes grotesque dimensions." In some areas, "where more than 10 percent of the population had a goitre, it was referred to as endemic goiter."[231] In 1918, the Swiss doctor Otto Bayard observed that the iodine deficiency causing this could simply be remedied with iodized salt and without any undesirable side effects. This likely soon gave rise to the phrase "superfluous as a goiter" for things that are completely avoidable and should be avoided. However, it wasn't until four years after Bayard's discovery that the Swiss doctor Hans Eggenberger, through a popular initiative, succeeded in having table salt iodized in the canton of Appenzell. The success of this measure broke the initial resistance from the medical community throughout Switzerland and Austria—but not the resistance in Germany. There, according to the two authors, "iodine had the reputation of being poisonous and was therefore still not used at all," despite about 30 percent of the German population at that time developing a completely unnecessary iodine deficiency goiter. Interestingly, the goiter became a symbol of the deficiency, rather than the much more problematic stunted mental development. According to the historical analysis titled *Introduction and Implementation of Iodine Deficiency Prophylaxis*, the beneficial iodized salt became available in Germany around 1959, several decades(!) after the groundbreaking findings from the two neighboring countries. But "initially exclusively as a dietary food" for treating existing brain damage caused by iodine deficiency. It then took a good two more decades, until 1981, before comprehensive iodine deficiency prophylaxis was finally introduced in Germany.[232] Although this made cretinism due to extreme iodine deficiency less common, the still suboptimal iodine supply in the general population continues to go unrecognized today. In Germany, for example, according to medical guidelines, neither the iodine level is examined nor iodine supplementation recommended when an underactive thyroid is diagnosed; instead, medication is always prescribed straight away.[233]

Is Mental Health Undesirable? Iodine deficiency is still considered the most common cause of preventable brain damage and mental disability.[234] However, despite the possibility of easily solving this enormous problem, at least 1.5 billion people worldwide were still living in iodine-deficient areas in 2017, including an estimated 300 million children who'll never be able to develop their full mental potential.[235] The victims of this global "health policy" are mainly found in Africa—a continent that primarily serves the countries of the global North as a source of raw materials. According to the results of a study published in 2018, up to two-thirds of the pregnant women examined there had suboptimal iodine levels, with tragic consequences for their offspring.[236] Researchers pointed out that the "method currently recommended by the WHO for assessing iodine supply status masks inadequate iodine supply during pregnancy."[237]

What does this teach us? Firstly: a deficiency of an essential trace element doesn't necessarily lead to death. Secondly, as long as it remains compatible with life, developmental disorders and health damage are sometimes tolerated. Thirdly: even the knowledge of an essential trace element deficiency doesn't automatically lead to its comprehensive correction—despite, or perhaps cynically because of, the proven harmful effects on mental performance. This brings us to the final, fourth point regarding adaptation options, which, however, requires radical rethinking—and a considerable degree of personal responsibility—due to this problem.

1. Recognizing and eliminating the causes of suboptimal health
   We can and must use the capacities of our mental immune system to develop a deeper understanding of these connections. As soon as we discard the belief in fate or the hope that pharmaceutical promises will one day be fulfilled and ultimately accept that we, as natural beings, have natural needs, we can eliminate the health-damaging deficiencies. This applies to a lack of iodine, aquatic omega-3 fatty acids, or lithium, which humanity has so far mainly obtained through a pescetarian (fish- and seafood-based) diet. This has been proven to be associated with excellent health and a high life expectancy. However, due to overfishing of the oceans and their toxic load, this diet is no longer recommended and isn't even possible for all people.[238] An omega-3 index (percentage of aquatic omega-3 fatty acids in the membranes of our cells) of less than 2 percent is not compatible with life. Ideally, it should be 11 percent, but the global average is already only around 5 percent—and falling, with catastrophic consequences for mental health in particular; I elaborate on this in detail in my book *The Algae Oil Revolution*, including what I believe is the only sensible

way out of this self-inflicted predicament.[239] We, therefore, don't need to return to the Stone Age or the coasts of Africa to revisit the conditions of our species-appropriate "Garden of Eden." Instead, what's needed is a "forward to our roots," where we acknowledge all deficiencies and find creative ways to remedy them—a future with conscious origins.

Seafood of Knowledge?

Over thousands of generations, the genetic makeup of the people who survived and evolved on the coasts of Africa adapted to the "Garden of Eden" of that time. This included not only the new living conditions but also the diet of seafood and root vegetables, which provided optimal conditions for the development of the mind. However, it's sometimes questioned whether it was actually the essential aquatic omega-3 fatty acids that played a decisive role in the development of our powerful brains. Physiologist Cunnane argues that these brain building blocks are also found in high concentrations in freshwater fish, for example. Natural scientist Crawford also believes that it was a combination of beneficial food components: "I believe that DHA was very important for our development and the health of our brain, but I don't think it was the magic bullet alone." Cunnane suspects that "a whole range of nutrients in seafood and fish (including iodine, iron, zinc, copper, and selenium) have contributed to our large brains," which are usually not found in physiologically relevant amounts in domestic foods.[240] He did not mention lithium, presumably because it's not yet considered an essential trace element, but it belongs on the list because—as we now know—it's not only essential and particularly relevant for the mental immune system, but it's also more concentrated in the sea than in fresh water.

The scarcity of many essential trace elements inland is geological. Over billions of years of Earth's history, rain has leached them from soils and groundwater, washing them into the sea. Consequently, inland foods often have very low concentrations of trace elements like iodine, selenium, or lithium. In the oceans, concentrations are generally ten to one hundred times higher than in lakes and rivers; seawater's iodine content averages 50 µg/l. Even with a salt content of 35 grams per liter, non-iodized sea salt still contains only about 1.4 mg of iodine per kg.[241] However, since fish and especially mussels accumulate iodine thirty to sixty times higher than its concentration in seawater, just a few hundred grams of these food sources are enough to meet an adult's daily iodine requirement of 150 to 250 µg (iodized table salt is an alternative).[242] This is how trace elements return to land dwellers—though, of course, only to those who can continue to eat from these sources. Otherwise, as we've seen with iodine, there's a risk of endemic iodine deficiency goiter, particularly in mountainous regions like the Alps.

The lithium concentration in seawater is about 100 times higher than in fresh water. With an average of approximately 175 µg/l of lithium at a salt content of around 35 g, 1 kg of sea salt therefore contains about 5 mg of lithium. Thus, with 10 g of sea salt, you consume around 50 μg of lithium. The edible soft parts of eight different mussel species collected in a coastal lagoon around Ria de Aveiro in Portugal had average lithium concentrations of 3.3 mg/kg.[243] This value corresponds to an accumulation in the tissue of about eleven times, assuming a lithium concentration in the lagoon of about 0.3 mg/l. Another experimental study on mussels calculated about 0.43 mg of lithium per kg of mussel tissue, which corresponds to an enrichment of only about 2.5 times.[244] "Shellfish may have been crucial to the survival of these early humans," write Marean and colleagues, who also list a large number of different species whose shells our ancestors left behind in large quantities in the coastal caves.[245] In the African "Garden of Eden," a wide variety of shellfish species formed the basic diet. Assuming our ancestors consumed about 1000 kcal daily from this primary food source, and given that mussel tissue contains about 1 kcal/g, this would equate to around 1000 g of mussel meat per day. With an average enrichment factor between 2.5 and 11, they would have ingested about 1 mg of lithium per day. Similarly high lithium concentrations of about 0.4 mg/kg were measured in sardines and anchovies.[246] Here, it certainly comes into play that lithium—as we have already seen in Argument 3—accumulates in nerve tissue. Indeed, in the head and spinal column (i.e., effectively in the brain and spinal column, including the spinal cord), values of up to 2.8 mg/kg were found (corresponding to up to a 20-fold lithium accumulation). It was, therefore, quite possible to obtain lithium levels of well over 0.4 mg per day from fish and seafood alone as staple foods. In view of the clinical study by Gerhard Schrauzer and colleagues mentioned above, we can also assume that this provided our ancestors with a higher degree of mental well-being and a tendency toward peacefulness.[247] In the Texas study discussed earlier under Argument 5, the lowest mortality was determined at drinking water concentrations of about 0.54 mg/l. Assuming that the people included in the study consumed, on average, one liter of lithium-rich water per day—we do not have precise information on this, as it is a retrospective correlation study that could not collect such details in retrospect—this minimum amount of lithium would also have been easily achieved by the diet of our ancestors, whose staple food was seafood.

We now know precisely when, where, and how our ancestors, during a crucial stage of our development into *Homo sapiens*, naturally obtained sufficient quantities of all vital substances still essential today—including essential lithium.

This also fulfills the last Argument 6 for the proof of lithium's essentiality:

Lithium was present in physiologically relevant quantities in food during human evolution's crucial phase.

Lithium Is Essential

It has become clear that there are not just one or two plausible arguments for lithium's essentiality, but six arguments that can be substantiated in detail. Here's a concentrated overview of the line of argument:

A lack of lithium has far-reaching consequences: **(1)** the greater the deficit, the more likely developmental disorders and illnesses become. The mental immune system, which is essential for resilience, cognitive ability, and social balance (including the capacity for peacefulness, is particularly affected.

A lithium deficiency thus leads to disorders of "higher" functions that are very noticeable in social communities. This is because **(2)** lithium controls the regulation of key proteins. If it's missing, malfunctions occur that trigger mental immunodeficiency syndrome. Lithium acts on specific target structures—the *Lithiome*—and regulates fundamental life processes crucial for the long-term maintenance of cognitive functions. To ensure this supply, lithium **(3)** is specifically enriched in the tissues that need it. Its concentration remains remarkably stable for a certain period, even in the event of a deficiency, as ion channels and transport systems compensate for natural fluctuations—a key factor for the mental immune system's functionality.

These mechanisms are deeply rooted in evolution: **(4)** Lithium targets are among the oldest cross-species key proteins of life. Their conserved regulation by lithium was even crucial for the development of complex organisms—without a functioning *Lithiome*, higher cognitive abilities would probably be inconceivable.

In addition to its importance for these higher functions, lithium's central significance is also demonstrated by the fact that **(5)** a serious deficiency is incompatible with life and reproduction. Animal studies show it leads to reduced fertility and premature death. Accordingly, lithium has a life-prolonging effect for many creatures, whereas a chronic deficiency, which is often present, shortens life expectancy. It's not a miracle cure but—like other essential vital substances—a necessary component of homeostasis. The deficiency, which is quite common nowadays, therefore directly impacts the health and lifespan of many people.

Lastly, lithium's key role in human development is also shown by the fact that **(6)** the functions dependent on lithium could only develop in dependence on lithium because it was present in sufficient quantities in the diet during the decisive phase of the evolution of *Homo sapiens* and the mental immune system.

CHAPTER 3

ESSENTIAL TRACES OF AN ESSENTIAL ACTIVE INGREDIENT

Anything that is against nature will not last in the long run.
—Charles Baudelaire (1821–1867)

Lithium Demand

Based on the six arguments discussed in Chapter 2, lithium can be classified as an essential trace element. As early as 2002—more than two decades ago—lithium expert Gerhard Schrauzer summarized the knowledge available at the time about this trace element in his article, "Lithium: Occurrence, Dietary Intake, Essential Importance for Nutrition" as follows: "The available experimental evidence now appears to be sufficient to recognize lithium as essential."[1]

As I was able to show, the scientific evidence regarding lithium's essentiality has even increased significantly since then. Nevertheless, in October 2024, the German Consumer Association, a nonprofit organization with a state mandate for consumer protection, announced: "Lithium is not an essential trace element, which is why the German Nutrition Society (Deutsche Gesellschaft für Ernährung, DGE) has not published a reference value for lithium."[2] A reference value—or rather a recommended dietary/daily allowance (RDA) for a particular nutrient—is the average daily intake that should be sufficient to meet the nutrient requirements of most healthy people, depending on age, body size, and gender. In contrast to the DGE, we already know at this point in the study that lithium is essential. Therefore, the question now arises as to what daily intake should be considered optimal—i.e., the recommended daily allowance (RDA). This is not an easy task, as to date there are no clinical studies that have determined the amount of lithium required for optimal development and maintenance of physical and mental health in controlled trials. Aside from the enormous costs, such a study also presents major challenges because, according to the law of minimums, the need for a systemically relevant factor like lithium can only be conclusively assessed in conjunction with all other equally systemically relevant factors. In other words: how much lithium is optimal in an individual case very probably also depends on the individual's health situation and supply of other vital substances. However, this ideal is fundamentally very difficult to implement in clinical studies, and it's possible that the daily requirement of other essential trace elements would also have to be reassessed if lithium supply is adequate.

If we accept this limitation, we can refer to clinical studies on microdosing lithium and epidemiological studies that also provide guidance on the approximate range for the RDA of essential lithium. As explained in the previous chapter, living in a region where the local drinking water happens to have a slightly higher lithium concentration is advantageous. There is a significantly measurably lower risk of developing mental illness, behaving antisocially or even criminally, committing suicide, or developing Alzheimer's disease in the long term. The lithium concentration in drinking water and the estimated average lithium intake, sufficient to minimize these risks as far as possible, can therefore provide an estimate of daily requirements.

So, how much lithium needs to be ingested via water to ensure mental health is no longer a matter of luck? To answer this question, I consulted a Texas study. It investigated the influence of different lithium concentrations in drinking water on the rate of admission to psychiatric clinics and on the murder rate. I plotted the data points collected over two years in a graph (see Fig. 21).[3]

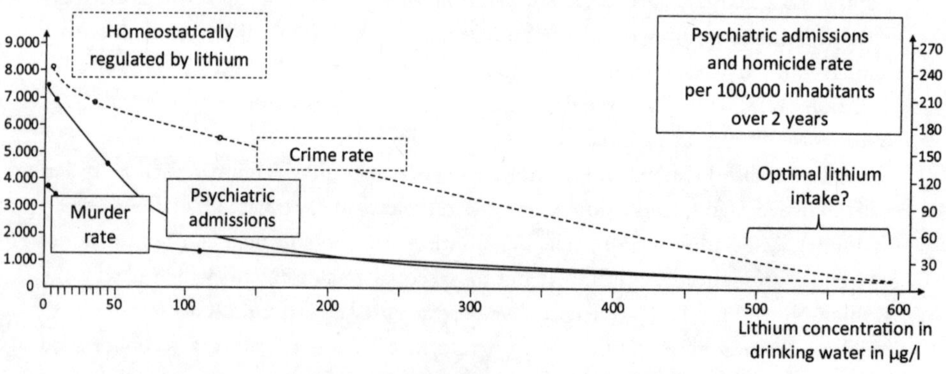

Figure 21

In addition, data points from another independent Texan study are shown; it determined the effect of lithium concentration in drinking water on the overall crime rate—even over a period of ten years.[4] If all three curves are projected into a concentration range where these lithium deficiency effects are minimized or reach a low point, they converge at a lithium concentration in drinking water of around 500 to 600 µg/l. Even if this is only a very rough estimate, the positive effect of about 500 to 600 µg/day (with a drinking quantity of 1 l/day) of lithium on this study population appears plausible. We recall: The causal link between lithium's effects within this dosage range was also demonstrated by Gerhard Schrauzer and his colleagues. Their controlled clinical intervention study showed that a 400 µg dose of lithium exerted a significant influence on mental health and well-being.[5] Moreover, a 2013 clinical study involving patients with early-stage Alzheimer's disease provides further corroborating evidence: Here, we're talking about 300 µg/day, which was able to completely halt their mental decline over the entire fifteen-month treatment period (more on this later in Chapter 4).[6]

The concentration range determined here also coincides with another study already discussed in the previous chapter titled: "Trace lithium in Texas tap water is negatively associated with all-cause mortality and premature death."[7] It not only confirms the results of a Japanese study that even comparatively low lithium concentrations of 59 µg/l in drinking water lead to life-lengthening effects, but also proves that this positive effect continues uninterrupted up to a value of 539 µg/l lithium in tap water. This indicates that lithium's life-prolonging effect may not have yet reached its maximum even at this intake level of around half a milligram (for a drinking volume of 1 liter per day). To prevent potential undersupply due to varying individual needs from an RDA set too low, I'd double the intake amount roughly determined from the graph. This would bring us to a lithium intake of about 1 mg per day. Based on these considerations, a daily intake or RDA of approximately 800 to 1,200 µg of lithium seems medically reasonable and justifiable for a healthy adult with an assumed average body weight of around 70 kg.

This value aligns with Gerhard Schrauzer's much-cited recommendation, who, back in 2002, proposed a provisional recommended daily dose of 1,000 µg for a 70 kg adult. However, I found no indication of how he determined this value.[8] He used the word "provisional" because lithium was not officially recognized as an essential micronutrient at the time (as of spring 2025), and still isn't. Consequently, his estimated RDA value hasn't been approved by authorities nor can it be integrated into nutritional practice. As mentioned earlier, the DGE does not commit to a specific value. Purely for legal considerations, I will therefore consistently refer to the RDA as *provisional*. This is the case even though the RDA estimated here falls within the range of natural basic lithium supply from fish and seafood, as was common during the evolution of the human mind in the "Garden of Eden." When I use the abbreviated term RDA for lithium going forward, it always refers to a provisional figure. However, authors like neuropharmacologist Timothy M. Marshall consider an RDA of 1 mg of lithium to be a conservative estimate, as it doesn't account for individual differences (see below: *What influences lithium levels*), which might necessitate an even higher lithium intake for optimal health.[9]

Lithium Supply for the General Population

What about the supply of essential lithium for Germany's general population? In 2020, an international research team, examining a large German study group with an average age of sixty-one (none of whom received high-dose lithium medication), determined a mean lithium level of about 0.96 µg/l.[10] This figure roughly aligns with a Berlin laboratory's reference value of 0.35 to 1.45 µg/l (with a mean of 0.9 µg/l) for whole blood, established from another German study group.[11] A clinical laboratory in Bremen provides a reference range of 0.3 to 2.2 µg/l, corresponding to a mean of 1.25 µg/l, collected from yet another German cohort.[12] The mean lithium level across all three studies, therefore, sits at approximately 1 µg/l, representing the statistical "normal value" typically achieved with a standard German diet. By incorporating the results of an experimental "mineral water study," as I'll refer to it, we can estimate the average lithium intake from food in Germany.[13] In this study, participants each drank 1.5 liters of mineral water containing a total of either 2.6 µg, 260 µg, or 2,600 µg of lithium. Blood lithium concentration rose rapidly, peaking after just thirty minutes, then halved with a six-hour half-life. The mean twenty-four-hour blood lithium content, depending on intake, was approximately 0.3 µg/l, 3 µg/l, or 30 µg/l. Since lithium intake levels and resulting average blood concentrations correlate well, a mean lithium level of about 10 µg/l can be calculated for a 1 mg/day intake (for lithium orotate, the expected lithium level is slightly higher due to its longer half-life; see below). This is roughly ten times the normal value of about 1 µg/l for the German population. From this, we can conclude that the total daily dietary intake of lithium from food and drink in Germany should be estimated at around 100 µg of lithium.

A Dutch study, which I'll refer to here as the "lithium chloride study," reached a very similar conclusion. Adult subjects took 1.74 mg (0.25 mmol) of lithium in the form of lithium chloride every day for several weeks. Their lithium level then rose from 0.97 µg/l to 27.0 µg/l—an increase of about twenty-eight times.[14] Since lithium is usually absorbed completely from food and drinking water and excreted evenly via the kidneys, it is also possible to determine the total dietary intake of the study group: If 1.74 mg of lithium additionally causes a twenty-eight-fold increase in the initial value, the daily intake before this dose increase must have been about 1/28 of 1.74 mg, or approximately 62 µg/day. This value is thus in the range of about 100 µg/day, which I was able to derive from the German mineral water study. This means that not only in Germany, but very probably also in the Netherlands (if the study group was representative), only one-tenth of the RDA of 1 mg lithium per day is consumed on average—or even less. A Dutch study from 2023 found an even lower intake of around 21 µg/day in a study group that, however, suffered from kidney problems.[15] In a study with Polish students, the value was as low as 10.7 µg/day.[16] To my knowledge, the lowest lithium intake was reported from Belgium; it was just 8.6 µg/day.[17]

Long-Term Lithium Memory—(Unfortunately) a Hairy Affair In 2002, Gerhard Schrauzer used hair analyses to calculate lithium levels in people from diverse backgrounds. He published these findings in his influential article on lithium's essentiality: results showed 406 μg/day for people from Munich, 494 μg/day for Potsdammers, and 348 μg/day for citizens of Gera.[18] Ten years prior, he and his colleagues had investigated how several months of additional lithium intake (1,000 or 2,000 μg/day) reflected in hair samples as a form of "lithium memory."[19] Combining these with lithium hair concentrations from a control group, they had three data points and inferred a linear relationship. On this basis, they developed a simple linear formula (lithium intake in mg/day = 11.6 x [lithium concentration in hair in μg/g] minus 0.43). They then used this formula worldwide to estimate average daily lithium intake by evaluating hair analyses. However, when I tried to apply the formula to very low hair concentrations, below the experimental daily doses of 1 and 2 mg, it yielded a naturally impossible daily lithium intake of less than zero. Problems with the formula also arose at the extreme upper end of the spectrum. For instance, based on hair concentrations in patients with bipolar disorder, Schrauzer and colleagues calculated a daily lithium intake of about 3 mg, instead of the actual 150 mg—a deviation of approximately fifty-fold. Therefore, a completely different curve must be assumed instead of a linear relationship between daily dose and hair concentration of lithium. Since hair growth requires a high rate of cell division, I assume that lithium is selectively accumulated there (similar to nerve cell-producing parts of the brain; see Chapter 2, Argument 3). When lithium concentration in hair is measured with a low daily intake, the actual intake is overestimated. Conversely, very high and completely unphysiological lithium intakes lead to only a relatively small increase in hair lithium concentration, as storage capacity may be limited. This results in a significant underestimation of daily lithium intake based on hair analyses. Therefore, the assumptions, the formula derived from them, and the lithium intakes calculated using them are incorrect. I assume this error was unintentionally caused by what I consider an inadmissible mathematical simplification. Consequently, the calculated and published intake quantities do not fit the overall picture for German cities—nor for many other cities and countries. This could have been used against my thesis, making it necessary to analyze this puzzling piece that distorts the overall picture. Hair analyses do not allow for statements about absolute intake quantities, but they do allow for statements about relative lithium intake from one hair sample to the next.

However, a massive undersupply must also exist in parts of the USA. For example, drinking water studies conducted in Texas and North Carolina found serious health effects even with small differences in lithium intake. These would not have surfaced if the population there had actually consumed between 650 and 3100 μg of lithium per day, as estimated by the US Environmental Protection Agency (EPA) in 1985. Incorrect assumptions and calculations are likely responsible for these values being ten to twenty times too high. About a decade later, typical dietary lithium intake in the USA was actually estimated at only 60 to 70 μg/day, according to a WHO report.[20] This value is much more realistic; otherwise, such clear and statistically significant effects would certainly not have been found in US drinking water studies (see below and also Argument 1): Anyone already adequately supplied with lithium certainly won't become healthier by consuming a few percent more. Those with a deficiency, however, benefit from even a slightly higher intake. According to the same report, the average lithium intake in Turkey was only about 102 μg/day, and in Finland, it was a mere 35 μg/day. In short, a global deficiency exists of which most people are unaware. In a 2019 article published in the *Ukrainian Journal of Ecology*, the authors, based on their analysis of actual lithium intake via food and drinking water, also concluded that "the actual consumption of the lithium with food and water indicates a low level of population provision with this trace element in most countries of the world."[21]

Lithium Levels: From "Normal" to "Natural" to "Toxic"

Normal values are merely a statistical measure. In medicine, they typically represent the average of clinical measurements taken within a population. However, they don't indicate what a species-appropriate value would be, and in most cases, they deviate significantly from this due to today's unnatural lifestyles. They are either too high or too low. For lithium, the normal value of about 1 μg/l shows that all individuals examined for that value were undersupplied with lithium. The total lithium intake at this blood level, as previously discussed, is approximately 50 to 100 μg/day, based on the aforementioned studies. However, using the results from the "mineral water" and "lithium chloride" studies, we can extrapolate what a "natural" or healthier lithium level would be if one consumes the recommended amount of 1 mg (or 1000 μg) per day. This value would then be about ten or twenty times higher than the average "normal" level, putting it at roughly 10 to 20 μg/l.

Lithium Levels: In Molar Quantities. Expressing quantities in moles offers several advantages over grams. Different atoms or molecules are equimolar if they contain the same number of particles. For instance, 1 mole of lithium and 1 mole of orotate yield exactly 1 mole of lithium orotate, even though lithium and orotate have different masses in grams. One mole of lithium weighs 7 grams, while 1 mole of orotate weighs 155 grams, making 1 mole of lithium orotate weigh 162 grams. By definition, one mole of a substance contains about 6.02×1023 or approximately 602 trillion particles. This number is known as the Avogadro constant and indicates the number of particles in a mole; it was named after the Italian physicist and chemist Amedeo Avogadro (1776–1856). Its function is to enable a direct conversion between the atomic mass unit ("u" for unit) and the macroscopic unit gram. This ensures that an atom or molecule's mass number in "u" corresponds exactly to the mass of one mole of that substance in grams. Since a lithium atom weighs about 7 u, 1 mol of lithium corresponds to about 7 g, 1 mmol to about 7 mg, and 1 µmol to about 7 µg of lithium. One mole of lithium per liter (blood) would thus be 7 g/l, which would not be compatible with life. The much lower lithium concentrations in whole blood and blood serum are therefore expressed either in µg/l, mmol/l, or µmol/l. These multiple ways of indicating quantities and concentrations are common for many vital substances. Additionally, some laboratories measure lithium in plasma, while others measure it in whole blood. The values are only comparable if the blood cells in the whole blood have approximately the same lithium concentration as the plasma (from which they were separated when determining the serum concentration). An average lithium level in the German population of 1 µg/l thus corresponds to 1/7 (one seventh) µmol/l, or approximately 0.14 µmol/l or 0.00014 mmol/l. For comparison: a lithium level of 1.2 mmol/l is targeted for treating acute mania. That's 8.4 mg/l, an incredible 8,400 times the average lithium level of about 1 µg/l in the German population that doesn't supplement lithium.

If you compare all of the results of our investigations and findings to date, the lithium level does not follow a linear but rather a sigmoid curve in relation to the intake quantity, which could be described as S-shaped (thick black curve in Fig. 22). You're likely familiar with sigmoid curves from other areas, such as exercise and sports: if you've done very little exercise for a while and then start again, you quickly experience a large increase in performance. If you assume that performance continues to increase linearly, you could calculate when a world record would be achievable. But reality quickly catches up: as performance progresses, it takes longer and longer to achieve a fixed percentage increase. If you're not careful, you'll end up overtraining, which could even lead to a drop in performance. If we transfer these findings to the relationship between lithium and health, the following picture emerges (see Fig. 22): According to animal experiment findings, a severe lithium deficiency reduces fertility and survival chances. (A complete deficiency wasn't studied, but it's highly likely to be completely incompatible with life. A complete deficiency is practically never reached in nature due to lithium's ubiquitous occurrence, at least in micro quantities). Starting from a severe undersupply ("severe deficiency"), different dose ranges can be distinguished along the sigmoid curve—from a moderate undersupply ("normal deficiency") to the natural requirement.

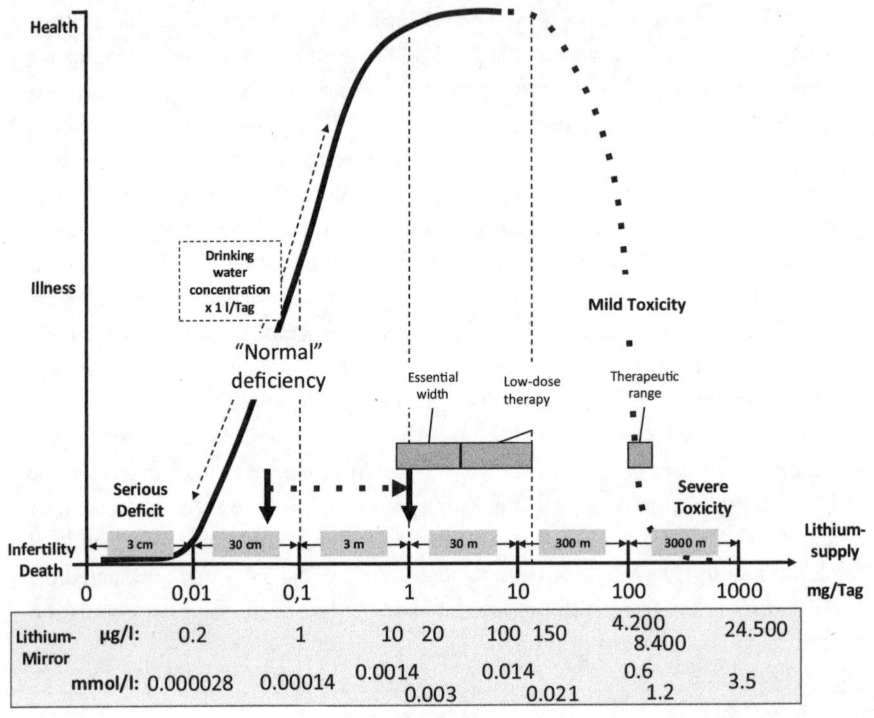

Figure 22

Since it was important to me to show you the entire spectrum of lithium levels at a glance, resulting from different intake quantities, I scaled the curve in Fig. 22 logarithmically. I've indicated the distances between intake quantities and the approximate resulting lithium levels in centimeters (cm) and meters (m). This clearly shows that the subsequent therapeutic range and the overlapping toxic range in the right part of the graph (dashed curve) are "kilometers" away from natural intake levels. If I had chosen a decimal scale, this page would have to be more than three kilometers across to fully display all areas. Let's look at the individual areas from left to right, i.e., in ascending lithium supply or increasing lithium levels, in a bit more detail:

Severe Deficit: An artificially induced severe deficit is associated with fertility disorders and neuropsychiatric developmental disorders, as we know from animal experiments (Chapter 2, Argument 5). Such a severe deficit is possible under natural conditions, defined here as an extreme deficiency with a lithium level below 0.2 µg/l.

Normal Deficiency: This reference range in Germany, considered normal at approximately 0.2 to 2.2 µg/l, could also be termed a "normal" deficiency. This range encompasses the entire steep rise of the sigmoid curve until it begins to flatten out to the right toward the "Essential Width." Most epidemiological drinking water studies were also conducted within this normal deficiency range. The curve makes it clear why even small differences in tap water's lithium concentration had significant health effects. The average value of "normal" deficiency in Germany is about ten to twenty times lower than what is achieved with a daily intake of 1 mg of lithium.

"Provisional" RDA: For a healthy adult with an average body weight of 70 kg (154 lbs), an intake of approximately 0.8 to 1.2 mg of lithium results in a new and, in my opinion, "natural" normal value of about 10 to 20 µg/l. According to test results from the Institute for Medical Diagnostics (IMD) in Berlin, the lithium level in adults, following the intake of 1 mg of lithium (in the form of 25.7 mg lithium orotate monohydrate), rose from approximately 0.5 µg/l to about 15.1 µg/l after three weeks of daily intake (see Fig. 22, black dashed arrow; IMD, personal communication).[22]

Essential Width: With the term "essential width," I want to highlight a fortunate circumstance for us: the human organism—thanks to its homeostatic regulatory mechanisms—can tolerate certain fluctuations around the optimal intake level determined here. Were this not the case, our ancestors would have had to precisely weigh how much and what they could eat or drink daily, using knowledge they couldn't possibly have acquired (e.g., lithium content in foods). Of course, they didn't have to do this, nor did they for other vital substances that naturally fluctuate. To give an example: a severe lack of vitamin C leads to scurvy, but just one apple could compensate for the deficiency and protect us from the deadly disease. It's completely harmless to our health to eat two, three, or even ten apples instead of one—regardless of potential digestive problems that might result. So, in the sense of the Law of the Minimum, we have low tolerance for deficiencies, but—in the sense of the Law of the Maximum—a relatively high tolerance for oversupply. Without advocating the false principle of "the more the better": we at least don't have to worry about the exact optimal intake amount. If a food happens to contain a little more lithium (we'll discuss specific cases shortly under *Lithium Sources—From Natural to Prescribed Supplementation*"), we can consume many times the RDA on some days. Furthermore, the exact requirement also depends on other factors. I know from people's reports that some feel much better mentally with just 1 mg more lithium per day, while others need to take 2, 3, or 5 mg to feel this positive effect. One can note that this upper, gently sloping part of the sigmoid curve is much wider than the graph suggests, at "30 meters." This means there's ample upward leeway regarding the daily lithium dose or lithium level, yet still remaining far from a toxic dose.

The Surprisingly Low Toxicity of Lithium. The minimum dose for long-term treatment of bipolar disorder with lithium is about 113 mg/day. A registration document on the safety of lithium carbonate with the European Chemicals Agency for the Evaluation and Authorization of Chemicals (ECHA) states: "Based on human data from routine long-term treatment of bipolar disorder with lithium carbonate, a NOAEL for long-term oral toxicity of 6.43 mg lithium carbonate/kg body weight per day was calculated."[23] NOAEL stands for *No Observed Adverse Effect Level*, which is the highest dose of a substance (in this case, lithium carbonate) at which no adverse effect was observed, even over long periods. This NOAEL corresponds to an intake of 1.22 mg lithium/kg body weight per day. An adult weighing 70 kg could therefore take 85 mg of lithium per day without having to expect adverse effects. For comparison: patients with bipolar disorder permanently receive around 120 mg/day. At this dosage, most patients already complain of side effects (dashed line in Figure 22). However, the safety and freedom from side effects of lithium, even with an 85-fold overdose measured against the RDA of 1 mg/day, is enormous. The resulting margin is significantly greater than for many other essential trace elements, such as zinc or selenium. Nevertheless, an article on the RDA and toxicity of zinc, for example, shows how differently trace elements are viewed. There you can read: "Zinc is considered relatively non-toxic, especially when taken orally."[24] According to the National Institutes of Health (NIH), an agency of the US Department of Health and Human Services, the recommended daily dose is 11 mg.[25] However, in contrast to the 85-fold tolerance factor for lithium, an increase in intake of twenty-five times the RDA of zinc can already lead to obvious toxicity. This manifests itself in nausea, vomiting, stomach pain, lethargy, and fatigue. Selenium deficiency also has health consequences; the requirement is given as 70 µg/day.[26] However, selenium has also been regarded as potentially highly toxic since at least the 1930s, as too much can lead to selenium poisoning (selenosis). At that time, it was discovered in South Dakota that livestock grazing in areas with high levels of selenium in the soil developed disorders known as alkali disease and "blind staggering."[27] In humans, selenium levels of over 900 µg/day cause symptoms of poisoning.[28] That is just about fifteen times the RDA! And even water is more dangerous than lithium in this respect. Increasing the RDA of 2 l/day by even ten times can be life-threatening. Water intoxication, resulting from drinking large amounts of water, leads to electrolyte imbalance and subsequent organ failure, which has even caused deaths.[29]

Low-Dose Lithium: The therapeutic low-dose range begins in the outer part of the essential width (from a dosage of 5 to 10 mg/day) and extends up to dosages of 20 mg/day. Low-dose lithium is, in many cases—and in my opinion, most cases, if a systemic approach is taken—an effective means of causally halting the progression of disease-causing mechanisms. This revolutionary application is something we'll discuss in detail later in Chapter 4. This approach could lead to many pharmaceuticals with significant side effects being either substantially reduced or even completely discontinued. It's important to note that the "low dose" in most clinical trials involving lithium is just below the high-dose treatment for bipolar disorder. This means it's still about ten to twenty times higher than the truly low dose of 5 to 20 mg/day of lithium that I'm proposing here. The reason for this typically very high "low dose" lies in the mono-therapeutic application of lithium. In contrast, I recommend a systemic approach (see Chapter 4), which should be effective with genuinely low doses.

High-Dose Pharmaceutical Treatment: If the dose is increased far enough, everything becomes a poison, according to the Law of the Maximum. Accordingly, at an unnaturally high lithium intake, as seen in the right part of the diagram (and already far beyond the sigmoid curve), there's a dose- or concentration-dependent deterioration in general condition (even if a positive effect is achieved in some patients, for example, regarding manic symptoms). The spectrum here ranges from potentially tolerable side effects to life-threatening consequences. In the treatment of bipolar disorder, the aim is for permanent lithium levels of 0.6 mmol/l and even up to 1.2 mmol/l in acute situations. This requires a permanent intake of around 120 mg/day of pure lithium (usually in the form of lithium carbonate). If an acute manic attack needs treatment, even two to three times this amount can be prescribed for a limited period. Most patients already experience more or less mild side effects at this dosage (dashed line in Fig. 22). As a result, many discontinue treatment on their own. Slightly toxic lithium levels of 1.5 to 2.0 mmol/l, which can be unintentionally reached during bipolar disorder treatment, are characterized by central nervous symptoms such as abnormally pronounced drowsiness, muscle tremors with muscle weakness, as well as diarrhea, nausea, and vomiting. This results in a very low therapeutic window of around 1.5.30.[30] This describes the relative distance between a drug's therapeutic blood concentration, where 50 percent of treated individuals (for lithium, this applies to bipolar disorder) experience a positive effect, and the blood concentration that produces a toxic effect in 50 percent of treated individuals.

Severe Toxicity: Toxic lithium levels of 2.0 to 2.5 mmol/l can be reached, for example, due to an accidental overdose or if acute renal insufficiency (a possible but rare side effect of high-dose lithium therapy) prevents excretion. The first scientifically documented case of an overdose comes from a doctor who undertook a self-experiment in 1913. He meticulously recorded the toxicological effects and published them in the *Journal of the American Medical Association*.[31] Within twenty-eight hours,

Dr. S. E. Cleaveland from the US Medical School of Case Western Reserve University ingested 2 g of lithium chloride four times. This amounted to 8 g of lithium chloride, or 1.32 g (1,320 mg) of pure lithium—a highly toxic dose. However, he knew what he was in for because, by his own account, he was familiar with the "hitherto described phenomena of lithium poisoning." Cleaveland reported:

> The symptoms appeared three to four hours after the first dose and consisted of slight dizziness and pressure in the head. [. . .] Shortly after the third dose, vision was so obscured that it was impossible to read anything smaller than the largest headlines in a newspaper. Dizziness and ringing in the ears were quite pronounced. There was also great general weakness and trembling. The fourth dose was taken at night, and the dizziness became so severe that it seemed as if the room was spinning all night, and it was almost impossible to sleep. The next morning the condition approached exhaustion. The symptoms from the night before were even stronger and the dizziness, weakness and trembling were so intense that you swayed and had to go to bed. The eye and ear symptoms persisted for about a day and a half after the last dose. The weakness and tremors lasted for five days.

A few months later, Cleaveland repeated the experiment with the same result.

In retrospect, it must be said that Cleaveland was lucky to have survived this insane self-experiment; with this dosage, he was in the potentially lethal range (far right in Fig. 22): At highly toxic lithium serum concentrations of 2.5 mmol/l and above, there is a risk of complete renal failure, impaired consciousness, even unconsciousness and coma, increased tendon reflexes, and seizures. There's also a risk of cardiac arrhythmia and a drop in blood pressure, potentially leading to shock. Such severe lithium intoxication of around 3 to 4 mmol/l typically leads to death without medical countermeasures. These are the kind of life-threatening effects that only occur at downright absurdly high concentrations, which have been perfidiously used to denigrate the trace element lithium. In the last chapter, we will delve into the associated procedure in detail.

Measuring Natural Lithium Levels

The greater a drug's therapeutic range, the safer it is. However, lithium's therapeutic range in bipolar disorder treatment is so small that there's only a fine line between treatment success and toxicity. Lithium levels must, therefore, be measured at short intervals. For this reason, most clinical laboratories specialize in determining lithium in the high concentration range. However, the measurement method they use isn't suitable for determining natural lithium levels, which are several orders of magnitude (powers of ten) lower. Since lithium isn't yet recognized as essential (as of spring 2025),

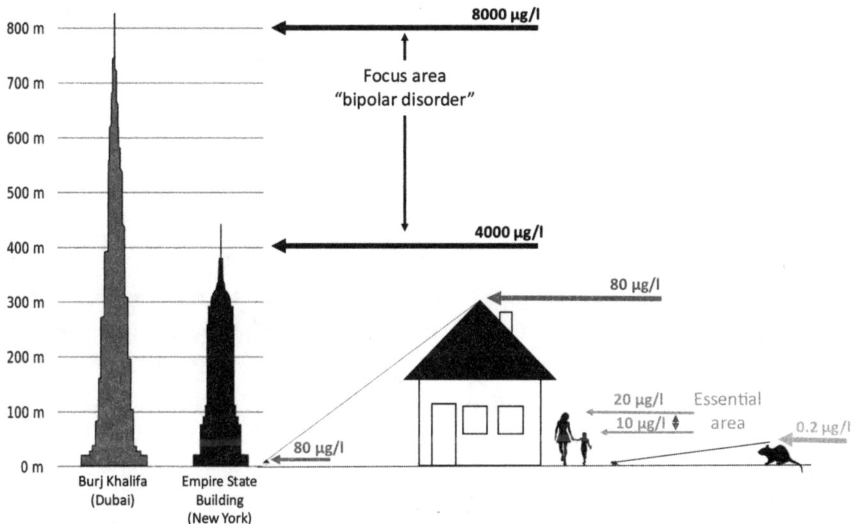

Figure 23

there's been little general or medical interest in measuring lithium concentration in the blood of people not on lithium treatment. Consequently, very few clinical laboratories in Germany, like the IMD in Berlin, specialize in measuring lithium levels in this comparatively low concentration range. This requires a completely different analytical method. If you could see lithium in the therapeutic range with the naked eye, you would need a microscope to see the quantities involved in the natural range. Fig. 23 below illustrates these relationships. The lithium level and concentration range for bipolar disorder treatment are shown here using two well-known skyscrapers.[32] The essential area, drawn to scale, isn't initially visible to the naked eye. It only becomes visible under magnification, represented by the two people in the diagram. The "low dose" range starts at intake levels from about 5 mg/day and should correspond to a lithium level of roughly 80 µg/l. In the diagram, this dosage is represented by a one-story house, which is barely recognizable compared to the maintenance dose (represented by the Empire State Building) from bipolar disorder therapy. If a doctor mistakenly sends your blood sample for lithium determination to a nonspecialized routine laboratory, the result will be below the laboratory's measurement range (for therapeutic doses). If you then take lithium in essential doses for a while and repeat the test, you'll be disappointed to get the same result. Doctors should, therefore, always ensure they commission a laboratory capable of determining lithium levels in the essential range. If you're taking a lithium supplement (see below), I generally recommend taking it in the morning. The measurement should then take place the next morning, on an empty stomach, and before your next dose. This avoids large lithium fluctuations during the measurement. If you opt for lithium orotate (we'll discuss its advantages shortly), which has a slightly longer half-life of about sixteen hours (IMD, personal communication) compared to, say, lithium from mineral water (about six hours, "mineral water study," see above), I would also wait ten to twenty days after starting intake before measuring, to ensure a steady-state value is obtained.

Essential Lithium Requirements for Children, Pregnancy, and Breastfeeding

If lithium is an essential trace element, it's logically essential for children too. After all, they belong to the same species as adults. In my opinion, the naturally increased accumulation of lithium in embryonic tissue and the brain is a clear indication of a naturally higher requirement before birth. Among many other essential functions, lithium activates the production of stem cells from which all human tissues develop.[33] The activation of neuronal growth factors and telomerase, necessary for maintaining cell division potential and viability, has already been discussed in detail in Chapter 2. Since not only the hippocampus but the entire brain, indeed the entire organism, still has to develop in children, their needs are very likely to be somewhat higher than an adult's. Gerhard Schrauzer also noted: "Special attention should be accorded to the potentially higher relative Li needs of children, adolescents and lactating mothers."[34] Once you've seen how children blossom when you compensate for their unnatural lithium deficiency, you're immediately convinced you've given them something vital. The comparison with a plant on the verge of wilting, which then flourishes again with a little water, seems very apt to me here. Potsdam-based pediatrician Dr. Christian Schellenberg, therefore, explains to parents of children with behavioral problems that administering lithium in essential quantities shouldn't be understood as drug therapy, but merely as correcting an unnatural deficiency. He repeatedly examines children who show clearly pronounced symptoms of deficiency.[35] However, he and I are certain that many children who don't show any conspicuous behavior would also benefit from compensating for a nutritional deficiency of essential lithium. Dr. Schellenberg also has a very illustrative comparison for this, which he gladly shares with parents:

> You can still cycle through life on a bicycle with almost flat tires which aren't yet down to the rim, but the ride is much more difficult than it should be. Even if you don't yet exhibit any pronounced clinical symptoms, you're less efficient: you tire more quickly and are at risk of chronic exhaustion. And you're in trouble if you drive into a small pothole or over a bump. With fully inflated tires, nothing would have happened, but now you have a flat tire and nothing works anymore. Figuratively speaking, you fall into a depression. You might even fall and injure yourself—not physically, but mentally. In worse cases, there's even a risk of post-traumatic stress disorder. In any case, the journey is over, and you ask yourself why this particular event has affected you so much.

When it comes to recommendations for children, many people's alarm bells ring—and rightly so. As Hippocrates once exacted with "*primum nihil nocere*" (first, do no harm). However, it shouldn't be overlooked that a lack of lithium would actually harm them, and in the dose range we're talking about, side effects are to be expected and weighed against the intended effect. Accordingly, doing nothing in the face of a deficiency—i.e., not supplementing lithium—would mean the actual harm. According to Hippocrates, instead of harming, the doctor should also "*secundum cavere, tertium sanare*," i.e. "second, be careful" and "third, heal." If a doctor isn't careful, they may overlook a problem, identify a false cause, or treat in a way that only creates new problems. Healing, on the other hand, is only possible by eliminating the true cause. Of course, if a deficiency of essential lithium is the real cause of a medical problem, then correcting that deficiency is exactly what needs to be done from the doctor's side.

Although the essential intake of lithium involves microdoses, also found in various types of medicinal water, it helps to know which macro doses are administered in other countries for children and adolescents with neuropsychiatric disorders such as bipolar disorder. The Mayo Clinic, for example, recommends the following dosages for adults and children aged seven and up for long-term treatment of mania, wherein the treating doctor can adjust the dose if necessary:

- For 30 kg body weight and up: 300 mg to 600 mg lithium carbonate two to three times daily; this corresponds to a minimum of 120 mg to a maximum of 360 mg pure lithium daily.
- For 20 to 30 kg body weight: 600 mg to 1,200 mg lithium carbonate per day, i.e., 120 to 240 mg pure lithium daily.
- If the body weight is less than 20 kg, the dcotor should determine the dosage.[36]

Interestingly, these recommendations do not differentiate between adults and children but are based solely on the weight classification. Despite these extremely high doses, which are well over a hundred times the requirement for this essential trace element, a meta-analysis (summarizing analysis of several studies) on the use of lithium in children and adolescents with bipolar disorder (BPD) and conduct disorder (CD), who are also frequently treated with lithium, found "no serious adverse events directly related to lithium [. . .]; the most common side effects were similar to those seen in adults."[37] Whatever these similar side effects may be, the high dose of lithium is fully recommended: "This systematic review supports the use of lithium in BPD and CD as an effective and generally well-tolerated treatment in pediatrics." If children with a body weight of 20 to 30 kg can be given "well tolerated" therapeutic doses of 120 to 360 mg of pure lithium per day, a more than 120- to 360-fold lower, essential dose of 1 mg of lithium can certainly be expected to have virtually no side effects—after all, one is far removed from the toxic range above these therapeutic doses. It has been shown that medicinal water containing the essential amount of lithium reduces the risk of serious negative effects of chronic deficiency on the mental development and mental health of children, such as the likelihood of developing mental disorders like AD(H)D, anxiety disorders, and depression (see Chapter 2, Argument 1). The pediatrician Dr. Schellenberg and I, therefore, assume that even children who don't show any mental abnormalities despite a lithium deficiency could be mentally more efficient and psychologically more resilient with a sufficient or natural supply of lithium. One should, therefore, not wait until a mental abnormality has been diagnosed to raise lithium levels to a healthy level for the mental immune system. Of course, this is always subject to the proviso that all other deficiencies in essential vital substances are also eliminated. Frequently, in children and adolescents, other serious deficiencies besides lithium are also found, e.g., in zinc, selenium, iodine, and especially aquatic omega-3 fatty acids, as Dr. Schellenberg told me in a public conversation.[38] To make it clear once again: if children have such deficits, including lithium, and these deficiencies are corrected, this isn't a drug therapy, but the correction of unnatural deficiencies of essential vital substances. In this case, lithium and the other vital substances do not become medication, even if therapeutic success is achieved or the psychologically abnormal symptoms are remedied by eliminating the deficiency. In fact, patients in Schellenberg's pediatric and adolescent practice in Potsdam often have lithium levels near the lower measurement limit in the range of 0.2 to 0.6 μg/l, which corresponds to a lithium intake of only about 10 μg/day (illustrated by the small mouse in Fig. 23).[39] For a teenager, this is just one hundredth of the lithium intake of 1 mg/day suggested here. Many of these children may be psychologically conspicuous precisely because—to return to the comparison with the bicycle—they simply have no more air in their tires. These children in particular often show astonishing improvements in their state of mind when their lithium levels are raised by daily supplementation of 1 mg lithium (in the form of about 25.7 mg lithium orotate, an organic lithium salt, the benefits of which we will discuss shortly). With such low initial values, lithium levels in the range of around 10 to often even 20 μg/l are measured after just a few days, which corresponds to an increase of 50 to 100 times. This in turn confirms that the daily lithium intake was previously only about one-fiftieth to one-hundredth of a milligram (10 to 20 μg) or even less.

Homeostasis with a Slightly Higher Requirement? Children's mental resilience is usually reduced as a result of a lithium deficiency. The resulting neuroinflammation (see next chapter) causes them to avoid potentially stressful situations, which often impacts their contact with other children. They risk becoming lonely.

With increasing lithium levels, they often truly blossom and can participate more actively in social life again. However, this is unfamiliar territory and therefore often associated with more stress, so that, according to Dr. Schellenberg, a dose adjustment may be necessary to achieve healthy homeostasis. Generally, the need for essential lithium may be somewhat higher in cases of preexisting neuroinflammatory conditions. In my opinion, the 1 mg/day dose should also be increased, at least for a certain period.

The ideal amount of lithium within the essential range for a particular person cannot be derived from the RDA, as this is a statistical value or estimate and thus only a rough guide. Dr. Schellenberg, therefore, always starts with a dosage in the lower range and increases it if necessary. For children weighing 10 to 20 kg (22 to 44 lbs), for example, 0.5 mg of lithium (equivalent to approx. 13 mg lithium orotate) per day could be appropriate based on the above-mentioned macro-dosages. For lower weights, correspondingly less, e.g., 0.25 mg of lithium (equivalent to approx. 6.5 mg lithium orotate) per day, directly after breastfeeding—I consider this dose to be appropriate according to current knowledge and would have given it to my own children if I had known then what I know now. Due to the current legal situation (non-recognized essentiality and prescription-only status) and the lack of experience with essential lithium, the pediatrician must be involved in deciding on these suggestions. But that should change in the not-too-distant future once lithium is recognized as essential and there's more experience with treating unnatural deficiencies. If your pediatrician is hesitant, in addition to reading this book, I recommend the related articles on my website.[40]

Lithium-Containing Tap Water and Fetal Growth. A 2015 study titled "Environmental exposure to lithium during pregnancy and fetal size: A longitudinal study in the Argentinean Andes" (hereafter, the "Andean Study") investigated the impact of environmental lithium on the development of unborn children. Researchers found that higher lithium levels in a mother's blood correlated with smaller fetal size, affecting the body, head, and thighs in the second trimester, as well as body length at birth.[41] For every 25 µg/l increase in blood lithium, a child's height decreased by 0.53 cm. A 100 µg/l increase in blood was linked to newborns being 2 cm shorter. The researchers concluded that "lithium exposure through drinking water was associated with impaired fetal size and this seemed to be initiated in early gestation." However, these findings contradict a large 2020 study involving pregnant women with bipolar disorder, who, on average, had lithium levels more than 120 times higher than those in the "Andean Study."[42] According to this "Bipolar Pregnancy Study" (as I'll call it), children of pregnant women with higher lithium levels were not born smaller; in fact, they were even slightly larger, a difference noticeable early in embryonic development. These findings also align with a larger prospective study from 1992.[43] But this higher birth weight is unnatural because unnaturally high amounts of lithium were administered. However, animal experiments have also shown that lithium, due to its function as an activator of stem cell proliferation, tends to stimulate rather than inhibit growth: As we've seen, lithium deficiency leads to reduced fetal growth, birth size, and birth weight. Therefore, there must be another explanation for the "Andean Study's" results of smaller birth size with higher lithium intake. The most plausible explanation is that the Andean water contained not only elevated lithium but also high levels of other substances, including up to 266 µg/l arsenic, 531 µg/l cesium, and 16,560 µg/l boron.[44] Exposure to cesium and boron during pregnancy alone has already been linked to significantly lower birth weight.[45] A meta-study also found that arsenic exposure was associated with a significant reduction in birth weight.[46] Although the authors of the "Andean Study" acknowledged these other contaminants, they focused solely on lithium in their explanation. We must also consider that other potentially toxic factors, not investigated, might have been elevated. For instance, high lithium content often correlates with increased uranium exposure.[47] Studies indicate that an increase in environmental or dietary uranium is also linked to significant fetal growth problems.[48] Until studies on microdosing lithium without the co-ingestion of multiple toxins are available, the "Andean Study"—whose drinking water was contaminated with toxins—must be regarded as a flawed piece of the puzzle. Its findings also contradict those of high-dose lithium therapy and the element's known natural effect profile.

Women during pregnancy and breastfeeding likely have a higher, not lower, lithium requirement, as Gerhard Schrauzer hypothesized. Whether this truly exceeds the proposed RDA of 1 mg/day to a relevant extent remains unclear at present (as of spring 2025) but might become evident with future lithium level determinations. Since we've already doubled the estimated essential requirement to be safe (as discussed above), I'd assume that, given the tolerances included in the 1 mg/day RDA, we're already very close to the optimum. Therefore, I wouldn't currently recommend higher doses to pregnant and breastfeeding mothers, though the essential range is certainly broad. A discussion of the surprisingly minimal negative impact of even hundredfold higher doses on a child's neurological development—observed in lithium treatment for pregnant women with bipolar disorder—can be found in Chapter 4 (see additional information: *Autism: A Danger from Lithium in Tap Water?*) Based on epidemiological drinking water studies and animal deficit studies, severe lithium deficiency is more likely to be a problem for a child's physical and, especially, mental development.

Factors Influencing Lithium Levels

In the previously mentioned lithium chloride study, participants' body weight and height were inversely correlated with serum lithium concentration: Specifically, starting from an average lithium level of 27 µg/l (with a 1.7 mg/day supplementation), a 10 cm decrease in height resulted in a 4 µg/l higher level, and a 10 kg decrease in weight led to about a 3 µg/l higher level.[49] In my opinion, these deviations of approximately 10 to 15 percent fall within the acceptable range or essential width of lithium. If your body weight exceeds the assumed 70 kg, I recommend checking your lithium level both before and, especially, after any supplementation to adjust your intake if needed. Daily fluid intake, total salt (NaCl) consumption, and the frequency and intensity of activities that cause sweating, such as saunas or exercise, can also influence lithium metabolism. If you know you significantly deviate from average values in these areas, it may also be worthwhile to monitor your levels. Certain medications, like nonsteroidal anti-inflammatory drugs (NSAIDs), i.e. anti-inflammatories and painkillers (like diclofenac, aspirin, and ibuprofen, etc.) can reduce kidney function, potentially leading to increased lithium levels. The same applies to some blood pressure medications and diuretics.[50] Individuals with kidney dysfunction should have their levels measured not only for lithium but for all other essential trace elements as well. I believe the risk of underdosing is significantly greater than that of overdosing; after all, with an RDA of 1 mg, we have a very wide essential range, allowing ample room for higher dosages.

Interactions of Lithium with Vital Substances and Medications

Since lithium, in essential concentrations, exerts beneficial biological effects across all vital bodily functions—otherwise it wouldn't be essential—interactions with other vital substances (potentially strengthening or weakening effects) are entirely natural. However, these interactions always occur within the framework of healthy homeostasis, as dosages and their effects should not exceed the bounds of healthy self-regulation. For example, essential concentrations of lithium can promote kidney function compared to a deficiency state (see Chapter 6). This could even contribute to a better excretion of many toxins or an undesirably faster excretion of medications. However, these are theoretical considerations. It would be presumptuous to attempt to comprehensively map the complexity of all conceivable interactions here, or to claim that it's possible to make a clear prediction in every individual case (for each toxin, medication, or other vital substance) for every individual. In this regard, only randomized, placebo-controlled, double-blind clinical trials can provide scientifically meaningful results. While some studies suggest adverse effects from low-dose lithium intake, they consistently exhibit systematic flaws and allow for more plausible alternative explanations.

Thyroid and Lithium. Like the brain, the thyroid gland can stabilize lithium concentration for a certain period, even in the event of a deficiency. This might explain why people who take unusually high amounts of lithium over extended periods for bipolar disorder treatment have approximately twice the risk of developing hypothyroidism compared to the general population.[51] While this increased risk isn't entirely negligible, it's remarkably small—especially when considering the potential drawbacks of low, essential lithium amounts. Compared to the extremely high doses used in these studies, the risk increase is barely significant. This makes the results of a study from the northern Andes all the more surprising: it claimed to find a 19 percent reduction in thyroid hormone levels for each additional 1 mg of lithium intake per day, starting from about 1 to 2 mg/day.[52] The findings of these two studies simply don't align: How can high doses of lithium pose a relatively minor thyroid problem, yet doses within the essential range present a comparatively larger one? As previously mentioned, people in the lithium-rich Andean region where that study was conducted consume not only more lithium but also a variety of highly toxic elements. For example, high boron intake can cause hypothyroidism.[53] The same thyroid-damaging effect is also known for arsenic, which is likewise highly concentrated with lithium in the northern Andean drinking water.[54] The observed thyroid dysfunction could very well be attributed to these contaminants, making it highly probable that lithium was not the cause of the effect seen. This is another reason why I must exclude this piece of the puzzle until studies using essential lithium are available that do not involve the ingestion of thyroid-damaging toxins.

As long as lithium is consumed in low doses, within the suggested RDA of about 1 mg of elemental lithium, to maintain natural blood serum values—which were also typical for *Homo sapiens* following a species-appropriate diet over millennia—all available evidence suggests there should be no negative interactions with other equally species-appropriate factors. Of course, to reiterate: no general statement can be made about interactions with unnatural, though potentially therapeutically beneficial, pharmaceuticals. Here are three examples:

Interactions with Antidiabetic Drugs: Lithium, by inhibiting GSK-3, can contribute to increased glycogen synthesis, which may lower blood glucose levels. Glycogen is the stored form of glucose, found primarily in the liver and muscle tissue. This is positive news, but it could also mean a significant reduction in insulin requirements. If you're a diabetic on medication or insulin, you should definitely speak with your doctor before starting lithium supplementation. If your doctor understands lithium's anti-inflammatory function and the importance of dampening chronic inflammation regarding the long-term complications of diabetes mellitus, they will hopefully work with you to determine your natural lithium needs and reduce your insulin dosage accordingly. However, this is a process that must be carried out under direct medical supervision. For this reason, I always recommend that any lithium supplementation, especially if you're taking other medications, be medically supervised. (I must also recommend this for legal reasons alone, as long as lithium isn't yet legally recognized as an essential trace element). The general principle is this: if medication addresses a problem caused by lithium deficiency, its dosage will likely need to be reduced or even discontinued entirely as the deficiency is corrected.

Interactions with Antidepressants: Lithium influences many systems that enhance our well-being, including increasing serotonin levels. It's not yet clear whether this effect only occurs at the lithium levels achieved in bipolar disorder treatment (though it's probable).[55] Unnaturally high serotonin levels can lead to serotonin syndrome, characterized by restlessness, confusion, muscle twitching, increased blood pressure, heart palpitations, fever, and in severe cases, seizures or unconsciousness. However, there's no risk of serotonin syndrome with essential or low-dose lithium. After all, there's no risk of serotonin syndrome when you increase your physical activity, even though this also leads to a physiological rise in serotonin (which, among other things, stimulates adult hippocampal neurogenesis). Antidepressants are typically the main triggers of serotonin syndrome.[56] However, alcohol consumption, which offers no physiological benefit, can increase this risk when combined with antidepressants. In my opinion, taking essential doses of lithium doesn't pose a problem but might necessitate reducing or even discontinuing antidepressant use. Physical and social activity, along with correcting all vital nutrient deficiencies, form the best combination against depressive moods.

Interactions with Other Medications Countless medications interact with various biological functions. Some interactions are more easily understood, while others may require further research. Due to lithium's blood sugar–lowering effect, antidiabetic medications might potentially be reduced (see Chapter 6). Similarly, antihypertensive medications might be reduced due to lithium's blood pressure–lowering effect (see Chapter 6). This should always be discussed with your treating physician. If you follow the systemic treatment concept I introduce in the next chapter, these (and many other medications) could very likely even be discontinued altogether. Fundamentally, I expect that as more people discover the benefits of essential lithium, we'll gain more experience with these desirable interactions and their effects. And since the consequence of these interactions is that unnatural, side-effect-laden medications become less important or entirely unnecessary, I view this as exclusively positive.

Natural Lithium Sources

Alongside hydrogen and helium, lithium is one of the three elements that formed as the first matter in the developing universe immediately after the Big Bang, condensing from pure energy.[57] Since then, it's been found worldwide. Lithium exists in the Earth's crust and all rocks at a low average concentration of about 0.006 weight percent.[58] Through slow erosion, this trace element is released, carried by rainwater into groundwater, and from there into the sea. This process, ongoing for billions of years since Earth's formation, has resulted in seawater generally having a hundredfold higher lithium concentration than groundwater, rivers, and lakes. It also accounts for the significant variability of lithium deposits in soils.[59] Consequently, higher lithium concentrations are usually found only in geologically younger soils, which are mostly of volcanic (magmatic) origin and from which lithium hasn't yet been largely leached out.[60]

The shift in human habitat from lithium-rich coastal regions, with their correspondingly lithium-rich food, to lithium-poor inland areas—or changes in dietary habits, with less fish and seafood as staples (now often even in coastal regions)—has led not only to widespread iodine deficiency, as explained above, but also to a lithium deficiency. Generally, I always recommend trying to meet your need for vital nutrients as completely as possible through a balanced diet before considering supplementation. This is because, for example, an apple contains much more than just vitamin C. However, since we no longer live in the real equivalent of the biblical Garden of Eden (see Chapter 2, Argument 6), this isn't always possible or advisable for all vital nutrients everywhere. For instance, fish and seafood are so contaminated with toxins like heavy metals and microplastics that these aquatic foods can no longer be recommended as staples[61]—to say nothing of the fact that global demand can't be met due to fish shortages caused by overfishing and environmental destruction. This eliminates a natural source of lithium, iodine, aquatic omega-3 fatty acids, and vitamin D3. Many people therefore supplement iodine with iodized table salt. I consider algae oil to be the best option for obtaining sufficient aquatic omega-3 fatty acids, as detailed in my book *The Algae Oil Revolution*.[62] I also believe supplementing with vitamin D3 is beneficial for protection against infectious diseases, as I explain in my book *Herd Health*.[63] But what options exist for essential lithium?

A large-scale study in Romania created a national database of lithium concentrations in 1,071 commercial foods and beverages, published in 2024.[64] According to its findings, consuming fruit, dairy products, eggs, and vegetables is a genuine, but generally insufficient, source of lithium. Furthermore, the study's authors note that the variability in the qualitative and quantitative composition of soil and water means that "vegetables, fruits, berries, cereals, livestock and poultry products grown in different ecological environments show significant differences in lithium content." Almonds and peanuts from the USA, Spain, and Turkey had lithium content up to about 10.46 mg/kg, though with enormous variability. Soy from Brazil and Poland contained between 8 and 11 mg/kg. Legumes, in general, seem to be a good source, especially when grown on volcanic soils: peas cultivated on the Canary Islands (Spain), known for their volcanic origin, showed a high lithium content of about 4 mg/kg, while beans had just under 3 mg/kg.[65] Canarian potatoes also registered quite high values at about 1 mg, compared to Italian potatoes with a lithium content of only 0.008 mg/kg.[66] The same applies to tomatoes: Canarian ones contained lithium at about 3 mg/kg, while tomatoes from Italy had only 0.002 mg/kg. Lithium concentrations in meat also varied tremendously by region. Pork from the Canary Islands contained about 4.4 mg/kg of lithium, whereas in Romania and Italy, it was only 0.002 mg/kg—a vast difference. According to the Romanian study, similarly large discrepancies were found in chicken eggs.

The authors of the Romanian study also point out "that low concentrations of lithium stimulate plant growth, while high concentrations, in varying doses depending on the species, individual, or organ, cause toxicity"—just as they do in humans (see also Additional Information *From Lettuce Heads to Human Heads—The Vital Function of Lithium,* below). For example, studies on biofortification (enriching the nutrient content of foods through plant breeding) as a method to combat dietary micronutrient deficiencies show that some plants can nonetheless accumulate considerable amounts of lithium in their edible tissues without any loss of yield or negative impact on quality.[67]

From Lettuce Heads to Human Heads—The Vital Function of Lithium. Lithium affects plants via the same lithium targets as it does in humans, and its impact is also dosage-dependent. Plants grow better with a small amount of lithium but fare worse with overly high concentrations. This has been demonstrated for lettuce, sunflower, and maize, among others.[68] Green amaranth has also been shown to benefit from lithium in the soil. A 2022 study revealed that low lithium concentrations stimulated plant growth.[69] Thus, in principle, lithium either stimulates or inhibits plant growth depending on its concentration, as a 2016 review summarizes.[70] While the authors state that "low lithium concentrations [. . .] improve plant productivity by increasing yield, accelerating maturation, and improving disease resistance," and that "lithium plays an important role in plant biochemistry," they reach a curious conclusion. Because lithium is toxic at higher concentrations (as in humans), they assert that there's no clear evidence "that lithium is essential or beneficial for plants [. . .] and that the importance of lithium for plants therefore remains controversial." Some highly intriguing, at times absurd, attempts to deny lithium's essentiality despite clear evidence are discussed in more detail in the final chapter.

For example, beets, rapeseed, or sunflowers can accumulate lithium in concentrations of up to 4,000 mg/kg, which suggests active ion transport in the plant kingdom as well (see Chapter 2, Argument 3). Accordingly, some scientists view *plant lithium mining*—the industrial extraction of lithium from plants (e.g., for battery production)—as a potentially viable future process.[71] If plants are being considered for industrial lithium extraction, then the targeted cultivation of lithium-accumulating crops in suitable regions with lithium-rich soils for human consumption could indeed become a sustainable option. This is especially true if the understanding that lithium should be an integral part of the human diet gains wider acceptance. In my opinion, the lithium concentration in edible plant tissue should be high enough that consuming 100–200g could cover the daily requirement. In Jordan, where the Jordan Rift Valley is a geologically active, volcanic region with lithium-rich soils, the authors of one study "recommend the consumption of 250-300g fresh weight of spinach per day per person to meet the daily lithium requirement [of 1 mg], which would have significant health and social benefits [!]."[72] I believe it would be possible to supply the world's population with essential amounts of lithium through food—particularly by cultivating lithium-enriching varieties in lithium-rich soils.

Brain Battery or Smartphone Battery? According to my theory, hippocampal growth, or adult hippocampal neurogenesis, generates the mental energy necessary for our mental immune system's functioning. It takes an intake of 1 mg of lithium per day to ensure this "brain battery" functions optimally. For comparison: a smartphone battery contains about 300 mg of lithium—nearly a full year's dose for our "brain battery." A laptop contains three annual doses of this vital element. An electric car holds enough lithium for approximately 90,000 years of human life. The world's total annual demand for lithium is around 2,880 tons, which is enough for about 96,000 electric cars. It will be fascinating to see how the raw material lithium is prioritized if it becomes scarce: for our "brain batteries" or for smartphone batteries? Who will—or should—be smart in the future?

Currently, the vast variability of lithium content in both domestic and imported foods, coupled with average dietary habits, means that hardly anyone in Germany, the USA, and many other countries (as noted above) consumes enough lithium. Otherwise, lithium blood levels wouldn't be as low as they are currently measured as "normal." However, if awareness of lithium's essential importance grows, its content in foods will likely be measured more frequently and, assuming official recognition, declared in the future. The first organic farms have already contacted me to share their plans for producing lithium-containing foods. It remains to be seen whether authorities will try to prevent this. Another natural source of lithium is lithium-containing tap or mineral water. To determine the lithium content in your local tap water, you can contact your water supplier. A list of 381 German mineral water sources and their lithium content was compiled in 2020.[73] Since some of these sources are in volcanic areas, they often contain many other dissolved minerals, some of which can be unhealthy in higher concentrations (see in the additional information: Lithium-Containing Tap Water and Fetal Growth). It's wise to check with the manufacturer beforehand, especially if you plan to drink mineral water daily over a longer period for its higher lithium content. Unfortunately, some lithium-containing brands are also relatively expensive, particularly for larger families. In many cases, supplementation offers a more affordable alternative.

(Prescribed) Lithium Supplementation

Given the widespread deficiency, one option would be to enrich tap water with lithium—indeed, a growing number of scientists advocate for this to protect mental health.[74] However, this goes against the principle of self-determination. An alternative would be to offer lithium-fortified table salt alongside iodized salt. Both options, aside from any potential drawbacks of the delivery method, would require lifting the absurd ban on lithium as a dietary supplement. Consequently, the only legal way to obtain essential lithium in the EU at present is through prescription medications.

In late 2022, I successfully convinced Sabine Bäumer, the pharmacist at the Eisbär pharmacy in Karlsruhe, to prepare 1 mg daily doses of lithium on prescription, and a year later, this became a reality. Many other pharmacies have since followed suit.[75] Beyond the unnecessarily high effort of finding a doctor who recognizes lithium's essentiality and is willing to issue a prescription,[76] this solution also incurs higher costs. This is because EU regulations prohibit pharmacists from prepackaging prescription substances into daily doses or commissioning third parties to do so, which drives up unit costs. My suggestion: Instead of one RDA dose per capsule, capsules could be filled with 10 or 20 RDA units, depending on family size. You could then dilute individual capsules yourself by a factor of 10 or 20 to achieve the daily RDA (see my instructions on YouTube).[77] For lithium orotate, this method reduces the cost per RDA to just a few cents. Despite all these disadvantages, the prescription requirement does offer at least one benefit: pharmacies guarantee GMP quality. GMP refers to the "Good Manufacturing Practice for Medicinal Products," which, according to EU regulations, aims to ensure that pharmacy products are of impeccably high quality. This is a guarantee you won't get on the current "black market" for lithium preparations.

Is Lithium Orotate the Better Lithium Salt?

If supplements are the most accessible source, the question arises: which of the various lithium salts is most suitable? Lithium doesn't occur naturally as a pure element; due to its chemical reactivity, it always exists as a positively charged ion (cation)—either dissolved in liquids, bound to a negatively charged ion (anion) as a salt, or embedded in rock. Neuropharmacologist Marshall noted that in general salts that easily dissolve (ionize) into their ions in an aqueous medium tend to irritate the gut when ingested, leading to side effects like nausea and diarrhea.[78] For example, low to moderate doses of easily ionizable magnesium salts like magnesium oxide or magnesium sulfate are poorly absorbed in the gastrointestinal tract. This creates an osmotic effect, clinically used as a laxative. Magnesium, however, forms a stable complex (chelate) with orotic acid, the salt of which is called orotate. Due to its low ionizability, this complex doesn't cause such irritations but boasts high bioavailability (over 60 to 70 percent absorbed in the gut).[79] Moreover, orotate appears capable of transporting magnesium ions into the cell, favoring intracellular magnesium concentration. When searching for the best "partner" for the lithium ion, orotic acid presented itself as an obvious choice. This is particularly true since, according to Marshall, lithium orotate is also absorbed very stably and intact in the gut, travels through the bloodstream to cells, and is efficiently taken up there via an orotate transport system. This means lower doses are needed for a good effect compared to other lithium salts.[80]

As early as the 1970s, German physician Hans Nieper (1928–1998), a pioneer in researching the orotic acid transport system, postulated that orotic acid acts as a mineral carrier, transporting ions of lithium, magnesium, or calcium across biological membranes.[81] Half a century later, in 2022, this orotate transporter was finally discovered.[82] The evolutionary biological explanation for its existence lies in the fact that orotic acid is particularly needed in highly bioactive cells (those with a high RNA synthesis rate) that divide frequently (high DNA synthesis rate). It is thus also the ideal transport system for lithium, which is also required in proliferating (growing and multiplying) cells. But if lithium orotate is so stable, how does it exert its effect as a lithium ion within cells? The explanation lies in the cell's demand for orotate. According to Nieper, the stable orotate-mineral complex breaks down upon entering the cell once the orotate is used for the biochemical synthesis of RNA and DNA building blocks, thereby releasing the lithium ion at its sites of action.

Thus, lithium and orotate are indispensable for tissue formation, cell regeneration, the regulation of the bodily immune system, and the growth and maintenance of the mental immune system.[83] For a long time, orotic acid was even referred to as vitamin B13 due to this essential function, until it was discovered that it can also be formed in the human body from the amino acids aspartic acid and glutamine. The high demand during body growth is perhaps the evolutionary biological explanation for why this molecule is primarily formed in the mammary gland, ensuring the infant receives sufficient amounts of this essential building material. Orotic acid was first isolated from cow's milk in 1904, giving the molecule its name: *orós* is the Greek word for whey. Depending on feed and season, cow's milk contains around 80 mg/l, sheep's milk 32 mg/l, and goat's milk 25 mg/l of orotic acid.[84] Human breast milk, however, contains relatively little orotic acid at about 1.7 mg/l.[85] Nevertheless, as a precursor for genetic building blocks, orotic acid is naturally beneficial for our microbiome and, via the well-known gut-brain axis, indirectly contributes to the mental immune system.[86]

Compared to lithium orotate, lithium carbonate, most commonly used in bipolar disorder treatment, ionizes readily in solution. This necessitates administering comparatively very high doses to drive lithium ions through the blood-brain barrier and into cells. Ultimately, the unphysiologically high concentrations needed for pharmacological effects are no longer naturally regulated by existing ion channels and transport systems. These combined disadvantages explain the high rate of side effects and the risk of toxicity associated with lithium carbonate. These pharmacokinetic differences between the two lithium salts (pharmacokinetics, from ancient Greek *phármakon* for remedy or poison and kinesis for movement, describes a drug's journey into, through, and out of the body) have also been demonstrated in animal models. When rats are injected with lithium orotate, the mineral's concentration in the brain after twenty-four hours is three times higher than with an equimolar amount of lithium in the form of lithium carbonate.[87] From these data, scientists concluded that therapeutic lithium concentrations in the brain can be achieved with significantly lower doses, meaning far fewer concerns about toxicity should exist for lithium orotate. This consideration was then tested in an animal model, the "manic mouse," whose abnormal behavior is induced by psychostimulants like D-amphetamine. Amphetamines, including speed and crystal meth, are known for their strongly stimulating or energizing effects, often accompanied by euphoria lasting several hours. Affected individuals feel highly capable, deeply focused, and literally feel the need to turn night into day. In mice, this leads to amphetamine-induced hyperlocomotion (AIH), an almost insatiable urge to move. Lithium orotate, at a ten-fold lower lithium intake compared to lithium carbonate, led to an almost complete blockade of AIH, indicating significantly improved efficacy and potency.[88] The authors also emphasized that despite its better efficacy, lithium orotate had no negative effects on kidney and thyroid function compared to lithium carbonate.[89]

Toxicology of Lithium Orotate Monohydrate. Although evidence suggesting increased brain availability and thus superiority over lithium carbonate was found as early as 1978, lithium orotate was not clinically adopted. This was because rat studies indicated it could cause greater kidney function impairment at the same concentration.[90] However, in these studies, lithium orotate was not administered orally, which would be natural, but injected into the abdominal cavity as a highly concentrated "slurry," as the authors termed it. This is a completely unnatural situation. Therefore, the experiment was repeated in 2021, this time using the natural oral route of administration. In the twenty-eight-day oral toxicity study, lithium orotate monohydrate showed no mutagenicity—meaning it didn't induce mutations, which is crucial for ruling out cancer risk, for example—in a series of genetic toxicity tests, even with repeated administration.[91] Furthermore, even at the highest intake level of 400 mg/kg body weight/day, *no* toxic effects were observed on the usual target organs for lithium toxicity. The fact that no adverse effects were seen even at this high dose (NOAEL) explains, according to the authors, why there have been no warnings of adverse effects to date, despite lithium orotate having been used as a dietary supplement in the USA for decades. If, to be cautious, we apply a safety factor of 10 for potential interspecies variability (accounting for possibly higher human sensitivity compared to the animal model), this yields a safe reference dose for humans for lithium orotate monohydrate of 40 mg/kg body weight per day. For an average adult weighing 70 kg, this translates to a safe long-term intake of 2,800 mg of lithium orotate monohydrate per day. This would correspond to a maximum intake of approximately 110 mg of pure lithium, which is roughly equivalent to the maintenance dose in high-dose lithium treatment. Given an RDA of 1 mg/day, this represents a safety factor of 110 up to the NOAEL—even higher and safer than the factor of 85 for lithium from lithium carbonate (see Additional Information above: *Lithium's Surprisingly Low Toxicity)*.

These advantages of lithium orotate (which, as of spring 2025, is not used pharmaceutically) over the disadvantages of lithium carbonate (which is primarily used pharmaceutically) are also evident in humans, for instance, in the treatment of alcohol dependence. Lithium orotate, at a dosage of 150 mg/day (approximately 6.4 mg pure lithium) administered over six months, proved very successful in reducing alcohol consumption in forty-two hospitalized patients.[92] Treatment with lithium orotate at this dosage was safe, and the observed side effects were minor, disappearing when this daily dose was given only four to five times per week. Lithium orotate was effective, whereas lithium carbonate showed little or no efficacy even at much higher doses of 600 mg/day (about 120 mg pure lithium), roughly twenty times the dose.[93] A prospective, double-blind, placebo-controlled study, conducted over six months, showed that lithium carbonate was no better than a placebo in achieving complete abstinence in alcoholic patients.[94] These studies on alcohol consumption demonstrate that lithium orotate, due to its entirely different pharmacokinetic properties, is effective not only in "manic mice" but also in humans at a significantly lower dosage than lithium carbonate. It's not just about how much lithium is administered, but how much actually reaches the cells and exerts its effect there. The lithium level in the blood, it must be understood, does not provide information about the function-critical concentration within the cells. Finally, the transport systems ensure that the proliferating cells, such as those in the hippocampus, maintain their lithium concentration even in the event of temporary insufficient supply—at least to a certain extent, as discussed in Chapter 2, Argument 3. Therefore, let's establish this: While lithium carbonate is more common than lithium orotate, in my opinion, it is less suitable—neither clinically for high-dose treatment of bipolar disorder nor as a vehicle for delivering the essential lithium dose. This is because it requires significantly larger amounts for the same effect, which in turn leads to a higher rate of side effects. This view is also shared by the two authors of the "Manic Mouse Study" in their article *"Lithium Orotate: A Better Option for Lithium Therapy?"*[95]

It is very important to me that these new findings regarding the disadvantages of lithium carbonate compared to lithium orotate gain acceptance in high-dose treatment. This is provided, of course, that high-dose treatment remains a valid strategy and is not successfully replaced by a systemic alternative, which we will address in Chapter 4. In addition to lithium carbonate and lithium orotate, there are other lithium salts:

What Quantity of Different Salts Contains 1 mg of Lithium?
23.4 mg Lithium Orotate
25.7 mg Lithium Orotate Monohydrate
5.3 mg Lithium Carbonate
6.1 mg Lithium Chloride
19.0 mg Lithium Aspartate
10 mg Lithium Citrate

Lithium Citrate: Offers no additional benefit over lithium carbonate, making it similarly disadvantageous compared to lithium orotate.[96]

Lithium Aspartate: There is limited literature on its efficacy and/or safety. While some studies explored its potential for curbing alcohol[97] and drug dependence,[98] they found little to no effect. This suggests it's less effective than lithium orotate, a point we'll further examine with the specific example of long COVID and post-vac syndrome in Chapter 5.

Lithium Chloride: When administered orally to rats, lithium chloride and lithium carbonate showed similar absorption and excretion patterns. However, lithium carbonate consistently resulted in higher overall plasma lithium levels across all tested concentrations.[99] Due to its high ionizability, lithium chloride doesn't appear to offer any advantages over lithium carbonate, and certainly none over lithium orotate.

Lithium Sulfate: Studies comparing the pharmacokinetics of lithium sulfate with lithium carbonate found no significant differences, even concerning the induction of side effects.[100] Therefore, there's no compelling reason to prefer lithium sulfate over lithium orotate.

Colloidal Lithium. Colloids (from ancient Greek *kólla* for glue and eidos for form) refer to particles finely dispersed in a liquid. The size of the individual particles is in the nanoparticle range. Colloidal silver, for instance, is defined as a mixture of silver ions and silver nanoparticles suspended in an aqueous medium.[101] Scientifically, very little is known about colloidal lithium, including its pharmacokinetics or pharmacodynamics. Fundamentally, lithium can only exert its biological function as a positively charged lithium ion. In my opinion, the amount of ionized lithium in colloidal lithium solutions remains unknown. A statement in the book *Colloidal Minerals & Trace Elements in Colloidal Form*, which promotes colloidal lithium for depression, anxiety, mania, bipolar disorder, etc., is perplexing: "Colloidal lithium has the advantage over pharmacological lithium agents that it has no side effects and you cannot overdose on it."[102] This claim that it fundamentally cannot be overdosed only makes sense to me if it has no effect per se—meaning it contains no biologically active ionized lithium. Until reliable studies on the composition, pharmacokinetics, and effects of colloidal lithium become available, I remain skeptical.

In summary, I believe there's currently no better form of lithium supplementation than lithium orotate. This supports the direction taken by orthomolecular medicine in the USA, where it has been used as an over-the-counter dietary supplement for years without complications. Though, as we'll see in the last chapter, the lifting of its ban there might not have been due to goodwill, but rather an administrative oversight (Chapter 7, Additional Info: *No Lithium Ban in the USA?*)[103]

CHAPTER 4

PSYCHONEUROIMMUNOLOGY—THE LITHIOME AS A LINK BETWEEN BODY, MIND, AND SOUL?

When the gods are set to ruin a man,
they instill in him wrong-headed thinking.
—Sophocles (497/496–406/405 BC)

The Vicious Cycle of Neuroinflammation

In Chapter 2, we were able to demonstrate lithium's essentiality by showing a direct causal chain between lithium, its various targets (the Lithiome), and its influence on central biochemical and physiological mechanisms, extending to the functionality of the human mental immune system. We were able to show that through these diverse regulatory mechanisms, lithium exerts a direct and indirect positive influence on adult hippocampal neurogenesis—the functional core of the mental immune system, yet also its Achilles' heel. This causal chain establishes lithium deficiency as a causative risk factor for the development of mental immunodeficiency syndrome. In this chapter, we'll take a step further. We'll explore how the mental and physical immune systems collaborate under natural conditions and in alignment with the evolutionary imperative during acute threat situations. We'll also investigate how this cooperation, when hampered by lithium deficiency, leads to a variety of neuropsychiatric developmental disorders, neurological dysfunctions, and chronic mental and neurodegenerative diseases.

In keeping with the evolutionary imperative, the physical and mental immune systems communicate via messenger substances. In dangerous situations, they can thus act as a cohesive unit, working as efficiently as possible for the benefit of the entire organism. For example, the physical immune system informs the mental immune system about an infection or tissue damage by releasing messenger substances (cytokines). The mental immune system then reduces its activity, a phenomenon we call sickness behavior. This allows the organism to dedicate more energy to overcoming the infection: you rest and withdraw to recover and reduce the risk of contagion for your immediate surroundings. This latter aspect is an impulse from the social immune system: through this cytokine-induced sickness behavior, we instinctively protect our environment. Conversely, if the mental immune system detects a threat, it triggers the release of stress hormones. These suppress the activity of the physical immune system, ensuring that almost all available energy is initially directed toward a fight-or-flight response, even in the event of injury. However, this "dialogue" between the two immune systems, mediated by messenger substances, can quickly escalate into a fierce "discussion" if there's a lithium deficiency. In such cases, neither the physical nor the mental immune system reacts adequately, risking entry into a pathological vicious cycle of chronic neuroinflammation, as illustrated in Fig. 24. Depending on the timing and duration of the triggering event, this can lead to hippocampal developmental disorders and even neurodegenerative diseases. This vicious cycle of neuroinflammation lies at the heart of a long list of neurological, psychiatric, and mental illnesses, which collectively represent the greatest medical and societal challenge of our time. Therefore, it's worthwhile to develop a deeper understanding of these neuropathological connections.

Microglial Cells and Their Danger Sensors

As can be seen in Fig. 24, the hippocampus is at the center of the mental immune system, with microglial cells playing a pivotal role. Glia is a collective term for all nonneuronal cells in nervous tissue, originally thought to simply hold nerve tissue together—hence their Greek name *glia*, meaning "glue." Microglial cells are the smallest of these nonneuronal brain cells, hence "micro," yet they are responsible for recognizing all forms of danger: pathogenic microorganisms (brain infections), physical and psychological brain damage (i.e., brain injuries and extreme stress damage). This recognition relies on a multitude of danger sensors, most of which are located in the outer and inner membranes of our body, particularly our immune cells. These danger sensors are also called Primitive Pattern Recognition Receptors. They're "primitive" because they're found in all animal and even plant organisms, having evolved long before adaptive immunity.[1] They're "Pattern Recognition Receptors" because, over a long evolutionary period, they've learned to recognize dangers or basic patterns of typical threats. This constitutes our first line of defense; the second is the adaptive immune system, which, for instance, develops highly specific antibodies against pathogens. However, that takes time, which the danger sensor defense system buys for us.

We possess several families of danger sensors.[2] One superfamily consists of Toll-like receptors (TLRs). In humans, this family includes over a dozen members that arose through gene duplications during evolution (much like IMPase or GSK-3 duplicated, as discussed in Chapter 2, Argument 4).

The TLR family owes its name to the discovery of a mutation of a family member in fruit flies. Its discoverer, German Nobel laureate Christiane Nüsslein-Volhard, spontaneously exclaimed "*Das ist ja toll!*" (That's great!") upon realizing the physical change. Following this "great" discovery, all other receptors in this large family were named Toll-like receptors. A commonality among the TLR superfamily and all other danger sensor families is their ability to recognize characteristic features of pathogens, known as Pathogen-Associated Molecular Patterns (PAMPs). PAMPs are conserved structures (molecular motifs) of pathogenic microorganisms like bacteria or viruses. For example, the spike protein of coronaviruses, present in all members of this large virus family, is a PAMP. In Fig. 24, PAMPs are depicted as small, lightning-like jagged lines. When PAMPs are recognized by the TLR-4 danger sensor, the corresponding microglial cell immediately goes on alert: viruses have invaded the brain, and life is in danger![3] After a PAMP—such as the spike protein of a coronavirus or the S1 subunit of SARS-CoV-2—docks onto TLR-4, an intracellular signaling cascade is activated. At the heart of this cascade lies the familiar lithium target, GSK-3.[4] GSK-3, activated by TLR-4 or by the TLR-4 complex (composed of various proteins), prompts the transcription factor NF-κB (pronounced: NF-kappaB), to migrate from the cytosol or cytoplasm (the outer cellular space distinct from the nucleus) into the cell nucleus. There, NF-κB specifically activates the transcription of genes into corresponding mRNAs, which then code for pro-inflammatory messenger substances (cytokines).[5] Outside the cell nucleus, the blueprints encoded in the mRNA are then translated into the corresponding cytokines. These pro-inflammatory cytokines, produced in large quantities, include Interleukin 6 (IL-6) and Tumor Necrosis Factor α (TNFα), which are directly released by the microglial cell after production, as well as Interleukin-1β (IL-1β) and Interleukin 18 (IL-18), which undergo activation in the inflammasome (a cytosolic multiprotein complex of the innate immune system) before their release.[6] Essential lithium prevents an overzealous neuroinflammatory response at multiple levels; thus, the greater the lithium deficiency, the more intense the activation of this pro-inflammatory signaling cascade. The trace element activates the signaling protein Akt (see Chapter 2, Argument 2, Additional Info: *Low Dose, Huge Effect?* Point 3) and thereby limits the number of TLR-4 proteins on immune cells.[7] Additionally, as thoroughly discussed in Chapter 2, it naturally inhibits GSK-3, as also shown in Fig. 24. This combination of lithium's functions ensures that not every stressful situation immediately leads to an inappropriate inflammatory response with adverse health consequences.

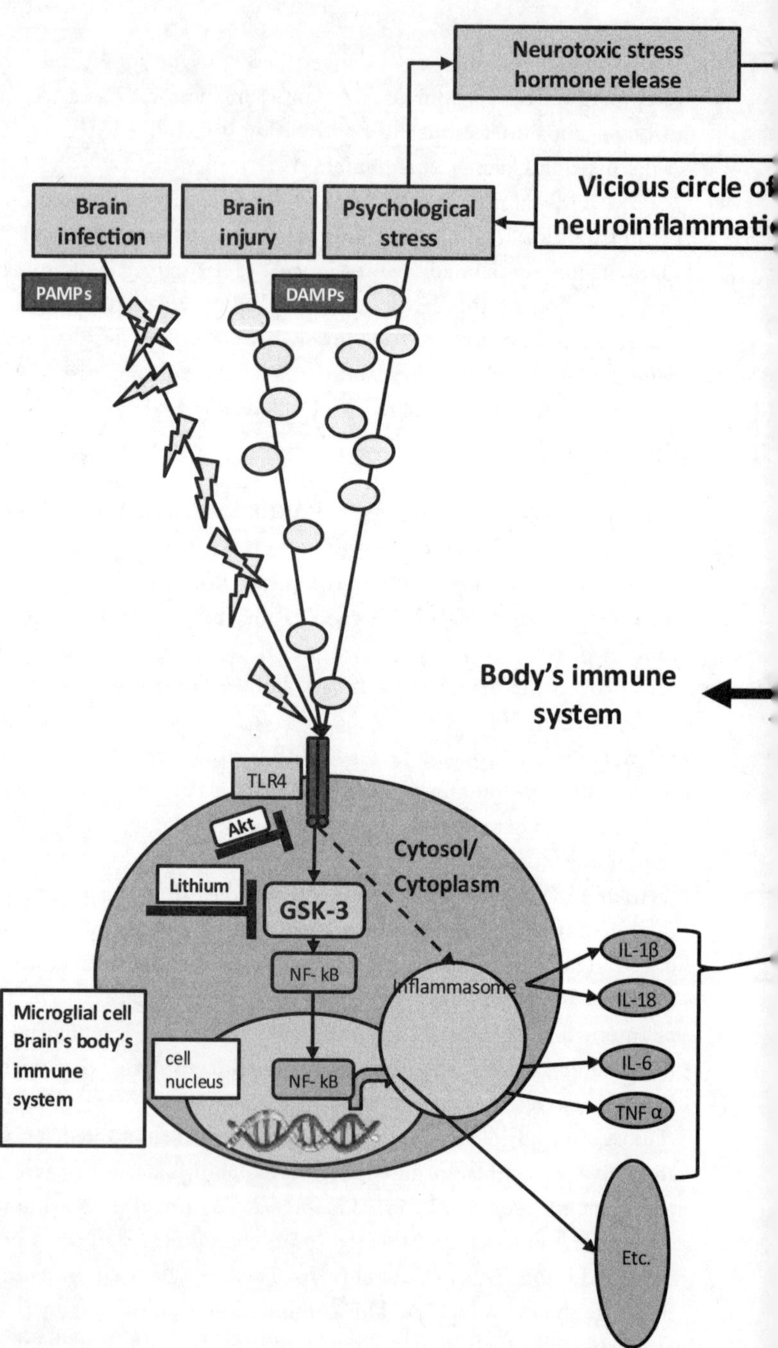

Neurotoxic stress hormone release
Vicious circle of
neuroinflammati
Brain infection
Brain injury
Psychological stress
PAMPs
DAMPs
Body's immune system
TLR4
Akt
Lithium
GSK-3
Cytosol/ Cytoplasm
NF- kB
Inflammasome
IL-1β
IL-18
IL-6
TNF α
Etc.
Microglial cell Brain's body's immune system
cell nucleus
NF- kB

Figure 24

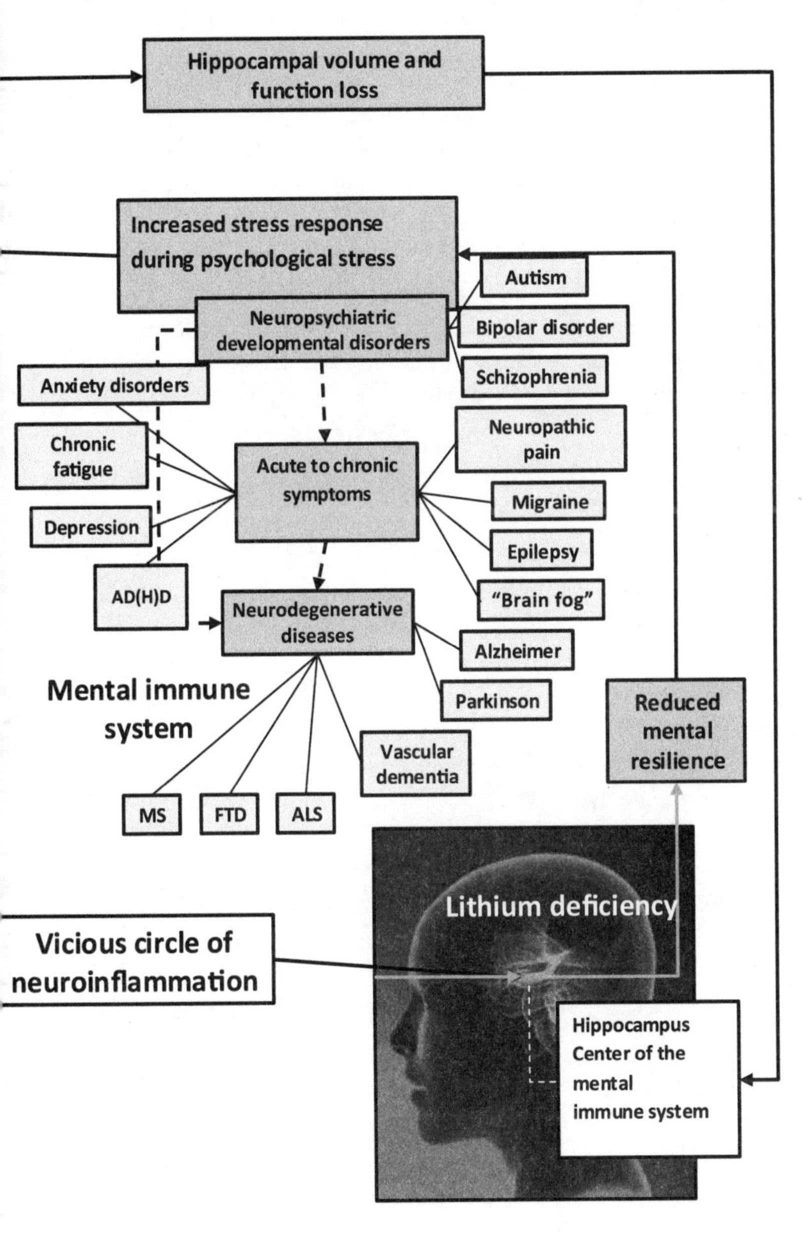

Hippocampal volume and function loss
Increased stress response during psychological stress
Autism
Neuropsychiatric developmental disorders
Bipolar disorder
Schizophrenia
Anxiety disorders
Chronic fatigue
Acute to chronic symptoms
Neuropathic pain
Migraine
Depression
Epilepsy
AD(H)D
"Brain fog"
Neurodegenerative diseases
Alzheimer
Mental immune system
Parkinson
Reduced mental resilience
Vascular dementia
MS
FTD
ALS
Lithium deficiency
Vicious circle of neuroinflammation
Hippocampus Center of the mental immune system

Natural Lithium or Unnatural Lithium Mimetics as Anti-Inflammatory Agents
The entire inflammatory activation cascade, from TLR-4 via GSK-3 to NF-κB, is central not only to all neuroinflammatory diseases but also to all chronic inflammatory conditions, such as rheumatoid arthritis and arteriosclerosis. Given the prevalence of these lucrative diseases, developing therapeutic strategies based on the pharmacological inhibition of the entire NF-κB signaling cascade is of significant pharmaceutical interest.[8] Inhibiting NF-κB activation by GSK-3 using lithium would be the simplest approach (see Fig. 24). However, this isn't patentable or marketable as an expensive drug (more on this in the last chapter).[9] Yet, this would be the natural and thus side-effect-free option. After all, lithium's essential braking function is to prevent even minor immune cell activation from leading to an excessive reaction and the release of pro-inflammatory messenger substances, which could otherwise spiral into a self-destructive vicious cycle.

The sickness behavior triggered by these and many other cytokines (such as IL-12, IL-13, and IL-23) is beneficial and evolutionarily explainable during an acute infection; it helps overcome the illness more effectively and protects those around you. However, it becomes problematic if there is no genuine infection—for instance, if genetically modified mRNA leads to artificial spike protein production following a "COVID-19 vaccination." (See Additional Info: *PAMPs as Bioweapons*) It's also problematic if an infection becomes chronic, like Lyme disease, which can even be accompanied by severe depression and suicidal thoughts due to neuroinflammatory damage to the mental immune system.[10] In both scenarios, there's a long-term risk of chronic spike production and actual chronic infections, potentially even leading to Alzheimer's disease.[11]

PAMPs as Bioweapons. Microglial cells can't distinguish whether their TLR-4 sensor has merely detected the brain-permeable spike protein of a coronavirus, or if the entire virus has actually invaded the brain; both scenarios trigger neuroinflammation. Therefore, if the goal is to cripple an opponent's mental immune system with a bioweapon, ensuring the highest and most sustained presence of the spike protein in the brain is sufficient. SARS-CoV-2 was precisely developed for this purpose, as I detail in *The Indoctrinated Brain.*[12] Through the genetic insertion of a furin cleavage site, a part of the spike protein, known as the S1 subunit, detaches from SARS-CoV-2. Unlike the virus itself, this S1 subunit can then cross the blood-brain barrier, where it activates TLR-4.[13] Even more dangerous are the genetically modified mRNAs declared as COVID-19 vaccines, which encode for the viral spike protein: These are: (1) genetically engineered to remain active in the human body for an extended period; (2) packaged in lipid nanoparticles designed for brain permeability, thereby inducing neuroinflammatory spike production in the brain; and (3) the spike protein itself, or its S1 subunit, is brain-permeable due to this bioweapon-like genetic modification, regardless of where the mRNA forces body cells to produce spikes. Chapter 5 will discuss therapeutic options for "spikeopathy" or brain fog as vaccine damage following mRNA injection using low-dose lithium orotate.

Before we delve deeper into the overall effect of acute and especially chronic infections on the mental immune system, let's look at how this pro-inflammatory cascade can be activated by other dangers. Because TLR-4, like other members of its family, isn't just a danger sensor for PAMPs, but also for DAMPs—which, particularly under lithium deficiency, can close the neuroinflammatory vicious cycle.

DAMPs Close the Neuroinflammatory Vicious Cycle

DAMP stands for damage- or danger-associated molecular pattern. It's a collective term for various intracellular cellular components that become DAMPs only when they are unnaturally located outside a cell, thus signaling cell or tissue damage via TLR-4.[14] As shown in Fig. 24, as soon as TLR-4 recognizes a DAMP, the same pro-inflammatory signaling cascade we've seen with PAMPs is activated. In contrast to PAMP activation, which is triggered by a pathogenic microorganism, DAMPs are referred to as sterile inflammation (in the case of *PAMPs as Bioweapons*, see additional information above, however, mRNA injections also constitute sterile inflammation). This activation of the innate immune response by DAMPs and PAMPs can occur anywhere in the body, even if we're focusing primarily on neuroinflammation here. Even an accidental cut on your finger releases DAMPs. The subsequent inflammatory reaction can be observed in real time (pain, swelling with impaired function, redness, etc.). Pro-inflammatory cytokines released outside the brain in such a case can, however, pass through an intact blood-brain barrier, indirectly causing neuroinflammation. It likely offers a survival advantage if a peripheral injury or infection also leads to messenger substances influencing behavior (via their impact on the mental immune system).[15] A direct way to release DAMPs in the brain is through a blow to the head, which injures brain cells. While the local inflammatory reaction can't be visually observed like a finger cut, you feel the pain and a sense of fogginess. If the blow or the inflammatory reaction is very severe, brain swelling caused by pro-inflammatory cytokines can become life-threatening. And just as an injured finger functions only to a limited extent, the brain also suffers a loss of function that can lead to coma or death. However, a look at Fig. 24 reveals that it doesn't always have to reach this point. For example, the inhibition of glycogen-synthase-kinase-3 is a new approach in the treatment of traumatic brain injuries, as per a 2016 scientific article titled "A new starting point for the treatment of traumatic brain injuries."[16] An adequate supply of essential lithium would be the natural way—instead of waiting for the development of pharmaceutical lithium mimetics—to reduce the risk of an excessive and thus life-threatening cytokine release. Other scientists have also suggested this, as evidenced by the title of a review published in 2014: "A New Avenue for Lithium: Intervention in Traumatic Brain Injury."[17]

To assess the extent of traumatic brain injury, in addition to clinical signs of functional loss, imaging techniques are employed. To determine the degree of neuroinflammation, blood and cerebrospinal fluid can be tested for specific inflammatory markers like IL-1α or IL-6. These pro-inflammatory cytokines, along with IL-12, IL-13, IL-18, IL-23, or TNFα, inhibit adult hippocampal neurogenesis, thereby temporarily crippling the mental immune system.[18]

Here, too, the evolutionary biological advantage of their release (as seen with acute infections) is that an acute craniocerebral trauma initially demands rest and additional energy reserves to repair the damage. However, if, for example, the neuroinflammatory vicious cycle depicted in Fig. 24 develops due to a lithium deficiency (or other deficits, as explored in *Systemic Prevention and Therapy of Multicausal Chronic Diseases*), even a mild concussion can lead to long-term cognitive deficits.[19] These long-term deficits encompass all effects of mental immunodeficiency: memory and attention impairments, changes in executive functions (problems with thinking and planning, decision-making, mental flexibility), and emotional instability. Furthermore, individuals can develop a variety of psychiatric disorders (e.g., depression, anxiety disorders, addictions) and even various forms of dementia after a single traumatic brain injury. Since our entire stress regulation system depends on productive hippocampal neurogenesis, a single injury to the mental immune system can become chronic and even lead to hippocampal shrinkage.

The hippocampus is the mental immune system's center and at the same time the central regulator of our stress reactions in situations that we perceive as threatening. This unique brain region is also the only one possessing the necessary prerequisites for this task. First, the nerve cells that mature daily through adult hippocampal neurogenesis and integrate into the hippocampal network receive all sensory impressions, as well as thoughts that signal or anticipate a potentially stressful situation. Second, these newly integrated nerve cells, via their neighbors, have indirect access to all our past life experiences. This enables these new hippocampal nerve cells to assess the danger potential of any new life situation and adjust our physical stress response accordingly, through regulating cortisol levels. Therefore, productive hippocampal neurogenesis is causally linked to our psychological resilience. As an animal study's title confirms: "Increasing Adult Hippocampal Neurogenesis is Sufficient to Reduce Anxiety and Depression-Like Behaviors."[20] In contrast, inhibition of hippocampal neurogenesis by pro-inflammatory cytokines triggered by DAMPs, i.e., physical or psychological trauma, or PAMPs, i.e., infection (or spike mRNA injection), has the opposite effect:

psychological resilience decreases, as illustrated in Fig. 24. Even a situation objectively not threatening is perceived as highly stressful. Under these conditions, the hippocampus not only stops growing, it even shrinks due to a lack of psychological resilience and thus chronically elevated neurotoxic and cytotoxic stress hormone levels, as also outlined in Fig. 24.[21] Essential lithium would counteract this fatal development in a natural way. It would inhibit excessive cytokine production. Simultaneously, as we know from Chapter 2 (see Argument 2), it would even stimulate adult hippocampal neurogenesis through multiple mechanisms and protect against apoptosis or neuronal death. Thus, a lack of essential lithium is highly problematic, even from a single blow to the head.

It's no surprise, then, that frequent concussions dramatically increase the risk of psychiatric and neurodegenerative diseases. For instance, the total duration of playing time in the US National Football League (NFL), along with the player's position (with head impacts being more frequent in certain positions), correlate closely with the decline of cognitive function and the development of chronic neuropsychiatric health problems such as depression and chronic anxiety.[22] A study examining the causes of death among former NFL players revealed an approximately threefold elevated risk of mortality from neurodegenerative diseases compared to the general US population.[23] The risk of developing and dying from Alzheimer's disease increases by a factor of 3.9. The risk of dying from amyotrophic lateral sclerosis was calculated to be 4.1 times higher. The risk of dying from Parkinson's disease also tripled.[24] Football is just one example here; similar patterns are seen in other sports, such as kickboxing, which also frequently lead to concussions and the aforementioned long-term neurodegenerative consequences.[25]

But words can also cause harm. Like physical blows, they can damage nerve cells if they trigger psychological stress, thereby releasing DAMPs. In their article "Stressed and Inflamed, Can GSK3 Be Blamed?" the authors explain: "Psychological stress has a pervasive influence on our lives. In many cases adapting to stress strengthens organisms, but chronic or severe stress is usually harmful. One surprising outcome of psychological stress is the activation of an inflammatory response that resembles inflammation caused by infection or trauma. Excessive psychological stress and the consequential inflammation in the brain can increase susceptibility to psychiatric diseases, such as depression, and impair learning and memory."[26] This has been repeatedly demonstrated in animal experiments. For example, the smell of a cat's urine for ten minutes a day over seven days is enough to increase the blood concentration of DAMPs by more than ten times in caged rats.[27] Indeed, any psychological overload of the nerve cells, particularly in the hippocampus, leads via this mechanism to the release of pro-inflammatory cytokines, which, as expected, triggers depressive behavior in animal models.[28] Likewise, in humans, all forms of distress such as social conflict, threat, isolation, and rejection cause neuroinflammation; This always occurs via this, one might almost say, "damp-ed" mechanism, which of course has its evolutionary biological justification in acute situations.[29] The concentrations of detectable inflammatory markers such as C-reactive protein (CRP), IL-1β, IL-6, and soluble TNF-α receptor in the blood, as well as NF-κB in white blood cells (leukocytes), correlate with the severity of neuroinflammation.[30] Thus, an increase in these inflammatory markers after an acute traumatic event can predict the development, severity of symptoms, and duration of mood disorders such as depression, anxiety, and posttraumatic stress disorder (PTSD).[31] This creates the same neuroinflammatory vicious cycle as with blows to the head. Due to reduced psychological resilience, there is an increased release of DAMPs as soon as the situation perceived as stressful is mentally relived. This further fuels the vicious circle and causes the hippocampus to shrink further via the mechanisms already discussed.

"Fear Eats the Soul"—Pre-Traumatic Stress Disorder. Rainer Werner Fassbinder (1945–1982) coined a famous phrase with the title of his 1973 social drama, *Fear Eats the Soul.* The sentence powerfully expresses fear's destructive force. Posttraumatic stress disorder (PTSD) is a severe depression combined with anxiety disorders, stemming from a traumatic experience. The hallmark of PTSD is a massively reduced hippocampus (especially the left one, see Chapter 2, Argument 3).[32] But it is enough to stir up fear: The hippocampus shrinks merely due to the perception of a great danger or potential trauma; neurobiologically, the perception is no different from the actual experience. Words are apparently powerful enough to trigger a *pre*-traumatic stress disorder in people.[33] The more the hippocampus shrinks due to growth inhibition (pro-inflammatory cytokines) and neurodegeneration (stress hormones such as cortisol), the more we lose access to our memories, until we finally lose ourselves and, metaphorically, the soul shrinks along with the hippocampus.[34] The severity of the symptoms is associated with accelerated cognitive decline—thus necessitating a systemic therapy approach (see: *Systemic Prevention and Therapy of Multicausal Chronic Diseases*).[35]

All the events exemplified here are, in essence, the "potholes" Christian Schellenberg described in the previous chapter, which many encounter with almost flat tires—i.e., a lithium deficiency. Consequently, one endures at least an unnecessarily arduous journey through life, constantly risking a breakdown (i.e., falling into neuropsychiatric emergencies) and an inability to progress. A single serious life event, like the death of a spouse, is enough to trigger an increase in inflammatory markers such as IL-1β and IL-6 in older adults, with the rise correlating with the severity of grief and depressive symptoms.[36] Clinical studies reveal how damaging latent chronic inflammation and chronic stress are for the brain: The higher the cortisol or IL-6 level in the blood, the smaller the hippocampal volume in middle-aged adults.[37] Mechanistically, lithium helps to interrupt the neuroinflammatory signaling cascade via GSK-3 and simultaneously (re-)activates adult hippocampal neurogenesis via the mechanisms discussed in Chapter 2 (Argument 2). However, sustained treatment success requires a systemic approach that helps people achieve healthy homeostasis. I'd like to present a draft of such an approach next.

Systemic Prevention and Therapy for Multicausal Chronic Diseases

Conventional medicine excels in advanced acute care, both pharmacologically and surgically. However, it often falls short with nearly all chronic diseases due to its noncausal approach; in these cases, its ineffective methods must be replaced by more impactful strategies. This applies equally to all neuropsychiatric developmental disorders, as well as to neurological, mental, and neurodegenerative diseases, which now have a very high prevalence and are consequently categorized as "diseases of civilization." Given the multicausality of these conditions, a systemic approach is always medically necessary for both prevention and causal therapy. While pharmacological intervention might offer acute relief or alleviate symptoms, it often fails to solve the long-term problem and can even worsen it. This is partly due to significant side effects, and partly because the underlying disease process continues unchecked while symptoms are merely masked. For example, strong painkillers ease the discomfort of an infected tooth root, but they don't, of course, address the root cause. Similarly, we must adopt a causal approach for diseases of civilization. Lithium deficiency frequently plays a crucial role, especially in neuroinflammatory diseases, which are graphically depicted in Fig. 24 as part of mental immunodeficiency syndrome. However, it's by no means the sole cause of dysfunctional adult hippocampal neurogenesis. While no other essential vital substance or aspect of our existence can replace lithium, conversely, lithium cannot compensate for a deficiency in another essential vital nutrient or human need—such as a species-appropriate diet, sleep, exercise, or social activities. Lithium also doesn't create a sense of purpose, which is equally vital for our mental health. Yet, correcting a lithium deficiency makes us more life-affirming, inquisitive, and mentally stable, which generally increases the likelihood that we'll consistently take on new challenges.

The Methuselah Formula and the Law of the Minimum

Building on this systemic approach and after approximately ten years of intensive analysis of lifestyles in the Blue Zones—regions worldwide where people not only live unusually long but also remain mentally sharp—I developed the Methuselah Formula in 2010 (Fig. 25). As you'll see, combined with the fundamental laws of life discussed in Chapter 1, this formula could form the basis for a species-appropriate medicine of the future, which we'll explore in the final chapter. The name "Methuselah" refers to the biblical figure from the Book of Genesis, described as the oldest human in history at an astounding 969 years.

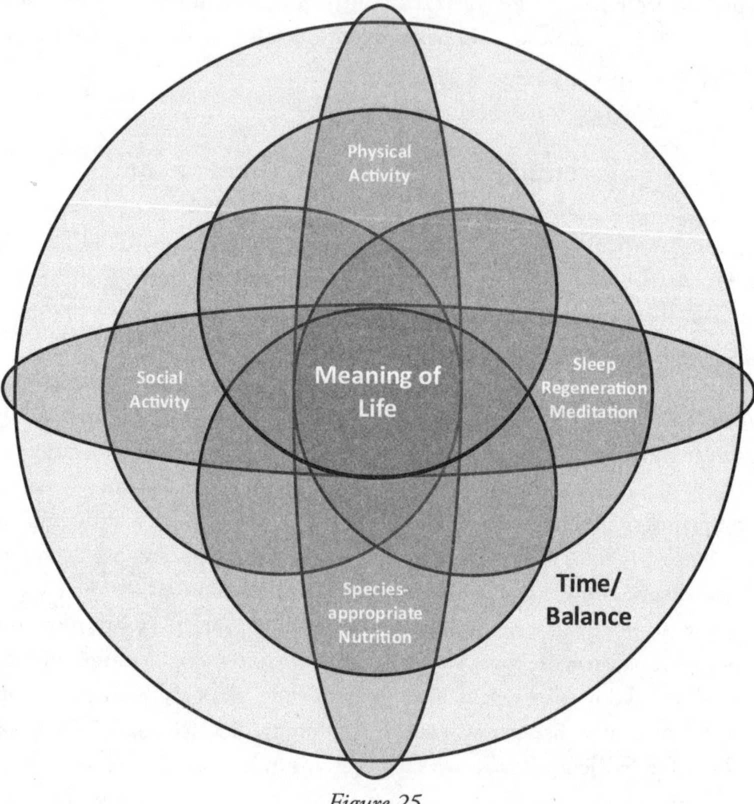

Figure 25

His name has thus become a symbol of longevity. Perhaps it's not a coincidence that the Methuselah Formula's graphic representation strikingly resembles the Flower of Life (see Introduction). Though I wasn't aware of it during the design process and created the formula independently, the depicted overlap of various human life areas was a conscious choice, as they mutually influence each other, a point we'll soon examine. My book *The Methuselah Strategy* explores leveraging the synergies of these overlapping life areas, considering the laws of minimum and maximum, with the goal of *aging healthily and becoming wiser*, as the subtitle reveals.[38] The Methuselah Formula, first introduced there, has been expanded in subsequent books due to its far-reaching implications for diverse aspects of life. Consequently, it became *The Formula Against Alzheimer's*[39] or the protective formula against life-threatening narratives in *The Indoctrinated Brain*,[40] all distinct, unnatural phenomena that would remain unobservable had humanity not strayed from a species-appropriate way of life.

In the following section, I'd like to delve deeper into these essential areas of life. I'll use illustrative examples of their interplay, particularly concerning this chapter's central themes: protection against neuroinflammation and the promotion of adult hippocampal neurogenesis, as well as the function of essential lithium. The aim is to demonstrate how this area influences both individual and societal mental and physical health. Building on these considerations, we'll then turn to the various clinical manifestations of the mental immune system. This way, we won't need to repeatedly cover these fundamental aspects, which are crucial for all these developmental disorders and diseases.

Meaning of Life: At the core of the Methuselah Formula lies the meaning of life. The evolutionary imperative "Be fruitful and multiply," coupled with the Evolution of the Grandmother (see Chapter 1), provides a lifelong range of tasks. This imperative motivates all living beings through various instincts. However, humans are almost certainly the only species on the planet capable of seeking their own meaning in life. Objectively, the question of whether life itself has meaning or serves a higher purpose cannot be answered. Therefore, this should be understood as a subjective meaning of life, not a higher destiny for humanity. What isn't good for us, though, is when we feel no meaning in our lives, perhaps even sensing we're living a meaningless existence. Meaning in life acts like a compass, providing direction. Without it, one feels aimlessly tossed by life's isolated events: without a consistent goal, the inevitable hurdles become impositions instead of challenges, resulting in a life that is often fraught with stress, as more and more studies are demonstrating.[41] Chronic distress causes chronic neuroinflammation and shrinkage of the hippocampus, as discussed in detail just now.[42] The results of a long-term study show that people who find meaning in their lives go through life with significantly lower inflammation levels than those who do not.[43] This is probably the reason why people with a sense of purpose in life have a comparatively lower risk of becoming chronically ill and dying prematurely.[44] And as already indicated, a healthy, functioning mental immune system, which requires, among other things, an adequate supply of the vital lithium, increases the likelihood of having the curiosity and courage to repeatedly devote oneself to new, meaningful tasks throughout one's life. Natural tasks include social engagement and investing in family and friendships. Crucially, spirituality and the belief in being part of something larger and more enduring have also been demonstrably linked to lower levels of inflammatory markers.[45]

Social Activity: Humans have always lived and survived solely within communities. This is why we instinctively know how dangerous social isolation is; even merely perceived loneliness causes psychological stress. For reasons of security and belonging, we desire to be a central part of a group, not on its fringes. Therefore, wherever there's significant socioeconomic inequality and "factors such as poverty, low levels of education, and neighborhood deprivation," there's an increased "risk for [developing] psychopathology," as a meta-study found, "including severe depression and schizophrenia."[46] Social isolation, which was imposed on a large portion of humanity during the orchestrated COVID-19 crisis, also led to an increase in neuroinflammatory diseases.[47] Perversely, the only way out of this isolation often involved accepting an mRNA injection. Since these introduce brain-toxic genetic material into the organism, this measure also contributed to the rise in neuroinflammatory diseases like Alzheimer's.[48]

But humans are not the only social beings; in animal models, social isolation also leads to neuroinflammation and inhibition of neurogenesis in the hippocampus, followed by depressive behavior.[49] Studies show that social activities are positively correlated with the overall volume of the brain, particularly with the number and density of neurons, the volume of the hippocampus, and the microstructural integrity of the brain network.[50] Hormonally, social activity leads to the release of the brain growth hormone BDNF and also oxytocin, both anti-inflammatory messengers that also highly potently activate adult hippocampal neurogenesis.[51] The fact that we instinctively feel more comfortable in company is due to these mechanisms. Oxytocin, famously known to trigger labor during pregnancy, even suppresses the sensation of pain.[52]

Conversely, a self-reinforcing "virtuous cycle" develops when our mental immune system is maintained and strengthened through social activity. This makes us more open to new experiences, more easily able to connect with others, and willing to broaden our horizons through deeper conversations. This way, we can also cultivate deeper friendships and partnerships. Another benefit of social interaction is that people are more likely to exercise together than alone. Vital lithium contributes to this in many ways, as I know from the practical experience of pediatrician Christian Schellenberg: children with lithium deficiency are more prone to neuroinflammation, which has the opposite effect—they avoid social interaction and become isolated.

Physical Activity: Our ancestors—and we don't need to go back to the Stone Age for this, just a few generations—didn't worry about whether they wanted to be physically active. They simply had no choice if they wanted to survive. We, however, have that choice, because our world is almost entirely mechanized. And from an evolutionary biology perspective, one might almost say: unfortunately. This is because humans are naturally "creatures of movement," as evidenced by over a dozen different hormones released from various hormonally active cell systems during physical activity, which synergistically stimulate adult hippocampal neurogenesis.[53] Furthermore, they even suppress the production of pro-inflammatory cytokines like IL-1β and TNF-α, and inhibit TLR signaling pathways.[54] Physical activity improves memory, thinking ability, sleep, and mood, and reduces psychiatric symptoms such as depression, stress, anxiety, and mood swings.[55]

Muscles, incidentally, are also highly active endocrine glands that produce many so-called myokines (messenger substances and hormones). For instance, the muscle hormone irisin is recognized for its role in preventing Alzheimer's disease, notably because it promotes adult hippocampal neurogenesis.[56] While it exhibits anti-inflammatory properties and stimulates adult hippocampal neurogenesis it should not be used indiscriminately in clinical practice.[57] The isolated use of a single substance is likely to fail, because it is against nature. There's a good reason why physical activity releases many hormones, not just one: each hormone has diverse additional functions, all of which would lead to severe side effects if a single hormone had to replace all others and therefore be active at much higher concentrations—exactly as is the case with a single-drug clinical intervention. For example, exercise also releases Erythropoietin (Epo). Epo stimulates the production of red blood cells (its name even hints at this) and also adult hippocampal neurogenesis. As a result, it makes us physically and mentally fitter. Yet, if EPO were to be used in isolation as a drug for preventing or treating, for example, Alzheimer's, the required amount would lead to a life-threateningly thick blood consistency due to excessive red blood cell production.[58]

This understanding of nature's complex workings and the natural multifaceted benefits of physical activity should form a compelling foundation for advocating more movement in an increasingly sedentary society, using rational arguments. This is all part of an overall species-appropriate lifestyle. Moderate exercise has been proven to make the hippocampus grow, without any side effects, of course. Simultaneously, it strengthens psychological resilience, which can be understood as a sign of a strengthened mental immune system—and this, in keeping with the Evolution of the Grandmother, even in advanced age.[59] Moderate physical activity is defined by a heart rate increase of no more than 50 to 60 percent above resting, such as walking or jogging; one should also be able to breathe in enough oxygen with a closed mouth. "Immediately after exercise, many positive outcomes are observed such as lower blood pressure, less stress and anxiety, better sleep and improved mood," as stated in the article "Healthy lifestyles and wellbeing reduce neuroinflammation and prevent neurodegenerative and psychiatric disorders." Anyone who exercises regularly will know this from personal experience.[60] Moderate physical activities that provide sufficient oxygen are also very effective in preserving telomeres, as a large meta-study has shown.[61]

There are close, mutually reinforcing interactions between physical activity and all other areas of life, such as social activity, sleep, and nutrition. Listing them all here would be beyond the scope of this discussion. However, a few examples should suffice. Physical activity stimulates the glymphatic system, which cleanses the brain of toxic metabolic products during sleep (more on this later).[62] Additionally, physical activity leads to an increased release of growth hormone during sleep, as well as the deep sleep hormone melatonin, both with numerous positive effects for our mental health (see below). I'd also like to mention that moderate physical activity can even improve the composition of the gut microbiome, which reduces both peripheral inflammation (i.e., outside the brain) and central nervous inflammation via microglia.[63] This insight, in turn, offers a possible explanation for the combined effect of a healthy, species-appropriate diet and moderate physical activity in preventing neuroinflammation and the resulting neurodegenerative and psychiatric disorders.

Interestingly, lithium and low-intensity endurance training in the rat animal model synergistically increase the production and release of BDNF, a neuroprotective factor, in the hippocampus.[64] The scientists interpreted this as a positive indication of the benefits of both factors (lithium and physical activity) in the prevention of neurodegenerative diseases. A further synergy exists between moderate physical activity and sleep due to the increased release of pro-neurogenic messengers during sleep, which also have an anti-inflammatory effect (and in turn increase physical fitness), such as melatonin and growth hormone.[65]

Sleep, Regeneration, and Meditation: A lack of exercise leads to poorer and thus less regenerative sleep, which, according to a Chinese study, explains the *link between lack of exercise and anxiety, depression, and suicidal thoughts in students*, as the work's title suggests.[66] Studies show that poor sleep increases circulating pro-inflammatory cytokine levels, which in turn contributes to neuroinflammation.[67] "Glymphatic" is a portmanteau coined from "glia" and "lymphatic" because, just like the rest of the body, the space between cells in the brain is bathed in a fluid called lymph.[68] When we sleep, hormonally activated oscillation of blood vessels cleanses the brain of waste and toxins, acting like lymphatic drainage. In this way, for example, so-called β-amyloid is removed from the brain, which is then broken down in the liver. β-amyloid is formed in the hippocampus during the day when we collect memories, in order to store them there. During deep sleep, memory contents are "uploaded" to the neocortical "hard drive," as I call our long-term memory storage. Meanwhile, the β-amyloid that has accumulated in the hippocampus but is no longer needed is also transported away via the glymphatic system so that new memories can be stored in the hippocampus the next day (more on this in *The Exhausted Brain*).[69] If this glymphatic removal does not take place or does not take place particularly effectively because we are not sleeping or sleep is often interrupted, the total concentration of β-amyloid gradually increases. This leads to neuroinflammatory neurotoxicity, which can, among other things, accelerate the development of Alzheimer's disease.[70]

Sleeping Pills Inhibit Glymphatic Brain Clearance and Promote Alzheimer's. When the brain transitions from wakefulness to sleep, its processing of external information decreases, while regenerative processes like the glymphatic removal of waste products are activated. During deep sleep, the brainstem releases the neurotransmitter noradrenaline approximately every 50 seconds. This causes the blood vessels to contract, "generating slow pulsations that create a rhythmic flow in the surrounding fluid to carry waste away," according to a January 8, 2025, publication of the scientific press *Cell* titled "How Deep Sleep Clears a Mouse's Mind, Literally."[71] Nevertheless, the related scientific study also demonstrates that the sleeping pill ingredient zolpidem suppressed noradrenaline oscillations, consequently impairing glymphatic flow and brain clearance and thereby elevating the risk of Alzheimer's disease.[72] This is not the first medication to exhibit brain-damaging side effects. Even the commonly prescribed (and mostly available over-the-counter) active ingredients in the benzodiazepine group, including the well-known Valium, nearly double the risk of Alzheimer's disease after a cumulative lifetime use of 180 days for sleep or anxiety.[73] Promoting sleep or psychological resilience should naturally therefore always receive priority.

Neurodegenerative diseases, including Alzheimer's, ALS, Parkinson's, Huntington's, and numerous others, are also known as "proteinopathies" because they are characterized by an accumulation of misfolded, aggregated intracellular, or extracellular proteins.[74] Poor sleep leads, on the one hand, to inadequate clearance of these potentially toxic protein conglomerates via the glymphatic system. On the other hand, overactive GSK-3, not sufficiently restrained by lithium, inhibits autophagy, a vital process for breaking down these toxins. "Sleep deprivation," an Indian research group writes, "is becoming an epidemic with detrimental effects ranging from immediate consequences such as traffic accidents to the most severe neurological disorders. All age groups are affected by sleep deprivation due to lifestyle or disease."[75] The result is always acute or long-term life-threatening neuroinflammation. Thus, good sleep hygiene is vital for better and, crucially, completely natural sleep. This is also promoted by essential lithium, as the trace element inhibits the negative influence of GSK-3 (which is otherwise hyperactive in lithium deficiency) on the internal clock, as we established in Chapter 2 (Argument 2): We sleep more deeply and, consequently, dream a bit more—a phenomenon more and more people report to me once they correct their essential lithium deficiency with supplements. "You do dream more," one essential lithium user commented on her experience, "but it's positive, as if someone is finally cleaning up upstairs." This "cleaning up" is indeed another essential function of sleep: we dream and, in doing so, recombine all our experiential knowledge, perhaps even waking up a little wiser. As Robert Stickgold, a professor of psychiatry and sleep researcher at Harvard Medical School, explains, "The brain will take information, integrate it with some old memory you have that you never would have thought of combining it with during your waking time."[76] According to Stickgold, "The difference between being smart and being wise is in the quality of their sleep."[77]

Beyond good sleep, psychological health and well-being can also be improved by becoming aware of our situation during stressful times and gaining new perspectives through reflection, allowing us to react more calmly or effectively. Mindfulness meditation and yoga can be used for this purpose, especially since they've been shown to reduce inflammatory processes.[78] A meta-analysis concluded that breathing exercises, meditation, yoga, and Tai Chi can downregulate the NF-κB signaling pathway and thus pro-inflammatory cytokines.[79] From an evolutionary biology perspective, it's not surprising that walks in nature also improve cognition, mood, and self-esteem, and, significantly, increase sleep duration and quality.[80]

Species-Appropriate Nutrition: "You are what you eat" is a truism. For optimal mental and physical health, a species-appropriate diet is a prerequisite. For humans, what constitutes a species-appropriate diet was determined by evolutionary biological adaptation processes to local food sources during the period that enabled the development of our mental immune system. This diet was predominantly pescatarian and, according to the *Adventist Health Study*, which compared various dietary styles, promises the best health and highest life expectancy.[81] Regarding a good lithium supply and the chance of healthy aging, it's interesting that the aforementioned Blue Zones, first conceptually presented by American author Dan Buettner in *National Geographic* in November 2005, are mostly islands of volcanic origin, like Sardinia (Italy) or Okinawa (Japan). In these areas, people not only consume a fish-rich diet but can also expect higher lithium levels in the soil.[82] It's also worth noting that lithium, as Gerhard Schrauzer reported, improves the transport of vitamins B9 and B12 into cells, thereby influencing mood-related parameters. According to Schrauzer, "the stimulation of the transport of these vitamins into brain cells by Li may be cited as yet another mechanism of the anti-depressive, mood-elevating and anti-aggressive actions of Li at nutritional dosage levels."[83]

The pescatarian diet, that of the prehistoric fisherman and gatherer, consists mainly of fish and seafood, along with a rich plant-based diet. This would ensure a supply of all essential vital substances. However, as discussed in Chapter 2, we currently lack sufficient fish, both in quantity and quality, to feed the entire world's population a pescatarian diet. Nevertheless, we can ensure an adequate supply of aquatic omega-3 fatty acids with algae oil. In my book *The Algae Oil Revolution*, I extensively describe what's needed to achieve a complete fish replacement through vegan supplementation with algae oil.[84] Plant-based foods contain a wealth of essential nutrients and many anti-inflammatory phytochemicals.[85] Consuming whole, organically grown foods therefore reduces chronic inflammation and neuroinflammation.[86] A healthy diet increases the chance of a longer and healthier life because, alongside physical activity (see above), it significantly influences the length of our telomeres, as a meta-study shows: "The main results suggest that a higher consumption of fish, nuts and seeds, fruits and vegetables, green leafy and cruciferous vegetables, olives, legumes, polyunsaturated fatty acids, and an antioxidant-rich diet might positively affect TL," the authors write. Conversely, "A higher intake of dairy products, simple sugar, sugar-sweetened beverages, cereals, especially white bread, and a diet high in glycaemic load were factors associated with TL shortening."[87]

However, longer daily breaks from eating, in the sense of intermittent fasting, are also quite natural. After all, our ancestors in the Paleolithic era had to leave the cave to find food and therefore probably had to leave the cave every day with an empty stomach. Evolutionary adaptation processes have most likely ensured that intermittent fasting combined with moderate physical activity, as it mimics Paleolithic foraging, can offer many health benefits. This is particularly true for the mental immune system, as our prehistoric ancestors had to be especially vigilant when foraging in the wild.[88] After just about twelve hours without eating, ketogenesis is activated.[89] As insulin levels fall, long-chain fatty acids can be mobilized from fat stores and transported via blood vessels to the liver, where they are broken down into ketone bodies. This is an important process for our brain function because, unlike saturated long-chain fatty acids, small ketone bodies easily cross the blood-brain barrier and are even a better energy source for brain cells than glucose. Furthermore, ketone bodies are also hormone-active messengers that, among other things, activate autophagy.[90] "Since protein homeostatic mechanisms such as autophagy are known to be activated by nutrient deprivation, it is not surprising," explains a US group of researchers on aging, "that evolutionary pressure promotes the elimination of pathogenic proteins during ketogenesis to improve cell health in organisms."[91] According to the authors, ketone bodies "are in a sense the janitors of our cells, removing damaged proteins or molecular waste so that organisms can function at maximum molecular fitness."

Daily intermittent fasting, aligning with natural evolutionary biology, is also less stressful and may therefore offer significant health benefits compared to extremely long fasting periods of several days or even weeks. While such longer fasts also stimulate autophagy, they come with increased physical stress.[92] Intermittent fasting has been shown to prolong lifespan, reduce free radicals, mitigate age-related diseases, and protect against declines in cognitive and motor function.[93] Intermittent fasting also reduces the risk of neuroinflammation at any point in life, thereby cumulatively reducing the risk of neurodegenerative diseases.[94] Combined with a Mediterranean diet, for example, it has proven beneficial in treating multiple sclerosis (MS).[95]

In a mouse model for Parkinson's disease, intermittent fasting slowed disease progression by upregulating protective factors for nerve cells and reducing neuroinflammation.[96] The anti-inflammatory ketone bodies mentioned earlier are responsible for this.[97] Sugar molecules like glucose, fructose, or galactose have the opposite effect: they are pro-inflammatory when they chemically react with molecules on the cell surface to form advanced glycation endproducts (AGEs).[98] These toxic foreign bodies are recognized by the immune system with the help of the receptor for AGE (RAGE). RAGE and TLR4 communicate directly with each other, and this key mechanism activates an inflammatory response responsible for the harmful effects of a high-sugar diet.[99] This doesn't mean carbohydrates should be avoided; rather, they should be complex and whole so they are digested slowly and blood sugar levels don't spike.

While insulin secretion, which regulates blood sugar levels, stops the release of ketogenic fatty acids from fat cell depots, neuroprotective ketogenesis can be maintained with ketogenic coconut oil.[100] You can find out more about this in my book *The Exhausted Brain.*[101]

The inhibition of neuroinflammation on one hand and the activation of adult hippocampal neurogenesis on the other are also governed by the gut-brain axis. The microbiome, or the microbial composition of our gut flora, is profoundly important here. In an animal model of MS, intermittent fasting increased bacterial diversity in the gut, thereby improving the clinical course of the disease, a finding also observed in MS patients.[102] The complexity of the gut flora is primarily determined by our diet's composition, which, though indirectly, massively influences brain health via the microbiome.[103] The more species-appropriate—meaning wholesome, varied, and nutrient-rich—our diet, the healthier the signals sent from the gut to the brain. This information reaches its destination via the blood, through microbial metabolic products. It happens even faster via the vagus nerve (from the Latin *vagari* for "to wander"), as this tenth cranial nerve runs from the brain to the intestines, sending signals in both directions. High activity of the vagus nerve (known as high vagal tone) is synonymous with inner calm. The vagus nerve is the main nerve of the parasympathetic system, which, unlike the sympathetic nervous system, lowers blood pressure and heart rate, for example.[104] Activation of the vagus nerve, which can also be supported by moderate physical activity, a calming social environment, and indeed, a healthy microbiome, also has an anti-inflammatory effect. Vagus stimulation is therefore used to treat various inflammation-related neurodegenerative diseases, as well as depression and PTSD.[105] In his brilliant book *My Grandmother's Hands* trauma therapist Resmaa Menakem refers to the vagus nerve as the soul nerve, as it connects the entire nervous system in many ways. In the author's view, its function can serve as a symbol of what it means to be human.[106] Thus, the vagus nerve—stimulated by a diverse and intact microbiome—also activates the release of BDNF in the hippocampus and thus adult hippocampal neurogenesis.[107]

Conversely, gut flora disrupted by a modern diet, low in vital substances and fiber, is associated with chronic diseases throughout the entire organism.[108] So-called gut dysbiosis is directly linked to neuroinflammation, leading to mood swings, anxiety, depression, pain, and cognitive impairment, as summarized in a review article titled "Mind-altering microorganisms: the impact of the gut microbiota on brain and behaviour."[109] The vagus nerve serves as the crucial link between the gut and the psyche.[110]

Time: Time is fundamental to our lives and simultaneously the final, all-integrating element of the Methuselah Formula (see Fig. 25), as it's intended to ensure homeostasis among the other areas. According to the Law of the Minimum, we must allocate sufficient time for all these areas. If we fail to do so, we risk falling into a pro-inflammatory state that makes us acutely unhappy and potentially chronically ill.

However, time itself is also subject to the Law of the Minimum. A lack of time, occurring when you've overcommitted, not only leads to neglecting many essential needs; above all, it leads to distress. This distress directly paralyzes adult hippocampal neurogenesis via high cortisol secretion. It's therefore wise to be mindful of this last element, because our time is limited and thus perhaps the most precious commodity we possess. It compels us to make smart decisions daily, which in turn requires a healthy mental immune system.

The Methusaleh Formula and the Law of the Maximum

It's clear that the combined benefits of all elements of the Methuselah Formula should be utilized for the prevention and causal therapy of all diseases. However, a systemic approach to prevention and therapy isn't just about the Law of the Minimum—that is, what's essential for our health and well-being—but also about the Law of the Maximum—what harms us if we overdo it. Generally, humans tolerate a violation of the Law of the Minimum less well than a violation of the Law of the Maximum, as we cannot compensate for deficiencies. Adhering to the Law of the Minimum akes us more tolerant of toxic influences, but avoiding toxic influences (insofar as this is de facto possible) doesn't make us more tolerant of deficiencies in vital needs. Unfortunately, humans are incredibly creative when it comes to self-harm. Therefore, discussing all the toxic influences on the human immune system that have developed during the industrial revolutions would far exceed this book's scope. I'll mention just a few here. A detailed discussion of other toxic influences can be found in my book *The Exhausted Brain.*[111] A few examples:

- Pesticides disrupt the blood-brain barrier, cause neuroinflammation, and have been shown to promote the development of neurodegenerative diseases (Parkinson's, Alzheimer's, MS, Huntington's disease, and ALS) and psychiatric disorders (depression, anxiety disorders, cognitive impairment, and autism).[112]

  Solution: Given this, it makes sense to obtain as much food as possible from organic farming.

- Electromagnetic radiation from cell phones and cell towers causes neuroinflammation and is particularly harmful to the young brain, which it can penetrate even more easily than an adult's.[113]

  Solution: Again, it's crucial to be aware of these dangers and educate your children. As with everything, the dose makes the poison—in this case, it's about daily exposure and its intensity. For example, holding a cell phone directly to the ear is very harmful to the brain.

- Aluminum is a causal risk factor for neuropsychiatric developmental disorders.

  Solution: Reduce exposure through food by cooking your own meals and avoiding corresponding food additives (see the last chapter on aluminum-containing additives and their E-numbers). Vaccinations are problematic due to the toxicity of their additives, though I won't delve into their effectiveness here. Generally, vaccinations with aluminum content should be avoided whenever possible due to the toxic burden (see Autism Spectrum Disorder below). Unfortunately, completely avoiding certain amounts of aluminum intake is no longer possible under current conditions. An effective measure that would reduce the harmful effects of the unavoidable aluminum content is a diet rich in silicate. These salts of orthosilicic acid possess the capability to bind aluminum, thus inhibiting its absorption, lessening its toxicity, and improving its excretion.[114] Good sources include particular mineral waters, alongside whole grains and carotene-rich vegetables.[115] As part of my Alzheimer's prevention program (see my book: *The Formula Against Alzheimer's*)[116] I recommend also a once-yearly, six-week regimen of 300 to 600 mg of natural dextrogyrated (r)-alpha-lipoic acid (ALA, α-lipoic acid). ALA forms stable chelates (complexes) with aluminum and many other brain-toxic metal ions like cadmium, mercury, or lead, thus promoting their excretion.[117] A major review article from 2019 summarizes the results of clinical trial evidence: "Analysis of data from clinical trials has shown that ALA is effective for certain diseases and conditions, including diabetic neuropathy, obesity, schizophrenia, multiple sclerosis, pregnancy issues [such as intrauterine bleeding often associated with threatened miscarriage, especially in the first trimester], and organ transplantation, with little or no side effects. ALA also appears to be a promising agent for improving quality of life, alleviating neuropathic symptoms, and even reducing the use of emergency medications often required by patients with diabetic neuropathy."[118] A systematic review also found that micro-amounts of lithium provide protection against the neuroinflammatory and neurotoxic effects of lead.[119] Lithium achieves this solely through its neuroprotective properties—supporting basic, universal cellular functions. This effect isn't due to a chemical reaction that neutralizes the toxic effect. Since these protective mechanisms are universal and not metal-specific, it stands to reason that lithium not only counteracts lead toxicity but is also effective against other neurotoxic metals like mercury or aluminum. The latter has been confirmed in animal experiments on rats.[120]

Prevention and Therapy of Mental Immunodeficiency Syndrome

I have often been asked whether *The Formula Against Alzheimer's*, since it is based on this systemic prevention and therapy approach to brain health, could not also serve as a formula against Parkinson's, ALS, or frontotemporal dementia (FTD), for example. As you can now easily see, the answer is a resounding yes. The formula is a road map to a species-appropriate lifestyle. However, what all these diseases (and many other neurological, mental, neuropsychiatric, and neurodegenerative conditions) have in common is neuroinflammation, which stems from an unnaturally dysregulated lithiome. Depending on the chosen perspective (e.g., focusing on causes or neuropathological mechanisms) and the weighting of specific symptoms, it would often be possible to group different clinical diagnoses under a single heading. Ultimately, the sole purpose of any classification is to identify common therapeutic approaches or causes. Even if the symptoms in the aforementioned diseases differ enough to distinguish them clinically, the pathological mechanisms and the root cause are often the same. Accordingly, from a systemic perspective, the same fundamental measures are always required to effectively prevent, reduce the effects of, or cure an existing disease. It doesn't need to be repeated for every clinical picture: Lithium dosages for prevention are in the essential range. Even in the case of illness, the dose would likely never need to exceed the low-dose range of 5 to 20 mg of pure lithium per day (corresponding to 128.6 to 514.3 mg in the form of lithium orotate monohydrate) due to the systemic approach. However, this must be decided by the attending physician. This comparatively low dosage could also apply to the treatment of bipolar disorder, which is currently sometimes treated with several hundred mg of lithium (in the form of lithium carbonate). A basic prerequisite, however, would be a systemic approach, as outlined above. This is the only way to eliminate as many disease-causing deficits as possible (law of the minimum), meaning significantly less lithium is needed to influence neuropathological processes. The same applies, of course, to eliminating harmful influences (Law of the Maximum). Furthermore, lithium orotate could be much more efficient than the conventionally used lithium carbonate (see Chapter 3).

Accordingly, the review article, "Potential application of lithium in Parkinson's and other neurodegenerative diseases," in *Frontiers in Neuroscience* published in 2015, states: "Lithium-only treatment may not be a suitable therapeutic option for neurodegenerative diseases due to inconsistent efficacy and potential side-effects, however, the use of low dose lithium in combination with other potential or existing therapeutic compounds may be a promising approach to reduce symptoms and disease progression in neurodegenerative diseases."[121] This highlights the serious problem of a mono-drug approach. However, for the authors, the solution doesn't lie in an obvious systemic approach that addresses the causes of the diseases but, unfortunately, only in a multidrug approach. Yet, a systemic therapeutic approach would not only be associated with fewer undesirable side effects, but it would also be the only effective—because truly causal—way forward. It could and should, therefore, revolutionize conventional medicine for chronic diseases.

In the following, some representatives of typical neurological, psychological, neuropsychiatric, and neurodegenerative disorders will be discussed in more detail as examples. For neuropsychiatric developmental disorders, I will focus on the autism spectrum. For pain syndromes, migraine, polyneuropathy, and neuropathic pain are grouped due to their similar origins and comparable pathomechanisms. The role of lithium, which we will discuss in this context, is almost identical in all three cases from the perspective of the systemic prevention and therapy approach. Regarding neurodegenerative diseases, the most recent discoveries in Alzheimer's research are noteworthy, as are, for example, those concerning epilepsy. In principle, however, these findings can be applied to all brain diseases listed in Fig. 24 regarding prevention and treatment. To keep this book within its scope, I will only mention them briefly. It is a certainty for me that only a systemic approach—as described above—can lead to a solution.

Autism Spectrum Disorder

Autism spectrum disorder (ASD) serves here as a representative for a range of other neuropsychiatric developmental disorders, including bipolar disorder, schizophrenia, and borderline personality disorder. ASD encompasses developmental disorders characterized by impaired social interaction and communication, repetitive and stereotypical behavioral patterns, and often accompanied by intellectual disability. Symptoms typically emerge in early childhood.

Asperger's—A Curse or a Blessing? The clinical presentation of ASD is highly variable, hence the "spectrum" designation. In contrast to medical literature, I do not classify what is known as Asperger's syndrome as part of the autism spectrum. After all, children with Asperger's syndrome show no delay in language development and typically even begin speaking early. They often possess unique talents in rational-analytical thinking and are generally above-average in intelligence. It's therefore likely that this isn't a delayed or disordered development, but simply a different kind of development. We owe many scientific, technological, and cultural advancements to the capacity to perceive the world differently, which is associated with Asperger's syndrome. Thus, from a sociocultural perspective, Asperger's syndrome is an invaluable part of the human developmental spectrum.[122]

In people with ASD, however, hippocampus-dependent learning, general memory, language skills, and emotion regulation are generally impaired.[123] Since symptoms usually appear between the ages of twelve and twenty-four months—precisely during the time window in which the hippocampus undergoes a crucial maturation process—more and more researchers are assuming that ASD is primarily a hippocampal developmental disorder.[124] For example, it has been found that a smaller hippocampal volume is closely associated with the severity of autistic symptoms and language skills in school-age children with ASD.[125] In fact, studies in children and adolescents with ASD even show a striking reduction in size, particularly of the left hippocampus (see Chapter 2, Argument 3).[126]

What could lead to neuroinflammation during this critical window, lasting until the end of the second year of life, thereby causally contributing to developmental disorder?[127] Any answer must also explain the dramatically high autism rate, which, for instance, in the USA was one in thirty-six among eight-year-olds in 2020.[128] According to the US Centers for Disease Control and Prevention (CDC), this represented a 4.2-fold increase compared to 2000.[129] Among the many factors that can cause neuroinflammation, researchers look for those that most closely align with the Bradford Hill criteria, also known as the Hill causality criteria. These were established in 1965 by the British epidemiologist Sir Austin Bradford Hill (1897–1991).[130] His nine criteria (e.g., strength of correlation in epidemiological studies, their reproducibility, plausibility, and not least, animal experimental verification) have since proven extremely useful in substantiating the assumption of a causal relationship behind an epidemiological correlation.

Of all the possible causes previously considered, aluminum has become the focus of investigations. This light metal is a known neurotoxin and causes strong inflammatory reactions, not only in the brain. Thus, the criterion of plausibility was met. Due to their pro-inflammatory effect, aluminum salts are added to many vaccines as adjuvants (from the Latin *adjuvare*, "to help") to enhance the immune response through inflammation. Applying the Hill criteria, researchers at the University of British Columbia, Vancouver, Canada, investigated whether exposure to aluminum from vaccines could contribute to the increase in autism rates in the Western world.[131] "Our results show," the authors wrote in their 2011 article, "that: (i) children from countries with the highest ASD prevalence appear to have the highest exposure to Al from vaccines." This fulfilled the criterion of reproducibility. Additionally, "Al exposure from vaccines in the US vaccination schedule from 1991 to 2008 shows a highly significant positive linear correlation with ASD prevalence at all three levels of exposure (Pearson $r = 0.92$, $p < 0.0001$ [that is an extremely high correlation!] . . .)." In summary, they concluded: "The application of the Hill's criteria to these data indicates that the correlation between Al in vaccines and ASD may be causal." Due to the fulfillment of most Hill criteria, the causal link became highly probable—and was indeed confirmed in animal experiments in 2020. Lambs were injected sixteen times within a year with either an aluminum hydroxide–containing vaccine, only aluminum hydroxide/the adjuvant, or just a buffer solution as a control. Compared to the control group, only the young sheep that received a vaccine with aluminum hydroxide or only aluminum hydroxide showed significant changes in individual and social behavior.[132] The animals in these two groups were easily excitable and prone to compulsive eating. Crucially, they exhibited far fewer natural social interactions, while showing significantly more aggressive behaviors and stereotypes—all typical symptoms of autism. Not surprisingly, elevated levels of stress biomarkers were also found in both groups. All symptoms in the animal experiment shared a common and causal factor: the injection of aluminum hydroxide. This fulfilled the last important Hill criterion.

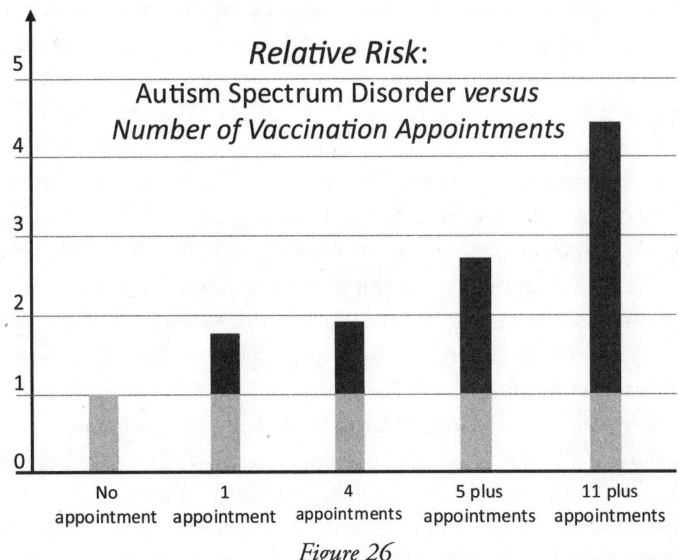

Figure 26

In January 2025, for the first time, a concrete examination of this suspected causal link in humans was undertaken (because official channels had previously avoided it; see "*Turtles All the Way Down*," Chapter 6). This involved comparing the health status of vaccinated children with that of unvaccinated children.[133] The study, which included over 47,000 nine-year-olds, unambiguously confirmed the suspected causal link: the risk of developing autism increases with each vaccination appointment (see Fig. 26; a relative risk of 1 corresponds to no vaccination appointment). "Children with only one vaccination appointment had a 1.7-fold increased risk for ASD compared to children with no vaccination appointments, while children with eleven or more vaccination appointments had a 4.4-fold increased risk for ASD compared to children with no vaccination appointments," wrote the authors from the nonprofit Chalfont Research Institute in Jackson, Mississippi.

However, vaccinations don't just causally increase the risk of developing ASD; they also heighten the risk for many other neurodevelopmental disorders. Fig. 27 illustrates the relative risk increase in vaccinated nine-year-old children, whether born prematurely or full-term. This is compared to children who were also born either prematurely or full-term but were unvaccinated. The risk for unvaccinated children was set at one in each case (for simplicity, only one bar is shown, though there should be two for each unvaccinated group), making the respective relative (average) vaccine-related risk increase for mental developmental disorders easy to discern. These are average values because some vaccinated children received only one vaccination, while others received more than eleven. Therefore, we are seeing mean values that, depending on vaccination status, contain even more drastic risk increases. If you compare Fig. 26 with Fig. 27, you'll notice that the average risk increase for ASD is higher than with just one vaccination but lower than with eleven or more. This means the risk for maximally vaccinated children is significantly higher than depicted here. Specifically, the risk of learning disabilities and encephalopathies (a collective term for pathological brain conditions) increases for both full-term and premature births (though even more dramatically for the latter), as the figure also shows. The risk of developing epilepsy or hyperkinetic syndrome like ADHD was also elevated. Even so-called tic disorders (not shown), including Tourette's syndrome (see below), are over six times higher in vaccinated children compared to their unvaccinated counterparts. These are serious disorders that severely impair quality of life.

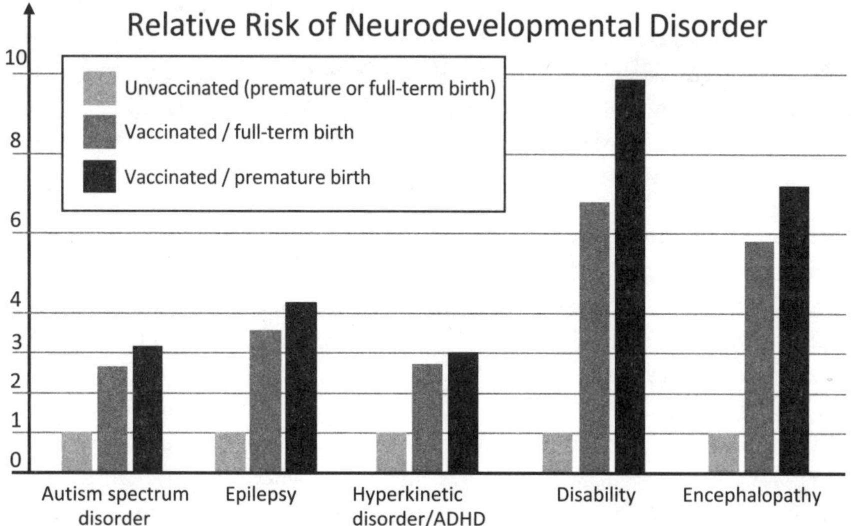

Figure 27

***Mahatma Gandhi on Vaccination in* A Guide to Health:** "Vaccination is a barbarous practice, and it is one of the most fatal of all the delusions current in our time, not to be found even among the so-called savage races of the world. Its supporters are not content with its adoption by those who have no objection to it, but seek to impose it with the aid of penal laws and rigorous punishments on all people alike. [. . .] No one can say that small-pox will necessarily attack those who have not been vaccinated. Dr. Jenner, the inventor of vaccination, originally supposed that perfect immunity could be secured by a single injection on a single arm; but when it was found to fail, it was asserted that vaccination on both the arms would serve the purpose; and when even this proved ineffectual, it came to be held that both the arms should be vaccinated at more than one place, and that it should also be renewed once in seven years. Finally, the period of immunity has further been reduced to three years! All this clearly shows that doctors themselves have no definite views on the matter. The truth is, as we have already said, that there is no saying that small-pox will not attack the vaccinated, or that all cases of immunity must needs be due to vaccination. [. . .] Those who are conscientious objectors to vaccination should, of course, have the courage to face all penalties or persecutions to which they may be subjected by law, and stand alone, if need be, against the whole world, in defence of their conviction. Those who object to it merely on the grounds of health should acquire a complete mastery of the subject, and should be able to convince others of the correctness of their views, and convert them into adopting those views in practice. But those who have neither definite views on the subject nor courage enough to stand up for their convictions should no doubt obey the laws of the state, and shape their conduct in deference to the opinions and practices of the world around them."[134]

Vaccinations, therefore, pose significant health risks. To provide another example: the increasing rate of infant and toddler vaccinations since the 1970s correlates with the dramatic rise in ASD.[135] Some authors even speculate that vaccinations during pregnancy can trigger neurodevelopmental disorders in the unborn child, as they lead to (1) activation of microglia, (2) oxidative stress, and (3) mitochondrial dysfunction. This creates a damaging neuroinflammatory trio in the brain. As the authors write, "Pro-inflammatory cytokines that cross the blood-brain barrier can trigger a neuroinflammatory cascade that begins with microglial activation." They conclude: "Inflammatory processes lead to oxidative stress and mitochondrial dysfunction, which in turn cause oxidative stress in a self-reinforcing vicious cycle, potentially leading to subsequent abnormalities in brain development and behavior."[136] Based on these findings, a harm-benefit analysis and an immediate rethinking of current health policy, which increasingly resembles a vaccination pandemic, are urgently needed. We should always ask ourselves what would be natural. For instance, it used to be common practice to breastfeed infants for their first two years, providing them with maternal passive immunity. This period was sufficient for the autobiographical memory center to mature. This center also serves as our spatial memory; from an evolutionary-biological perspective, it likely needs to begin functioning with the first separation from the mother during the weaning phase. Furthermore, studies show that, assuming sufficient hygiene measures and good wound care when necessary (the toxin-producing bacterium *Clostridium botulinum* can only thrive in oxygen-deprived conditions, not in clean, well-perfused wound tissue), even tetanus vaccinations in this early life phase may not hold the indispensable importance generally attributed to them (see further literature such as *Vaccination Pros & Cons: The Handbook for Individual Vaccination Decisions*).[137]

Pseudo-vaccines, such as the genetic engineering interventions against COVID-19, do not contain aluminum hydroxide. However, they cause brain-damaging neuroinflammation in entirely different and potentially even more effective ways (See additional information above: *PAMPs and Bioweapons*). For instance, severe post-vac symptoms[138] or accelerated Alzheimer's disease development[139] indicate the brain-damaging effect. Ultimately, all individuals—especially parents—must inform themselves as long as authorities can be accused of either not having the best intentions or of being misguided due to structural dependence on industrial or political interests. For example, the US CDC, the equivalent of the German RKI, promotes three mRNA injections against COVID-19 for infants up to the end of their first year of life.[140]

Prenatal Disruption of Neurogenesis—With Chronic Effects. Experimental animal studies prove that when TLR-4-activated production of proinflammatory cytokines like IL-6 occurs in pregnant rats, these can cross the placenta and directly affect fetal brain development.[141] Moreover, IL-6 can not only cross the placental barrier but also the blood-brain barrier.[142] According to a 2021 review article on autism, immune activation during pregnancy "leads to activation of microglia, oxidative stress, and mitochondrial dysfunction—a damaging trio in the brain that can lead to neuroinflammation and neurodevelopmental disorders in the offspring."[143] The authors see the danger of a self-reinforcing vicious cycle that can lead to abnormalities in brain development and behavior. An inhibition of hippocampal neurogenesis caused during pregnancy can persist into adulthood.[144] Since not all TLR-4 activation can be prevented by preventive behavior, it's crucial, in my opinion, especially during pregnancy—when there's typically an increased need for essential vital substances—to ensure no deficiency arises. This also applies to a lithium deficiency, which, if left unaddressed, leads to an inflammatory reaction that is more intense and prolonged than necessary, with negative consequences for both the mother and the developing child.

Instead of blindly following vaccination schedules that are geared more toward available products than actual need, it's wise to carefully avoid any vaccinations whose potential benefit to the individual is outweighed by their risk of harm. As we've seen, vaccinations are highly risky. This is evident even though, according to research presented in *Turtles All the Way Down*,[145] not a single currently recommended childhood vaccination has been adequately studied for health concerns—another compelling reason to make informed decisions. Anyone who chooses to forgo as many vaccinations as possible should then strive for a lifestyle aligned with natural human needs, especially a healthy diet. Since vaccines, when effective, primarily reduce morbidity (the severity of infection) but hardly mortality (as most infections inherently carry a low risk of death), this can usually also be achieved through an adequate supply of essential vital substances (see my book *The Corona Syndrome*).[146] This approach would not only serve as an alternative to many vaccinations but also aid in preventing neurological developmental disorders: While vaccinations are a significant causal risk factor, they don't trigger ASD with absolute certainty, "only" with a significantly increased probability—otherwise, the rates would be much higher. In fact, considering the relative risk increases in Figures 26 and 27, the decades-long "vaccine pandemic" cannot be the sole cause of the enormous increase in disease incidence by more than a hundredfold (!) since the 1970s.[147] There are certainly other causal factors that promote the development of ASD via neuroinflammatory damage to the developing hippocampus. These could contribute to vaccine side effects or even synergistically multiply their effects in interaction with them. Such factors include toxic influences like radio radiation,[148] toxins in food like glyphosate[149] or microplastics,[150] to name just a few introduced by our modern, industrialized world. Beyond reducing these toxic influences, especially during the vulnerable developmental phase of the child's brain, it's crucial that no essential deficiencies make the growing brain even more susceptible to these toxic influences or directly impair its development, such as a lack of aquatic omega-3 fatty acids or lithium. According to the Law of the Minimum, sufficient lithium intake through food or drinking water, along with a relatively good supply of all vital substances that counteract neuroinflammation and promote adult hippocampal neurogenesis, is critical.

Indeed, a study conducted in Arizona, analyzing the hair of autistic children and their mothers for thirty-nine different toxic and essential trace elements, found that mothers of young children with ASD had 56 percent lower lithium levels than mothers of healthy children. In toddlers with ASD aged three to six years, lithium levels were about 30 percent lower than in healthy peers.[151] Essential amounts of lithium (see Chapter 3) during pregnancy and in the first years of life could prevent or reduce the risk of hippocampal-damaging neuroinflammation due to its anti-inflammatory effect. This assumption is further supported by a 2022 study of ninety-two autistic children. The authors wrote: "The present study shows that the degree of decrease in lithium levels was related to the severity of some autistic symptoms."[152] Thus, patients with ASD are found to have increased concentrations of inflammatory mediators like interleukin 1β (IL1β), tumor necrosis factor (TNF), and interleukin 6 (IL-6) in serum, brain, and cerebrospinal fluid.[153] It's therefore unsurprising that there's increasing evidence suggesting that dysregulation of GSK-3 might be key to understanding the development of ASD and related disorders.[154] Interestingly, in an animal model, administering valproic acid—which is used to treat bipolar disorder in humans—causes ASD, and then the effect of small amounts of lithium on clinical symptoms is investigated. The authors conclude: "The results show a protective effect of environmental lithium exposure doses [i.e., levels commonly found in the environment or drinking water] on neurobehavioral deficits in the valproic acid model of autism in rats, suggesting that it may be a potential drug for the treatment of autism."[155] Small amounts of lithium are thus capable of preventing autism-like neurological symptoms in animal models. Essential lithium could therefore be effective in prevention and therapy. Conversely, a deficiency could increase the likelihood of autism.

Autism: A Danger from Lithium in Tap Water? According to a 2023 Danish study, there's a positive correlation between the regional lithium concentration in drinking water, or its possible intake during pregnancy, and the risk of a child developing ASD.[156] We're discussing regional differences in extremely low lithium concentrations in tap water, ranging from 0.6 to 30.7 μg/l. For the study in question, this range was divided into four equally sized concentration brackets. In the three higher brackets, an approximately 23 percent increased autism rate was found compared to the bracket immediately below it. Let's assume pregnant women drink one liter of tap water daily. If we compare the lithium intake of an additional 30.7 μg/day in the highest group with the lithium intake of 200,000 μg/day during the treatment of pregnant women with bipolar disorder, it becomes clear that there can be no causality for ASD development here. This has been confirmed elsewhere. For example, a 2012 Dutch study found no neuropathological abnormalities in the children of patients, even with these enormous amounts of lithium. The authors wrote: "The continuation of lithium therapy during pregnancy had no negative effects on the growth, neurological and cognitive development and behavior of the exposed children."[157] A larger follow-up study from 2022, also examining the consequences of lithium treatment in pregnant women, similarly found "no evidence of significantly altered neuropsychological functioning in lithium-exposed children aged 6 to 14 years compared to non-lithium-exposed controls."[158] However, faster fetal growth and higher birth weight were observed, which can be explained by lithium's function as an activator of stem cell proliferation.[159] If lithium amounts more than six thousand times higher don't cause ASD, there must inevitably be another explanation for the results of the Danish drinking water study. Of course, correlation isn't proof of causation. Additionally, statistical biases can easily lead to perceiving connections where none exist; a similar criticism has been voiced by other scientists.[160] Yet, even if there were a causal relationship between the ASD rate and tap water, and this clearly cannot be explained by lithium, then the water would need closer examination (see also Chapter 3, Additional Information: *Lithium-Containing Tap Water and Fetal Growth*). The reason: Lithium concentration in water is generally somewhat higher when it originates from volcanic areas. Such water often contains other minerals, such as lead, uranium, thallium, cadmium, mercury, tin, and tungsten, all of which are suspected of genuinely increasing the risk of ASD. However, unlike the Arizona study mentioned earlier, these were not examined in the Danish study.[161] In contrast to these toxic elements, which are poisonous even in very low concentrations, lithium is not harmful in essential doses and, in extremely low quantities, is definitely not a cause of autism—quite the opposite. Nevertheless, the untenable results of the Danish study, which don't stand up to even superficial scrutiny, have been used and disseminated in mainstream media, possibly to incite entirely unjustified fear of even the smallest amounts of lithium and, by extension, lithium-containing medicinal springs.[162]

In the treatment of ASD, a systemic approach is also indispensable. The goal must be to ensure that the hippocampus is adequately supplied with all essential vital substances. Concurrently, all measures to reduce GSK-3 activity should be employed to alleviate the resulting autistic behaviors.[163] Remarkably, a study involving thirty children and adolescents with autism demonstrated improvement in 43 percent of patients with lithium treatment alone. This was particularly true for those who exhibited two or more symptoms of an affective disorder.[164] A retrospective study of sixty children with autism also showed a significant reduction in symptoms of social and emotional disorders with lithium administration.[165] This further suggests a causal link between autism and lithium deficiency, especially since lithium, through its targets, reduces the likelihood of neuroinflammation—and thus the risk of an autistic developmental disorder during the susceptible time window. This susceptible period precisely overlaps with the time when most young children must endure a significant number of high-risk vaccinations.

The Trisomy-Autism Connection. Down syndrome (DS), or trisomy 21, offers additional insight into the cause and function of lithium. It is one of the most common genetically determined causes of intellectual disability and often co-occurs with numerous other conditions. According to meta-analyses, between 16 and 18 percent of children with DS also develop ASD, with some estimates from individual studies, using different criteria, reaching as high as 39 percent.[166] Experiments on animals that developed Down syndrome due to genetic manipulation (referred to as DS mice) have shown that hippocampal neurogenesis is impaired in Trisomy 21. Lithium therapy reactivated the hippocampal formation of new neurons, thereby improving the performance of DS mice in cognitive tests.[167] "These results suggest," the authors of this paper state, "that restoring a functional population of hippocampal neurons in adult DS mice preserves hippocampal plasticity and memory, implying adult neurogenesis as a promising therapeutic target for alleviating cognitive deficits in DS patients." Although ASD and Down syndrome have different origins, both largely converge in impaired hippocampal neurogenesis and developmental abnormalities. This is precisely why lithium could exert a therapeutic effect in both cases.

I don't wish to create false hope for a potential cure. However, three aspects make attempting a systemic therapeutic approach appear worthwhile. Firstly, there's the circumstance that—as other scientists and doctors, myself included, infer from existing data—ASD is likely a hippocampal maldevelopment. Secondly, there are indications that this is attributable to a culturally increasingly unnatural lifestyle (which, regrettably, includes most vaccinations). This implies that it's a multicausal developmental disorder. Since we know that the hippocampus is in principle capable of regeneration due to its ability to undergo neurogenesis, it bears examining whether the simultaneous correction of all factors contributing to the development of the disorder is not significantly more effective than a single measure. However, a corresponding multicausal study that largely eliminates all currently known causes doesn't yet exist, either for prevention or therapy. There are ample good reasons to attempt this; just consider what essential lithium alone can achieve. Lithium expert James M. Greenblatt writes:

> While much attention from the traditional medical community has been paid to the development of novel drugs for ASD, there are few who include lithium in their considerations. I consider this a glaring omission, particularly as lithium's calming, mood-stabilizing properties have been known to western psychiatry for nearly a century. In our clinic, we have found that irritability can effectively be treated with lithium. This natural mineral, when administered

at low, nutritional doses in the form of lithium orotate, can safely and reliably alleviate ASD-associated irritability, aggression, and agitation. Unburdened by such symptoms, ASD patients may be able to better engage not only with their care providers but also their loved ones, opening the door to more significant and longer-lasting therapeutic gains.[168]

Bipolar Disorder

John Cade had an early suspicion that a lithium deficiency might be the cause of bipolar disorder, concluding from his observations that it could be an essential trace element.[169] Both of his hypotheses proved correct. Not only is lithium essential, but in 2018, a significant correlation was found between lithium levels in groundwater or tap water and the likelihood of developing bipolar disorder. A lithium concentration below 40 µg/L, compared to a level above 40 µg/L, more than doubles the risk (without statistical manipulation of the data).[170] One of the consequences and hallmarks of bipolar disorder is excessive GSK-3 activity due to lithium deficiency.[171] This hyperactivity triggers a pro-inflammatory state, circadian rhythm dysregulation with sleep disturbances, and diminished neurogenesis alongside an elevated susceptibility to apoptosis (neurodegeneration) in the central nervous system, resulting in accelerated hippocampal volume loss.[172] Genetic variants in the GSK-3 gene's regulatory region that inhibit its activity mirror the effects of lithium and lower disease risk, further substantiating this link.[173] Thus, substantial evidence suggests that bipolar disorder, in both its origin and ultimate outcome, is a disease or developmental disorder of the hippocampus.[174]

For example, lower hippocampal volume is associated with structural abnormalities of the hippocampus, including reduced cell density and biomarkers indicative of neuronal dysfunction. In contrast, permanent lithium treatment prevents this neuropathological degradation mechanism and even leads to an increase in hippocampal volume.[175] An optimal supply of essential amounts of lithium during development could potentially even prevent these structural developmental disorders. "The dysfunction of the GSK3 signalling pathway is involved in all the aforementioned 'biological causes' of BD," write two scientists from Canada's University of Saskatchewan in their article, "GSK3 Signalling and Redox Status: in Bipolar Disorder: Evidence from Lithium Efficacy." They conclude that in this complex scenario, dysregulated GSK-3 can be seen as the common denominator linking them all: "Inhibition of GSK3 [but also of IMPase and many other lithium targets] by lithium may at least partially explain the beneficial effect on these biological dysfunctions and the superiority in clinical efficacy."[176] Lithium thus appears to not only halt the disease process of bipolar disorder but perhaps even reverse it through hippocampal growth promotion. Consequently, numerous imaging studies now highlight lithium's neuroprotective (brain-protecting) and neurotrophic (brain-growth-promoting) effects, suggesting it can slow the degradation of brain matter in bipolar disorder or, notably, even promote the normalization of hippocampal volume.[177] However, this outcome isn't consistent. Remarkably, contrary to expectations, a further decrease was found in some patients with bipolar disorder despite lithium therapy—but always only in those in whom no symptom improvement was observed as a result of the administration.[178] It would be interesting to use the new imaging techniques for investigating lithium distribution in the brain (see Chapter 2, Argument 3) to find out whether lithium accumulation in the left hippocampus could be fundamentally disturbed in individuals who respond to lithium neither symptomatically nor through measurable hippocampal growth. As far as I know, a multicausal prevention therapy has not yet been carried out or published in people with bipolar disorder, aside from one study involving alcoholics, some of whom also suffered from bipolar disorder in addition to many other illnesses. Although only a fraction of the program for a lifestyle aligned with natural human needs was implemented, the success was remarkable in many respects, as described by the treating physician:

> Other benefits of lithium therapy [5 mg pure lithium (in the form of lithium orotate) 4–5 times per week] included improvement in liver and cardiovascular function, reduction (and in some cases complete disappearance) of migraine headaches, relief of symptoms of Meniere's disease [an inflammation of the inner ear resulting in vertigo, nausea, fluctuating sensorineural hearing loss, and tinnitus], and improvement in seizure freedom in epilepsy. In patients with leukopenia [lack of white blood cells] as a result of chemotherapy, there was an increase in white blood cells; in patients with liver cirrhosis, there was a decrease in edema and ascites [fluid in the abdomen]; and in patients with lung cancer, there was a decrease in pleural effusion [fluid between the lungs and pleura] and lymph node swelling. Three patients with this affective [bipolar] disorder no longer experienced manic episodes during treatment with lithium orotate. Hyperthyroidism also improved in four patients.[179]

A whole range of positive effects on many different disease symptoms (migraine, epilepsy, etc.) are described here within the context of a controlled clinical study with low-dose lithium orotate. Therefore, the assumption is justified: Bipolar disorder—which to date has only been treated with high-dose lithium purely through medication, with a modest success rate of only 30 percent and a nearly 100 percent side effect rate—could very probably be successfully treated within a systemic approach using significantly lower doses of lithium orotate. And without undesirable (side) effects!

Schizophrenia

"There is reason to believe that hippocampus disorders are responsible for the development of psychotic symptoms in 'schizophrenia,'" concludes the review article "Hippocampus in Health and Disease."[180] Schizophrenia is both a neurodevelopmental disorder and a progressive neurodegenerative disease causing irreversible loss of cognitive function.[181] It often emerges between late adolescence and early adulthood and is characterized by a range of neuropsychiatric abnormalities, largely indicative of hippocampal dysfunction. "The hippocampus undergoes changes that are characteristic of schizophrenia, and these can be observed in both early and advanced stages of the disease," a group of neuroscientists from Ruhr University Bochum concluded in a review article. "Hippocampal development therefore plays an important role in the manifestation of schizophrenia," they further explain.[182] This is known to be due to a lithium deficiency disorder, which negatively affects hippocampal development and function in a variety of ways. Scientists at the Neuroscience Institute of Psychiatry, King's College London, even consider schizophrenia to be primarily a disorder of GSK-3 regulation: "Converging evidence suggests that the regulation of glycogen synthase kinase 3 (GSK-3) might be important in schizophrenia. [. . .] We propose a variant on the neurodevelopment and dopamine hypotheses of schizophrenia, whereby (i) an early dysfunction in GSK-3 regulation has neurodevelopmental consequences that predispose to disease and (ii) dysfunction in GSK-3 regulation in the adult brain alters dopamine signalling events, causing psychotic symptoms and cognitive dysfunction."[183] A lithium deficiency during brain development could therefore favor the development of schizophrenia via a dysregulated Lithiome, where GSK-3 plays a crucial role. This explains the correlation observed in drinking water studies between reduced psychiatric hospital admission rates and higher lithium concentrations in drinking water. Vaccinations are, consequently, also specifically suspected of causing schizophrenia—this includes COVID-19 mRNA injections. This suggests that these cause medically induced neuroinflammation in children and adolescents, resulting in a neuropsychiatric developmental disorder.[184] Thus, the same recommendations apply here as for ASD discussed above.

The multicausal origin of hippocampal developmental disorders is also evident in the example of schizophrenia, where a deficiency of aquatic omega-3 fatty acids, common in adolescents, significantly influences the risk of developing the disease. In a groundbreaking study, young patients at risk of schizophrenia, aged thirteen to twenty-five, received either aquatic omega-3 fatty acids or a placebo for just three months.[185] Over the subsequent nine months, the likelihood of a new psychotic episode was almost six times lower in the omega-3 group than in the placebo group. This result was so spectacular that it was decided to extend the observation period for another six years. The expectations were indeed confirmed: while 40 percent of the sham-treated subjects developed manifest schizophrenia, the figure was less than 10 percent in the omega-3 group—which was highly unexpected, given that the three-month therapy had concluded several years prior.[186] If all deficiencies and toxic influences were eliminated or at least significantly reduced, neuropsychiatric developmental disorders like schizophrenia would be virtually nonexistent. Of course, this applies to all other similar conditions as well. Nevertheless, neither causal prevention nor causal therapy is currently the norm.

Tourette's Syndrome

Tourette's syndrome (TS) encompasses a group of chronic neuropsychiatric developmental disorders that typically emerge in childhood. These disorders are characterized by tics—sudden, repetitive movements or vocalizations that, while consciously experienced, are difficult to control. Vaccinations may play a role (see Fig. 27). TS is frequently accompanied by disorders falling under the umbrella of mental immunodeficiency syndrome, such as AD(H)D (see below), obsessive-compulsive disorder, anxiety disorders, or depression. When these co-occurring conditions are present, TS patients tend to have poorer developmental outlooks and less favorable treatment outcomes than those without them. The precise mechanism of TS remains unclear. However, the potential connection between abnormal immune activation, neuroinflammation, and neuropsychiatric disorders has garnered significant attention over the past two decades. It appears that dopamine dysregulation and immune system dysfunction play a crucial role in the underlying causes and progression of the disease.[187] In TS, microglial inflammatory activity is heightened, while microglia-mediated clearance mechanisms are impaired.[188] Similar to migraine, TS involves reduced Akt activity alongside elevated GSK-3 activity, leading to neuroinflammatory and dopaminergic dysregulation (see Figs. 12 and 24).[189] Since the regulation of GSK-3, Akt, and dopamine are all under the control of the Lithiome, it's anticipated that correcting a lithium deficiency or administering low-dose lithium could causally improve symptoms. In fact, I've personally been contacted by a TS patient who experienced a marked improvement simply by taking approximately twice the recommended daily allowance (RDA) of lithium (in the form of lithium orotate), as determined in Chapter 3. Similar observations were documented in several case reports in 1977 and 1983: Lithium led to the complete disappearance of symptoms without side effects, unlike the antipsychotic haloperidol.[190] This strongly suggests that lithium deficiency could be the primary cause of TS.

Borderline Personality Disorder

Individuals with borderline personality disorder (BPD) often display impulsive and aggressive behavior. The hippocampus (along with the amygdalas) is part of the limbic system, which plays a central role in controlling these manifestations of emotional reactivity. In patients with BPD, significantly smaller volumes were found in both the left hippocampus and the amygdalas compared to healthy controls. This suggests that a reduced volume of both hippocampi (and the amygdalas, also as an indication of dysfunction) is the neuronal correlate for BPD development.[191] Another study concluded that memory problems in BPD can be explained in part by a disturbed relationship between hippocampal activation and successful memory coding.[192] In addition, a clinical study provides evidence that that structural changes in the hippocampus promote aggressive behavior typical of BPD.[193] Clinical researchers have also suggested that obsessive-compulsive disorder (OCD), which is often associated with borderline personality disorder, should be well treated by low-dose treatment with lithium—after all, it is also due to the same pathomechanisms (neuroinflammation, etc.)[194] It is therefore obvious that in the case of BPD and closely related symptoms such as obsessive-compulsive disorder or anxiety disorder, a systemic approach as outlined above should be promising.

Addictive Behavior

As we saw in Chapter 3 on the benefits of lithium orotate, lithium helps individuals with alcohol dependence maintain abstinence. But how does addictive behavior arise, and what role does the Lithiome play in it? Addiction is the pursuit of something that promises feelings of happiness. Dopamine is well-known as the central "happiness hormone." However, it also has a darker side, appearing to play a central role in the development of addictive behavior.[195] This is because the dopamine system helps our brain distinguish between important and unimportant things. According to German addiction specialist Prof. Dr. Falk Kiefer, this includes "on the one hand, dangers, but on the other hand, reward-associated stimuli, otherwise we would walk past food when hungry."[196] A "dopamine high" therefore leads to selective attention: "Things that have been marked with dopamine," says Kiefer, "become increasingly important." The adolescent brain, in particular, is "sensitive to new imprints, meaning it's geared towards new reward stimuli." According to Kiefer, this "makes a lot of evolutionary sense because this is the phase where detachment from former reward stimuli like the mother and family occurs, and young people need a new orientation." A particularly large number of new preferences are set during adolescence. Excessive dopamine release at this time is harmful if it leads to uncontrollable addictive behavior. And this is where the Lithiome comes into play again, as the trace element counteracts "dopamine-dependent behavior mediated by the [. . . GSK-3] signaling cascade," as stated in a scientific paper's title. In it, the authors point out that, according to their data, hyperactive GSK-3 (or GSK-3 not inhibited by essential lithium) is an important signaling mediator of dopamine and lithium's effects. This leads them to conclude that lithium could be relevant not only in the prevention and treatment of addictive behavior but also for the treatment of dopamine-related disorders such as attention deficit hyperactivity disorder and schizophrenia.[197]

The Center for Recovery and Wellness (CRW), an inpatient and outpatient health center for individuals with addiction in New York's East Village, employs what they call a "triad approach" to address the three most common causes of addiction: physical pain, emotional pain (trauma), and attention deficit disorder. The doctors successfully use low-dose lithium in the form of 150 mg lithium carbonate, which is roughly equivalent to 28.7 mg of pure lithium. Due to the higher efficacy of lithium in the form of lithium orotate (see Chapter 3: *Is Lithium Orotate the Better Lithium?)* I assume that 5 to 10 mg of pure lithium in the form of lithium orotate could achieve the same, possibly even better, effect, which is quite remarkable. "The introduction of low-dose Lithium in an addiction treatment setting where trauma, untreated ADHD and medical conditions are common was useful in helping patients achieve and maintain progress in their lives," writes Lead Physician Sudhir Gadh.[198] "Nearly all our clients are diagnosed in accordance with the DSM-5 with a singular or polysubstance dependence. [*Diagnostic and Statistical Manual of Mental Disorders,* 5th Edition). Greater than 75 percent are also diagnosed to have an underlying psychiatric condition such as Major Depression, Post Traumatic Stress Disorder, Attention Deficit Disorder, or a substance induced condition. Furthermore, the majority have been in multiple addiction treatment facilities indicating that the standard treatments were not effective enough." According to CRW's experience, any patient with a history of trauma, addiction, incarceration, and/or relapse can be successfully treated with low-dose lithium for "emotional pain" as part of a holistic-systemic approach. Notably, the use of many allopathic medications could sometimes be completely eliminated as a result.

Comparing blood values before lithium administration and three months after the start of low-dose lithium therapy revealed no negative changes in laboratory values for kidney function, leukocytes (white blood cells), or thyroid function. Taking lithium at this dosage, therefore, presents no health problem, in stark contrast to many medications that numerous CRW patients regularly took before therapy. These include benzodiazepines, most commonly prescribed in the USA for anxiety and sleep disorders. However, due to their many side effects—which, grotesquely, are identical to what they are supposed to treat—such as "increased excitability, nightmares, anxiety, insomnia, panic attacks, depression, hallucinations, irritability, paranoid thoughts, social phobia, poor memory, poor concentration, delirium, and even psychosis" —the CRW adheres to a "Zero Benzo" strategy.[199] The use of the sleeping pill zolpidem is also stopped at the CRW (see above, Additional Information: *Sleeping Pills Inhibit Glymphatic Brain Clearance and Promotes Alzheimer's*). However, according to the CRW, these withdrawal measures were successful only because of the low-dose lithium in the addiction treatment. The same applies to the use of antipsychotics, often administered as tranquilizers. These could also be reduced by 72 percent with lithium. Moreover, according to lead physician Gadh, there was a significant reduction in polypharmacy, such as the simultaneous use of several antipsychotics. While taking these "led to control of potential self-harm and agitation, it came with a significant burden of side effects." Among these, Gadh lists "weight gain, sedation, cognitive impairment, and hypercholesterolemia." However, the use of low-dose lithium made these drugs superfluous. These successes should serve as an important indication that lithium in essential (preventive) and low (therapeutic) dosages makes many medications unnecessary—very probably not only in the case of individuals with addiction.

AD(H)D

Attention-deficit/hyperactivity disorder (ADHD) describes a behavioral disorder in children, adolescents, or adults characterized by abnormalities in three core areas: severe attention and concentration deficits (ADD), and often pronounced impulsivity and significant physical hyperactivity (H), hence AD(H)D in its full manifestation. It's now well-documented that high impulsivity could be caused by a lithium deficiency.[200] This deficiency might also be responsible for the dysregulation of the so-called dopaminergic system, which is under the control of the Lithiome and considered one of the causes of ADHD.[201] Neuroinflammation is regarded as a decisive cause.[202] However, neuroinflammation is also a consequence of AD(H)D, as constantly failing to meet the demands of the social environment is extremely stressful.[203] An increased release of DAMPs is the inevitable consequence, perpetuating the vicious cycle of neuroinflammation (see Fig. 24). As expected, a 2023 study therefore reports a significant shrinkage of about 23 percent of both the left and right hippocampus in children with ADHD compared to control subjects. As with neuroinflammation, this can be both a cause and a consequence of AD(H)D.[204] All functional disorders—lack of attention and hyperactivity—are thus under the control of the Lithiome. This means that lithium deficiency is causally related here as well; addressing it therefore appears sensible for both prevention and therapy. It's not surprising that pharmacological inhibition of GSK-3 is also recommended in such cases.[205] But why search for lithium mimetics, as the pharmaceutical industry does based on the same understanding? The lack of GSK-3 regulation can be remedied directly and entirely naturally by essential amounts of lithium. This should obviously be done as part of a systemic approach that also corrects other deficiencies (in aquatic omega-3 fatty acids, zinc, selenium, vitamin D, etc.) that are unfortunately far too common in children. The clinical successes achieved solely by correcting these deficiencies—as reported by Dr. Christian Schellenberg, among others (see Chapter 3)—are simply spectacular. The common alternative to this natural approach includes "medications" like Ritalin® and Medikinet®, whose effects frighteningly resemble amphetamines such as "speed," methamphetamines like "crystal meth," and their derivatives like "ecstasy." I therefore assume (and wish for all affected individuals) that standard medication with side-effect-rich methylphenidate stimulants will give way to causal therapy with essential active ingredients. However, the term "therapy" in the context of essential lithium administration is also somewhat misleading in that no therapeutic doses are administered; rather, existing deficits are merely corrected in order to stop neuroinflammation and, in concert with all other systemic measures, allow the hippocampus to naturally grow again.

Depression

Stress-induced neuroinflammation is mediated by the TLR4 receptor and GSK-3-driven signaling, which increases susceptibility to depression-like behavior.[206] The inhibition of GSK-3 activity in microglia is common to almost all types of antidepressants.[207] Ultimately, this inhibition directly and indirectly leads to an increase in hippocampal neurogenesis, which represents the actual antidepressant effect by reactivating psychological resilience. Indeed, the central mechanism of all clinically available antidepressants is likely the indirect activation of adult hippocampal neurogenesis.[208] This was first observed in animal models and then confirmed for human application.[209] Although depression itself isn't *directly* caused by a disturbance in adult hippocampal neurogenesis, the resulting reduced psychological resilience *indirectly* makes individuals more susceptible to developing depression.[210] All the Lithiome functionalities listed in Chapter 2 directly or indirectly influence productive adult hippocampal neurogenesis. This, in turn, is linked to a functional mental immune system that protects us from developing depression. For this reason, a Lithiome dysregulated by lithium deficiency increases the risk of depression. It's therefore not surprising that hyperactive GSK-3 is considered a key factor both in the development of depression (A Master Player in Depressive Disorder Pathogenesis) and in the effectiveness of medicinal treatments (A Master Player in Treatment Responsiveness).[211] Even if GSK-3 is a central modulator of mood regulation, only a systemic approach that encompasses all aspects of life as outlined above, and adheres to the Law of the Minimum and the Maximum, can be promising in the long term, especially in prevention and therapy.[212]

Suicidal Tendency

Independent and purely random mutations in five different genes, including the lithium targets GSK-3 and IMPase, share three grave effects. First, these respective mutations measurably increase the suicide risk (which is how they were identified). Second, the dysregulation triggered by the mutation directly or indirectly activates GSK.[213] Third, this leads to the development of neuroinflammation, which, according to Swiss pharmacologist Hans O. Kalkman, "is causally linked to dysphoria and anger, two factors relevant for suicidal thoughts and suicide attempts."[214] Even if such gene variants are very rare and thus cannot be solely blamed for the high global suicide rate, these findings show that dysregulation of the Lithiome has a causal influence on an increased probability of suicide. In these rare cases, GSK-3 becomes hyperactive due to random genetic variants, functionally mirroring a lithium deficiency. This, in turn, provides the causal link and explains why lithium in drinking water contributes to lowering suicide rates through its regulatory influence on the entire Lithiome, particularly GSK-3. However, lithium not only directly influences neuroinflammation but also all functions associated with a healthier psyche, including telomerase activity. In fact, comparatively short telomeres can predict an increased risk of suicide or suicidal ideation in patients with schizophrenia and bipolar disorder.[215] This closes the causal chain between lithium deficiency and suicide: it's already well-established that lithium deficiency increases the risk for mental and psychiatric disorders, which in turn are associated with an increased suicide risk. The cause is neuroinflammation, promoted by lithium deficiency and triggered by GSK-3 hyperactivity, which in turn blocks adult hippocampal neurogenesis. As a result, one of the most common mental disorders, depression (Major Depression), develops, which is itself associated with a drastically increased suicide risk.[216]

"Evidence from both basic and clinical researches support that lithium may decrease impulsivity and may at least partially, exert its antisuicidal effect via reinforcing 'top-down brakes' of impulsive action," writes Greek scientist Orestis Giotakos. According to him, based on the current state of research, "we may suggest that even natural lithium level intake can influence impulsivity, a possible core factor that mediate to the manifestation of both suicidality and aggressiveness, or even criminality. Moreover, we may suggest that a lithium deficiency state may precipitate these situations."[217] Based on these findings, I believe it's imperative from both a medical and ethical perspective to offer low-dose lithium therapy, especially to individuals with suicidal tendencies, to break the life-threatening vicious cycle of neuroinflammation.

Conduct Disorder

Conduct disorder (CD), the clinical picture of aggressive, antisocial, and rule-breaking behavior during childhood, is associated with a noticeably smaller left hippocampus, similar to many other mental immune system dysfunctions (see Chapter 2, Argument 3).[218] Childhood maltreatment or abuse, the main risk factor for mood swings, anxiety, substance abuse, psychotic disorders, and personality disorders, is also associated with a reduction in the size of the hippocampus, primarily affecting the left hippocampus, suggesting a link.[219] CD is the main risk factor for the development of antisocial personality disorder (APD) in adulthood—although not necessarily. In the search for an explanation, a finding emerged: The larger the volume of the left hippocampus or a specific region of the left hippocampus, the lower the likelihood that SSV will progress to ASP.[220] Anti-inflammatory, proneurogenetic, and thus systemic therapy is clearly the necessary approach, especially since it has been shown in animal models that hyperactive GSK-3 promotes aggressive behavior via neuroinflammation.[221]

"Mounting data have demonstrated," writes US physician Austin Perlmutter in *Psychology Today*, "that inflammation changes the way we think, altering our mood and even our decision-making. This powerful information has significant implications for both individual and public health."[222] According to Perlmutter, it is obvious "that the ability to make good, long-term oriented decisions is imperative for our well-being. It's what keeps us from spending all our money on a whim, from yelling at our boss, and from eating junk food at every meal. On the other hand, when we engage in impulsive, quick-fix solutions, the results can be catastrophic. That's why it's both illuminating and frightening to see that higher levels of blood inflammation predict decisions characterized by impulsivity and an inability to delay gratification." He is convinced that neuroinflammation changes our perception of the world. But this is an unnatural view, clouded by a lack of essential lithium. I am therefore convinced that our view of the world will be profoundly different in a positive way once the global lithium deficiency is remedied.

Anxiety Disorders

Neuroinflammation and resulting changes in the brain (especially in the hippocampus) lead to anxiety disorders. They are also associated with chronic stress, with chronic stressful situations accelerating the development of anxiety disorders—a vicious circle illustrated in Fig. 24.[223] Among other things, the left hippocampus is again most affected, suggesting lithium deficiency as a direct or indirect cause (see Chapter 2, Argument 3).[224] However, the tendency to develop into a more anxious person could already be established during pregnancy, as animal research suggests. Neuroscientists from various New York universities explain their findings: "Since maternal exercise/activity is known to reduce both brain TNFα [225] and offspring innate fear [226], while maternal stress was reported to increase brain-TNFα [227] and offspring fear and anxiety [228] maternal brain-TNFα may report environmental conditions to promote offspring behavioral adaptation to their anticipated postnatal environment."[229] Because any stressful situation during pregnancy can lead to neuroinflammation, which also influences the child's future, it's crucial here to ensure an adequate supply of vital substances, especially anti-inflammatory lithium. Concurrently, injections of genetic material responsible for neuroinflammatory spike proteins in the brain, for example, should be viewed very critically. We risk raising an entire generation of children already prone to excessive anxiety from birth—and who are easier to control with fear propaganda. If this were indeed the intention, as I suspect (see Chapter 7 on this issue), it would offer another explanation for why lithium is banned as a dietary supplement in the EU, Switzerland, and many other countries, despite its essentiality.

Fundamentally, and I'm happy to reiterate this, a systemic approach should always be pursued in both the prevention and treatment of anxiety disorders. Humane medicine must, without exception, account for all basic human needs if it aims to foster an open, reflective, and not purely anxiety-driven humanity.

Pain Syndromes

Pain is a warning signal essential for life and survival. However, it can also shorten life if it becomes self-perpetuating, chronic, and impairs daily living. The fundamental distinction between physical and emotional pain isn't as clear-cut as the terms suggest—after all, body and soul form a unified whole. Emotional pain makes us more susceptible to physical pain, even leading to allodynia, where even light touches can trigger pain. By contrast, physical pain, in turn, burdens the soul.[230] Accordingly, pain can easily initiate the neuroinflammatory vicious cycle (see Fig. 24). Whatever the cause of a chronic pain condition—be it physical or psychological—the common pathological denominator for the chronification of pain is hyperactive GSK-3 or a dysregulated immune system.

Polyneuropathy and Neuropathic Pain

The neuroinflammatory vicious cycle of chronic pain is triggered by a stress-related release of DAMPs in the brain. However, it can also be initiated by a neuropathic process distant from the brain, which in the affected tissue triggers the release of pro-inflammatory messengers that reach the brain via the bloodstream and across the blood-brain barrier. Neuropathic pain originates from pathologically altered sensory nerves. The pain, usually very severe, typically occurs in episodes and is often described as burning, stabbing, or dull. Polyneuropathy (also known clinically as multiple peripheral neuropathy) describes the malfunction of several peripheral nerves simultaneously. Symptoms of polyneuropathy include pain in affected body areas (often the legs), as well as sensory disturbances, abnormal sensations like tingling, burning, and a "crawling" feeling. Additionally, disturbances in touch, pain, or temperature sensation can occur. Causes of neuropathic pain include spinal cord injuries or damage from the body's own immune system, as seen in MS. Frequently, however, it also stems from metabolic disorders (diabetes mellitus), arteriosclerosis, or herpes zoster (shingles).[231] Cancer can also cause neuropathic pain, either directly through nerve compression by tumor tissue or indirectly through the release of cytokines like TNFα. The latter is released by the immune system due to local necrosis (dying tissue) of a rapidly growing tumor—hence the name TNF: tumor necrosis factor.[232] Radiation or chemotherapy can also damage nerve tissue.

The neuroinflammatory processes in the hippocampus triggered by nerve cell damage therefore play a crucial role in developing depressive disorders as a result of chronic pain.[233] The local release of pro-inflammatory messengers such as TNFα and IL-1β, which is caused by GSK-3 hyperactivation, interrupts hippocampal neurogenesis, causing even peripheral nerve injuries to lead to changes in the structure, volume, and function of the hippocampus This has been shown in an animal model.[234] This mechanism contributes to the development of depression in most pain patients.[235] In the long term, the chronic blockade of adult hippocampal neurogenesis even threatens the development of Alzheimer's dementia, as I was able to describe in detail in a review article.[236]

The activation of neuroinflammation via dysregulated GSK-3 is considered a crucial mechanism for the development and maintenance of many types of pathological pain.[237] Since chronic pain severely alters the hippocampus in the long term, the vicious circle must be broken quickly. Direct inhibition of GSK-3, for instance, leads to a reduction in neuropathic pain, as shown in animal models using lithium mimetics.[238] Even though these pharmaceutical products aim to replace inexpensive lithium, they still demonstrate the benefits of lithium, which I would always prefer due to its natural multifunctionality within a systemic therapy. As a natural GSK-3 inhibitor, lithium could even play a crucial role in developing comprehensive treatments for spinal cord injury. This insight comes from a 2023 review article that attributes to synthetic GSK-3 inhibitors (lithium mimetics) the ability to "promote neurogenesis," thereby indirectly proving the therapeutic benefits of this trace element.[239] "Furthermore," the authors write, "signaling pathways associated with GSK-3 also participate in the pathological process of neuropathic pain that remains following spinal cord injury." Lithium could also be effective against neuropathic pain because it releases β-endorphin in the brain—an endogenous, pain-reducing morphine-like opioid peptide (a small protein with opiate effect), as demonstrated in animal models: "Our results provide evidence that lithium induces a long-lasting analgesia in neuropathic mice presumably through elevated brain levels of beta-endorphin and the activation of mu opioid receptors."[240] Interestingly, lithium simultaneously appears to naturally inhibit the mechanism of tolerance development to opiates like morphine. It could, therefore, reduce the risk of addiction if opiates are used in addition to lithium for pain treatment.[241]

Migraine

A migraine is characterized by recurrent, typically severe, unilateral headaches that usually last between four to seventy-two hours and are accompanied by other symptoms such as nausea and sensitivity to light and sound. In about a third of cases, migraine attacks are preceded by a so-called aura. An aura is generally a five- to sixty-minute period of discomfort with slowly spreading visual disturbances like visual field defects, flickering, zigzag lines, or flashes. This can, however, precede not only a classic migraine attack but also an epileptic seizure—suggesting a similar disease mechanism. Sensory disturbances such as tingling or numbness in the extremities are also common. Some individuals also experience speech problems during this phase, such as difficulty finding words or speaking. In the worst cases, there may even be temporary complete hemiplegia, resembling a stroke. Migraine affects more than 15 percent of the general population, making it one of the most common chronic diseases with significant disabling effects in terms of prevalence and productivity loss due to workplace absences.[242] Women are three times more likely to be affected by migraine than men.

Some researchers consider migraine a special form of neuropathic pain due to some fundamental neuropathological mechanisms.[243] Thus, as with other acute and chronic pain conditions, neuroinflammation very likely plays a key role as a trigger for migraine attacks.[244] Neuroinflammatory signaling pathways (see Fig. 24) could, for these researchers, represent therapeutic targets and perhaps even biomarkers for migraine. Besides IL-1β, TNFα is also a focus in the hypersensitization of pain-sensitive cranial nerves. In migraine, this particularly involves the trigeminal nerve, also known as the fifth cranial nerve. Among other functions, it's responsible for facial sensation and, like other sensory nerves, reacts to TNFα. A migraine attack can be triggered by an injection of TNFα, while injecting TNFα antibodies has been shown to relieve pain in humans.[245] TNFα levels in the blood rise at the onset of migraine pain and gradually decrease after the attack. In addition to TNFα levels, IL-6 levels are also significantly increased in migraine patients compared to healthy controls between attacks. It's therefore not surprising that recent research indicates GSK-3 and its activation of NF-κB represent a central signaling pathway for the release of IL-1β and TNFα in the initiation of a migraine (see Fig. 24). Dr. Bianca Raffaelli from the Headache Center of the Department of Neurology with Experimental Neurology at Charité in Berlin commented on why migraines often occur during menstruation in a press release dated February 23, 2023: "Animal models suggest that fluctuations in female hormones, especially estrogen, lead to an increased release of CGRP, an inflammatory neurotransmitter, in the brain."[246] DNF-κB also regulates the messenger CGRP (Calcitonin Gene-Related Peptide).[247] Raffaelli points out that CGRP dilates blood vessels in the brain, causing an inflammatory reaction there; which could be "one of the reasons behind the severe headaches people experience with migraine" (blood vessels have pain receptors, the brain does not). She and her team confirmed this connection in a study.[248] Interestingly, the increased release of this key molecule for migraine is regulated by the activation of GSK-3β and the inactivation of Akt.[249] Like GSK-3, Akt comprises a small family of kinases belonging to the Lithiome. However, unlike GSK-3, Akt is activated, not inactivated, by lithium. When Akt is activated by lithium, it, in turn, inactivates GSK-3, thereby enhancing lithium's anti-inflammatory and pain-relieving effect (see Chapter 2, Fig. 12, lithium target D). Under lithium deficiency, however, GSK-3 is more easily activated, and Akt is more easily inactivated. This could explain why, with lithium deficiency, migraine patients are more likely to experience psychological (PAMPs) or physical (DAMPs) stressors and/or more intense attacks (see Fig. 24). Also, cycle-dependent CGRP release is not attenuated from the outset in the presence of lithium deficiency. Because the lithiome controls both neuroinflammation and CGRP release, lithium would be an effective means of preventing and treating migraine. This was confirmed as early as 1982 in a study by the US neurologist Jose L. Medina. He writes: "The drug that most effectively controls cyclic migraine is lithium carbonate."[250] According to Medina, the patient typically needs a maximum of 300 mg of lithium carbonate three times a day: "It is not necessary to reach high blood levels to achieve relief of the headache. In 19 of 22 patients who were treated for cyclic migraine with an average dose of 900 mg lithium carbonate per day, there was an improvement. Five of them experienced complete remission, five experienced a 75 percent reduction in headache duration, and nine patients experienced a 50 to 75 percent reduction in headache duration." Taking 300 mg of lithium carbonate three times daily totals 900 mg, which is approximately 180 mg of pure lithium. Since there are indications that lithium orotate could be about ten times more efficient (see Chapter 3), just three doses of 6 mg lithium daily (154.3 mg lithium orotate monohydrate each) could be enough to reduce or even completely prevent migraine symptoms—a low-dose lithium treatment might thus suffice to prevent attacks or at least alleviate pain intensity. Perhaps even less is needed, as Medina adds that "patients usually self-adjust their dose after the first two months, and many of them only need 300 mg of lithium carbonate daily or every other day" to keep their headaches under control. This dose reduction suggests that lithium could have a causal therapeutic effect rather than merely a symptomatic one: once the vicious cycle is broken, the essential dose might be sufficient.

If treatment with low-dose lithium is so promising, why has no one in the medical world seemed interested in this spectacular result since then? There are no follow-up studies or mentions of Medina's groundbreaking work by other researchers; it seems as if it never existed. Given that the multiple signaling pathways involved in migraine are highly lucrative pharmaceutical targets, a familiar suspicion repeatedly arises as to why lithium continues to be overlooked as a simple and natural solution.

Cluster Headache

As with migraines, cluster headaches are characterized by frequent attacks of severe, always unilateral headaches, usually in the area of the eye or temple. Cluster headaches differ from other headaches due to accompanying symptoms such as tearing of the eyes, a runny nose, or a slightly drooping eyelid. Paralysis symptoms and hypersensitivity to noise and light can also be observed, similar to migraines. Another parallel is their cyclical occurrence, suggesting that chronobiological rhythms or the signaling pathways responsible for them may be dysregulated in both pain syndromes or have an indirect influence on the pain attacks.[251] Chronobiological rhythms, like the circadian rhythm, are regulated by lithium via the Lithiome (see Chapter 2). It is therefore not surprising that in the article "Management of Chronic Cluster Headache," published in 2011, the authors consider "lithium to be the drug of first choice for the prevention of chronic cluster headache alongside verapamil (a calcium antagonist with many side effects, which can itself cause headaches)."[252] Two Brazilian neurologists also recommended lithium carbonate in their 2010 review article as the first choice for lithium-responsive headaches that occur cyclically, particularly at night or upon waking: "Nocturnal Migraine, Cluster Headache, and Hypnic Headache." According to the two authors, "All three types show a positive therapeutic response to lithium and the average dose is from 300 to 900 mg at bedtime."[253] Here too, as already discussed for the treatment of migraine, a significantly lower amount of lithium could be sufficient with lithium orotate, which could possibly be reduced even further as part of a systemic prevention and treatment approach.

Epilepsy

Globally, approximately 65 million people are affected by epilepsy.[254] This neurological disorder, characterized by seizures, can stem from many causes. Rarely, it's due to genetic changes, for instance, in ion channels, which can lead to a "short circuit" in the nervous system.[255] While nerve cells themselves only very rarely cause their own dysregulation, the more than twenty antiepileptic drugs certainly influence their function. This is why these medications only achieve symptomatic seizure control in a minority of patients, often with severe side effects. Crucially, they don't alter the origin or progression of epilepsy. Therefore, to act causally, we need to understand what causes epilepsy.

Much more frequently than genetic factors, environmental factors are the cause of epilepsy's development, such as traumatic brain injury. Epileptic seizures were first described in connection with a "gaping wound on the head" in the Edwin Smith Papyrus from Babylon, dating from around 1700 BC.[256] Approximately 20 percent of all epilepsies are primarily caused by traumatic brain damage.[257] The neuroinflammatory process caused by the damage leads to chronic, faulty wiring of nerve cells, which increases the likelihood of seizures.[258] A developing hippocampus reacts even more sensitively to harmful environmental influences than that of an adult, as is already known from schizophrenia research.[259] Neuroinflammation during brain development is therefore particularly problematic, as the vaccination studies previously discussed under the topic of autism have shown. Routine childhood vaccinations increase the likelihood of developing epilepsy by an average of about four times (see Fig. 27).[260]

"It is believed that hippocampus has an inhibitory effect on seizure threshold (i.e., it keeps it elevated)," explain two Indian neurologists in a review article published in 2012.[261] If one of the two hippocampi is damaged during its development, seizures occur more frequently. Each seizure causes further damage (e.g., due to oxygen deprivation during the seizure), which, in turn, leads to even more seizures; a vicious cycle that makes treatment increasingly difficult, causing cause and effect to merge and become self-perpetuating. The two authors summarize the data as follows: "In up to 50 to 75 percent of epilepsy patients who died from temporal lobe epilepsy that was not treatable with medication, hippocampal sclerosis [loss of nerve cells and their "replacement" by glial cells] can be detected at autopsy. However, it is not yet clear whether the epilepsy is a consequence of hippocampal sclerosis or whether repeated seizures damage the hippocampus." However, the two doctors assume that "the mechanism of hippocampal sclerosis in epilepsy could be related to the development of uncontrolled local inflammation of the hippocampus," which in turn explains the increased risk associated with vaccination.

Interestingly, in most cases, the epileptic seizure begins in the left temporal lobe and is accompanied by left-sided hippocampal atrophy.[262] What could be the reason for this directional lateralization? As we saw in Chapter 2 under Argument 3, the left hippocampus accumulates significantly more lithium than the right. This selective accumulation can be interpreted as an indicator that it naturally has a higher demand than the right. This, in turn, could provide an explanation for why the left hippocampus is more sensitive to a neuroinflammatory stimulus or environmental factors (such as an armada of vaccinations in childhood) than the right one—at least under conditions of lithium deficiency, which is often severe in our society. This also explains why the left side of the hippocampus is more frequently affected by epilepsy than the right.

Neuroinflammation is both the cause and driver of the epileptic disease process. It's therefore not surprising that clinical studies have shown that specific anti-inflammatory or drug interventions in the TLR4 signaling cascade up to IL-1β (see Fig. 24) reduce seizure frequency in epilepsy patients, even in those who don't respond to current antiepileptic drugs.[263] However, the most natural way to dampen this neuroinflammatory and thus epileptogenic signaling cascade would be to administer essential amounts of lithium.[264] Studies have demonstrated the neuroprotective effect of low-dose lithium on epilepsy in animal models. While high-dose lithium, as used in bipolar disorder, can trigger seizures, low-dose lithium reduces the likelihood of an epileptic seizure.[265] These dose-dependent, opposing effects of lithium are typical, as we'll see with further examples in Chapter 6; however, essential to low doses are always health-promoting. Evidence that low-dose lithium orotate leads to an "improvement in seizure freedom in epilepsy" has been known since 1986; we've already discussed the corresponding clinical study concerning bipolar disorder.[266] As with all neuroinflammatory diseases, a systemic prevention approach is not only the most promising for epilepsy but could also manage with even lower doses of lithium. Given the hippocampus's regenerative potential, this also represents a causal therapy option.

Neurodegenerative Diseases

As we saw in Chapter 2, the Lithiome is not only crucial for brain development, but also for its maintenance of function into old age—in keeping with the Evolution of the Grandmother. Due to their transgenerational effect (more surviving grandchildren, as discussed in Chapter 1), grandparents need to remain physically and mentally fit. The development of a neurodegenerative disease contradicts this evolutionary imperative and, under conditions of a truly species-appropriate lifestyle, would not occur. Among the many causal risk factors for neurodegenerative diseases that stem from disregarding the Laws of Minimum and Maximum is a lithium deficiency and, consequently, a dysregulation of the entire Lithiome. This trace element is so fundamental that epidemiological studies have even found an increased risk for Alzheimer's dementia with reduced lithium intake via tap or other potable water, as we observed in Chapter 2 (Argument 1).

The numerous disease-promoting mechanisms associated with a dysregulated Lithiome—including chronic neuroinflammation, impaired nerve cell regeneration due to insufficient autophagy, and so on—are key factors not only in Alzheimer's disease but also in all neurodegenerative diseases like Parkinson's, Huntington's, and many others. If a lithium deficiency causally leads to an increased risk of disease, this provides justifiable hope that this essential trace element (within a systemic approach) will exert not only preventive but also therapeutic effects. This could happen by lithium supporting the regenerative potential of our extremely plastic brain. However, here too, the majority of industrially or state-funded research doesn't view lithium as an essential trace element whose deficiency promotes neurodegenerative diseases, but merely as a tool for identifying pharmacological targets.[267]

Alzheimer's Disease

Alzheimer's disease serves as a sad prime example of the ultimate outcome of an increasingly brain-damaging cultural trend in Global North countries like the USA and Germany, where this dramatic neurodegenerative condition is a leading widespread illness.[268] Increasingly, aging signifies senility rather than seniority, despite our innate potential for lifelong mental growth, making Alzheimer's entirely preventable.[269] In addition to the ever-increasing toxin load, which also includes the simultaneous and permanent intake of five or more products from the pharmaceutical industry, a prevalent chronic deficiency in vital substances, for which our bodies cannot compensate, is particularly critical. As extensively demonstrated, lithium, even in small doses, acts as a neuroprotective agent (shielding nerve cells) by restoring the cellular activity of GSK-3 in the brain to a natural level once a deficiency is addressed. Findings from an animal study "contribute to a better understanding of the ability of low-dose lithium," write the scientists responsible for the article, "to influence GSK3 activity in the brain and its potential to prevent Alzheimer's Disease."[270] Insulin resistance in the hippocampus also contributes to the development of Alzheimer's disease because it causes the nerve cells in the memory center to literally starve. As previously discussed, a ketogenic diet can counteract this. Nevertheless, studies on hippocampal neurons in an animal model of Alzheimer's disease have also revealed that lithium increases glucose uptake, glucose metabolism, and energy production, thereby eliminating insulin resistance, a critical factor in the Alzheimer's disease process.[271]

Early clinical indications that lithium offers a causally protective effect against the development of Alzheimer's disease emerged from studies involving patients suffering from another disease within the mental immunodeficiency syndrome group. According to the results of a 2007 study, older people chronically treated with lithium for bipolar disorder showed an Alzheimer's disease risk of about 5 percent—roughly equivalent to that of the general population of the same age. However, the risk in a comparison group not treated with lithium, also suffering from bipolar disorder, was over six times higher, at around 33 percent.[272] A 2022 study further compared the dementia risk of patients with lithium-treated bipolar disorder directly with an otherwise similar control group that did not have bipolar disorder (and thus did not receive lithium): "After accounting for sociodemographic factors, smoking status, other medications, other mental and physical illnesses, lithium use was associated with a lower risk of dementia [-44 percent], including Alzheimer's disease [-45 percent] and vascular dementia [-64 percent]," the authors write.[273] Lithium treatment of bipolar disorder leads to an increase in hippocampal volume in both hemispheres, an effect that became apparent after a short treatment period of about four weeks.[274] This is astonishing; after all, one would expect an accelerated loss of hippocampal volume in bipolar disorder, with corresponding functional effects up to an equally accelerated development of Alzheimer's disease.[275] Since the hippocampal volume or its relative shrinkage is considered a prognostic marker for impending Alzheimer's disease, this trend or risk was reversed by the addition of lithium (albeit with the considerable side effects of high-dose treatment).[276]

Lithium therefore has properties that could have a positive effect on the progression of Alzheimer's disease. These properties could be particularly important for the prevention of Alzheimer's disease. In 2011, this was more thoroughly investigated in a randomized, controlled long-term study. The study involved people who suffering from memory problems—more specifically, amnestic mild cognitive impairment (MCI), a precursor to Alzheimer's disease. However, the dosage used here was approximately 25 to 50 percent lower than that used for bipolar disorder.[277] Despite the generally positive outcomes, a placebo-controlled study conducted two years later demonstrated that microdoses of merely 300 µg of lithium daily were even more effective—suggesting that more is not always better.[278] This considerably lower dose, within the RDA range, significantly altered the progression of Alzheimer's disease (see Fig. 28). At the study's outset and every three months up to the fifteenth month, participants in both the lithium and placebo groups underwent a Mini-Mental State Examination (MMSE), consisting of 30 questions and tasks, with one point awarded for each correct answer. Scores of 30 to 27 indicate no dementia, 26 to 20 indicate mild dementia, 19 to 10 indicate moderate dementia, and below 10 indicates severe dementia. As shown in Fig. 28, participants began with an average of 19 points, marking the transition from mild to moderate dementia. This score stabilized only in the lithium group, while cognitive performance declined as expected in the control group treated solely with placebo. The fact that these microdoses of lithium, even below the RDA of 1 mg/day, also have a preventive effect was successfully demonstrated in animal experiments on mice: Two mutated genes were implanted into their genomes, which are known to accelerate the Alzheimer's disease process in the animals kept in cages under completely unnatural conditions. However, when given lithium, their memory function remained intact.[279]

These findings, obtained with lithium levels even slightly below the essential range (Chapter 3), provide solid evidence that a lithium deficiency is a causal factor in the development of Alzheimer's disease. This impressively confirms the findings from the drinking water studies, which indicate an elevated risk of Alzheimer's disease associated with lithium deficiency (Chapter 2, Argument 1). Another study from Texas even found that the risk of dying from Alzheimer's also decreases significantly with higher amounts of lithium in drinking water (again, we're referring to only a few hundred micrograms, roughly equivalent to those in the clinical study mentioned).[280] At this point, it is important for me to emphasize that lithium was not used as a medication in the microdose study, but due to the very small amount, rather, due to the minute quantity, it likely only compensated for a preexisting lithium deficiency.

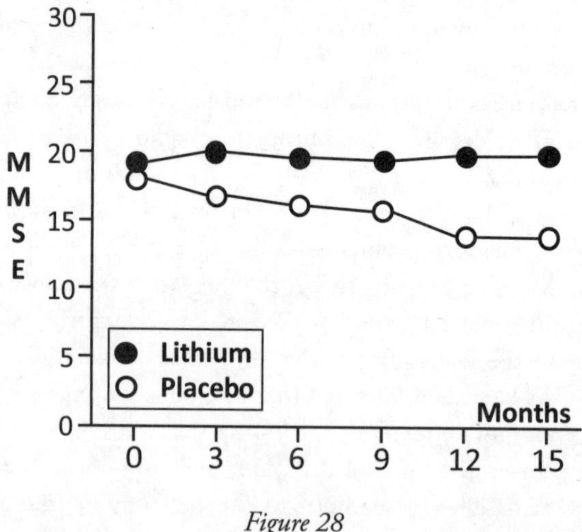

Figure 28

These results, using lithium levels even slightly below the essential range for lithium (Chapter 3), provide concrete evidence that lithium deficiency is a causal factor in the development of Alzheimer's disease. Studies in mice where GSK-3 was genetically, rather than pharmacologically, inactivated, showed that this central regulator is required for hippocampal memory.[281] The authors noted: "Based on these observations, we propose that GSK-3β may contribute to help maintain brain function during aging. Our results may explain the poor efficacy of GSK-3β inhibitors in preserving memory capacity in AD patients." As the saying goes, the dose makes the poison. In a systemic review article published in August 2024, which analyzed five independent studies, the authors accordingly stated:

> The reviewed evidence shows that trace-Li levels in the water are sufficient to lower the incidence or mortality from dementia. Considering the lack of options for the prevention or treatment of dementia, we should not ignore these findings.[282]

Nevertheless, these findings are ignored by authorities, while the pharmaceutical industry pushes a vaccination strategy. In this case, it's an "immunization" using medicinal antibodies against the body's own β-amyloid. This approach rests on the misguided assumption that β-amyloid is the sole cause of Alzheimer's disease and that, therefore, one only needs to monocausally eliminate this small protein to halt neurodegeneration. However, the β-amyloid hypothesis is not only unproven but a veritable dogma. Even when it became clear that successful removal of β-amyloid from the brains of Alzheimer's patients could not stop cognitive decline—thereby questioning the entire established Alzheimer's research—German biochemist Dr. Christian Haass, who had garnered numerous scientific awards for his Alzheimer's research, warned in the November 23, 2016, FAZ article "Alzheimer's Antibodies: A great beacon of hope flops": "Under no circumstances should the unspeakable debate that amyloid is the wrong target molecule be started again."[283] By this point, it would have been appropriate to fundamentally question the amyloid hypothesis, especially since it had, until then, blocked every other explanation for the steady increase in Alzheimer's disease, including lifestyle-related causes. The prevailing hypothesis not only harmed patients but also the cause, as it diverted attention from the actual solution, which I had published and put up for discussion a few months earlier, on July 15, 2016.[284] As previously discussed, β-amyloid is necessary for memory formation, which is why its medicinal removal is clearly not only largely useless for patients but even dangerous. Mice from which the precursor of this protein was removed from their genome using molecular genetic methods—a fact known since 1999—showed reduced learning ability, a decreased density of hippocampal synapses (through which nerve cells communicate and store memories), and increasingly cognitive deficits.[285] In short, they developed Alzheimer's-like symptoms. But, mind you, not because they produced too much β-amyloid, according to the β-amyloid hypothesis based on falsified studies (see below), but because they had too little of it. The β-amyloid-deficient mice also displayed conspicuously reduced exploratory behavior.[286] The reduced curiosity points to decreased psychological resilience and damage to the mental immune system.

In the USA, the drug Aduhelm, containing the antibody Aducanumab, was approved for the treatment of Alzheimer's disease in June 2021 through an accelerated procedure and a controversial FDA decision. The Mayo Clinic website reveals the astonishing profile of the (un)desired (one inevitably begins to doubt) effects of this "therapy": visual disturbances, confusion about time, place, or people, dizziness and falls, hallucinations, headaches, depression or anxiety, nausea, nightmares or unusually vivid dreams, difficulty moving, walking or speaking, seizures, drowsiness or unusual fatigue. As if that weren't enough, here's my personal highlight: "holding false beliefs that cannot be changed by fact."[287] You may have noticed that these are almost all symptoms of Alzheimer's—a not-to-be-overlooked indicator that the disease is not halted but even worsened by this method, as a meta-analysis explained in its very title: "Accelerated Brain Volume Loss Caused by Anti–β-Amyloid Drugs."[288] In other words, patients are getting significantly worse, which is why responsible scientists and doctors called for an equally accelerated withdrawal of approval in May 2022—a plea that, to the chagrin of countless patients, had not yet been fulfilled by the time of this writing in Spring 2025.[289]

In my book *The Alzheimer's Lie*, I had already warned of this drug attack on the brain in 2014.[290] The cognitive decline of Alzheimer's patients used as guinea pigs accelerated; this was already evident from the results of the first antibody studies against β-amyloid, published in 2003.[291] "Later, signs were found in deceased patients that successful immunization not only had serious side effects but also had not truly stopped the disease process,"[292] I wrote then, adding: "Meanwhile, under the impression of the experimental results in humans, active immunization against β-amyloid was repeated in healthy laboratory mice [. . .].[[293]] The immunized mice lost their natural curiosity and suffered from severe memory impairment [due to TLR-4-dependent or activated neuroinflammation.] Exactly the opposite of what was intended happened: an immune system gone wild [attacking the brain] caused Alzheimer's-like symptoms, from which the preventive therapeutic measure should actually have protected. Nevertheless, research continues on the preventive therapeutic strategy of autoimmunization." The deadly folly of this "vaccination strategy" against an endogenous protein was therefore well known, and, following the pharmaceutical industry's business model, the research simply continued. Indeed, the entire "β-amyloid-is-the-cause-of-Alzheimer's-concept" has been based on scientific lies and deception for decades, as two articles in the renowned journal *Science* pointed out.[294] Even the health authorities, who should have protected us from the pharmaceutical industry pursuing its goals without regard for people's well-being, did not intervene—and still do not.

Is Alzheimer's a Lie? In my book *The Alzheimer's Lie*, I use countless studies to demonstrate that age is not a causal risk factor for Alzheimer's disease, rather, violations of the Laws of Minimum and Maximum are causally responsible.[295] I've also scientifically published these findings.[296] The probability of developing the disease only correlates with age because the illness typically follows many decades of a brain-damaging lifestyle before becoming symptomatic. The more inadequate, unnatural, or unhealthy the lifestyle, the faster the neuroinflammatory destruction of the brain progresses, which is why more and more people are falling ill at younger ages. *The Alzheimer's Lie* posits that nearly all scientific works declare the correlation with age as causality. This renders any shift toward a healthier lifestyle absurd and thus hinders prevention: Why change your lifestyle if you can't alter your Alzheimer's fate anyway? It's interesting to see how, for example, the Alzheimer's Registered Society in Munich reacted to my book. Its then-chairman, Dr. Christian Haass, described in a statement as "one of Germany's leading dementia researchers," posed the curious question, "Well, is Alzheimer's a lie?" which he immediately answered himself: "If you look at the number of patients, 1.4 million people in Germany alone, it's probably hard to speak of a lie here," as if he hadn't grasped the specific problem the article addressed. Elsewhere, he eventually touches on the topic: "We are getting older and older, and age is known to be the greatest risk factor for developing dementia."[297] According to Haass, "Every one of us has the potential to develop this disease because everyone produces the toxic, Alzheimer's-triggering amyloid throughout their lives." By this logic, every person would inevitably become diabetic because our bodies can produce glucose, or contract gout because they can produce uric acid. This is, of course, nonsense. As I demonstrate in *The Alzheimer's Lie*, β-amyloid only becomes toxic if, due to an unnatural lifestyle, we don't break it down sufficiently, leading to its accumulation and clumping in the brain (Greek *amylon* means "starch" or "glue," hence amyloid means "starch- or glue-like.") I also show that β-amyloid, when clumped into so-called oligomers, is by no means the sole disease-causing mechanism, and that neuroinflammation, impaired autophagy, and many other pathological processes are paramount. These, as also shown in my book, are influenced by our lifestyle. However, according to Haass, this lifestyle has no real preventive influence on Alzheimer's disease: "The disease cannot be prevented in this way at all, but it can of course be concealed to a certain extent, the onset of the disease can be 'delayed,' but the changes in the brain will still progress and inevitably lead to dementia." And then I'm even accused of serving only my vanity and harming people through my educational work: "Preventing Alzheimer's solely through the right lifestyle is too much of a pipe dream and certainly doesn't do justice to the

1.2 million affected people in Germany. On the contrary, the author's treatise harms those affected and their families, as it raises completely false hope." In the *Science* article about the revelation of a gigantic number of falsified β-amyloid articles, published on September 26, 2024, Haass was quoted: "People will of course be shocked, as I was. I sort of fell off my chair."[298] Whether the "true lies" about Alzheimer's will ever all be uncovered remains to be seen—hope dies last.

In a review article, three researchers from the Department of Psychiatry at Tokyo Medical University in Japan compared the research on whether low-dose lithium or the pharmaceutical industry's antibody strategy shows better efficacy—using Aducanumab as an example. The result was unequivocal: "A network meta-analysis showed that lithium was significantly more effective than Aducanumab in the primary endpoint [protection against mental decline in patients with MCI or Alzheimer's disease]."[299] According to the authors, effective treatment with lithium costs only around $40/year, whereas Alzheimer's-accelerating treatment with Aducanumab costs USD $28,000/year (plus the exorbitant costs for treating the numerous side effects, a truly profitable situation for Big Pharma). Meanwhile, further Aducanumab-like "vaccines" have been approved, but these also invariably lost out when compared to lithium. This finding was brought to light by a meta-study that evaluated 8 randomized, placebo-controlled trials with 6,547 participants: "In the [MMSE], lithium performed significantly better than *Donanemab*, *Aducanumab*, and Placebo."[300] As of this writing (spring 2025), passive vaccination is approved for all three anti-β-amyloid antibodies in the US, while Donanemab is still under review in the EU. Lecanemab, however—despite devastating side effects—has been approved for certain Alzheimer's patients since November 14, 2024.[301] But low-dose lithium not only works better than all current Alzheimer's drugs, but also—due to its essentiality and adherence to the Law of the Minimum—better than future preparations are likely to ever achieve, as no pharmaceutical can remedy a causal vital substance deficiency. Even the anti-inflammatory and neurogenesis-promoting deep sleep hormone melatonin outperforms the "Alzheimer's vaccine" or daily physical activity, as a meta-analysis revealed.[302] What if, instead of waging a hopeless war against an endogenous protein (with the brain paradoxically as collateral damage) that is fraught with massive side effects and costs, everything that actually promotes health were implemented at once? Why isn't this being attempted? Lithium, melatonin (via sleep hygiene), some exercise as part of a systemic prevention and therapy concept—along with the elimination of all other causes of this multicausal disease process? What if the money were not spent on pointless and highly health-damaging medication, but on comprehensive health advice? How much quality of life would people regain? How much money would be saved because the industry would have to forgo astronomical profits?

Does Lithium Protect Against Blindness? Glaucoma is a progressive eye disease characterized by damage to the optic nerve. It also involves the destruction of the macula (the yellow spot at the back of the eye with the highest density of photoreceptors for sharp vision). Glaucoma is classified among the neurodegenerative diseases and is the leading cause of irreversible visual field loss, including complete blindness, worldwide.[303] Currently, there are no clinical studies on the effect of essential lithium on the risk of developing glaucoma. However, given the known anti-neuroinflammatory and neuroprotective signaling cascades controlled by lithium, which are causally involved in glaucoma development, a preventive effect is highly probable.

At least in animal models, it has already been shown that the inhibition of GSK-3 protects against age-related macular degeneration.[304] A hyperactive GSK-3 due to lithium deficiency is most likely a causal risk factor, meaning correcting the deficiency with essential amounts of lithium could reduce the risk of glaucoma. This argument is also presented in the review article "The Actions of Lithium on Glaucoma and Other Senile Neurodegenerative Diseases Through GSK-3 Inhibition: A Narrative Review."[305]

Parkinson's Disease

Parkinson's disease is a progressive neurodegenerative disorder. While Alzheimer's disease originates in the hippocampus, Parkinson's disease centers on the *substantia nigra.* This is a complex of nerve nuclei in the midbrain rich in iron and melanin (hence nigra for black). The "black substance" is a crucial motor center and plays a decisive role in initiating and controlling movements. Dopamine serves as the most important neurotransmitter within the *substantia nigra.* Parkinson's disease is the third most common neurodegenerative disease worldwide, after Alzheimer's disease and vascular dementia (see below). It primarily affects older individuals, making lifestyle factors a likely cumulative cause over time (see above, e.g., *Football*). It was first described by James Parkinson in 1817 as "shaking palsy." It's a progressive disease characterized by both motor and non-motor symptoms. Motor symptoms include slowed movement, stiffness, and tremor. Non-motor symptoms include uncontrolled salivation, a mask-like facial expression, gait disturbances, sleep disturbances, and restless legs. The estimated probability of developing dementia is 27 percent after ten years of illness, 50 percent after fifteen years, and 74 percent after twenty years.[306] If left untreated, patients become stiff and disoriented during physical activity, which can lead to life-threatening respiratory complications such as aspiration pneumonia and pulmonary embolism. These often prove fatal.

In the 2023 review article "GSK-3β: An Exuberating Neuroinflammatory Mediator in Parkinson's Disease," the authors highlight the crucial role of GSK-3 in the disease process of Parkinson's disease.[307] In the 2024 review article "Therapeutic Potential Effect of Glycogen Synthase Kinase 3 Beta (GSK-3β) Inhibitors in Parkinson Disease: Exploring an Overlooked Avenue," the authors investigate what they describe as previously overlooked therapeutic approach, namely, "an overactivity of GSK-3β that drives the neuropathology of Parkinson's disease by triggering mitochondrial dysfunction and neuroinflammation."[308]

In the Parkinson's animal model, lithium inhibits damage to dopaminergic nerve cells via GSK-3.[309] In humans, however—due to the multiple causes and associated inconsistent clinical results—sole and thus often high-dose treatment with lithium is not a suitable therapeutic option due to side effects. However, the administration of low-dose lithium (similar to Alzheimer's disease), in combination with a systemic, multicausal approach, would be a promising strategy. This also applies to frontotemporal dementia, ALS, and all other neurodegenerative diseases.

Frontotemporal Dementia

Frontotemporal dementia (FTD) encompasses a diverse group of neurodegenerative syndromes characterized by progressive changes in behavior, personality, executive function, language, and motor skills.[310] Only about 20 percent of FTD cases have a known genetic cause.[311] Nevertheless, as with other forms of dementia, there's increasing evidence that—regardless of a possible genetic cause—neuroinflammation is involved in the clinical progression of FTD.[312] "The clinical syndromes of frontotemporal dementia are clinically and neuropathologically heterogeneous, but processes such as neuroinflammation may be common across the disease spectrum. These *in vivo* findings indicate a close association between neuroinflammation and protein aggregation in frontotemporal dementia," writes an international group of researchers.[313] "GSK-3β activity has been strictly related to neuroinflammation and neurodegeneration," explains a group of scientists from the University of Trieste.[314] According to these authors, GSK-3β plays a significant role not only in Alzheimer's disease but in almost all neurodegenerative diseases, including Parkinson's disease, amyotrophic lateral sclerosis, frontotemporal dementia, Huntington's disease, and the autoimmune disease multiple sclerosis. Since the fine regulation and networking of all these GSK-3-dependent signaling pathways form the basis for the meaningful use of GSK-β inhibitors in neuroinflammation and neurodegeneration, this should also apply to FTD. There are some case reports on the use of lithium in the treatment of FTD that report some success in managing agitation, mood swings, and other behavioral disturbances.[315] However, neither lithium orotate was used, nor was a systemic approach followed, which would very likely have made the treatment more successful.

Amyotrophic Lateral Sclerosis (ALS)

One of the most famous ALS patients was British theoretical physicist Stephen W. Hawking (1942–2018), diagnosed at age twenty-one. During the course of the disease, there's a selective destruction of motor neurons—the nerve cells that control skeletal muscles. This loss causes the muscles to waste away, as if they were no longer nourished (Greek *amyotroph*, literally "muscles are not nourished."). As with many other neurodegenerative diseases, GSK-3–dependent chronic neuroinflammation is prominent, along with an intracellular accumulation of conglomerates of dysfunctional proteins. This suggests that impaired autophagy is another important factor in nerve cell loss—after all, cells "cluttered" with dysfunctional organelles and proteins are less viable.[316] Mitochondria, in particular, are massively affected, meaning that overall, a functionally impaired Lithiome could be the common denominator of all forms of ALS. A range of data suggests that lithium could increase the survival chances of ALS patients. This is also stated in the review article "Lithium and its Effects," published at the end of 2024. Simultaneously, it raises the important question, based on contradictory clinical and animal studies: *Does dose matter?*[317] As with Alzheimer's, the assumption here is that purely monotherapeutic trials with lithium in ALS are possibly not particularly successful (and in some animal experiments, even harmful) because the doses used were consistently far too high. Regarding the failed high-dose therapy trials, the authors write: "Conversely, lithium's ability to counteract cognitive decline appears to be exerted at subtherapeutic doses, possibly corresponding to its molecular neuroprotective effects," and "Indeed, lithium can reduce inflammation and provide neuroprotection, even at doses many times lower than those commonly used in clinical trials."

According to the Law of the Minimum, a deficiency of essential lithium has serious consequences, making compensation for the deficiency necessary to prevent illness. However, according to the Law of the Maximum, everything becomes toxic if the dosage is too high. Monotherapeutic interventions usually require extremely high dosages and are therefore often harmful; many other deficits that also endanger the organism continue to be ignored. I am convinced that for chronically progressive diseases like ALS (and all others), only a systemic approach has a real chance of success—even if, in individual cases, this only means slowing down the disease process as much as possible.

Huntington's Disease

Huntington's disease is a psychiatric disorder characterized by so-called choreatic (erratic, wavelike, involuntary, and dance-like) movements, associated with a progressive loss of movement-related nerve cells. Stress and neuroinflammation accelerate the disease's progression, while a healthy lifestyle can slow it down. Approximately one-third of patients are still alive fifty years after diagnosis, with death usually resulting from a respiratory illness. Although Huntington's disease is genetic, caused by a single defective gene (autosomal dominant inheritance), lithium has been successfully used in its treatment due to its neuroprotective effect.[318] Lithium prevented the progression of chorea and also helped stabilize mood. Beyond inhibiting neuroinflammation and apoptosis, the activation of autophagy appears particularly crucial. This is because the mutated gene leads to the production of altered huntingtin protein, which forms nerve cell–toxic aggregates that negatively impact motor coordination and cognitive functions.[319] A systemic approach that supports the vital process of autophagy could be promising for significantly slowing down the disease process.

Vascular Dementia

Vascular dementia results from impaired blood supply to the brain, stemming from dysfunction of the blood vessels (Latin: *vasculae*). This dysfunction can be caused by diabetes mellitus, lipid metabolism disorders, often insidious and silent chronic inflammation, and high blood pressure. Smoking and many other aspects of an unnatural lifestyle, such as a chronic lack of essential vital substances, ultimately lead to chronic arteriosclerotic reduced blood flow to the brain; strokes accelerate the dementing process. A lithium deficiency is also a factor, as this trace element influences practically all of these causes in multiple ways, as we'll discuss in more detail in Chapter 6. In brief: lithium also offers multifaceted protection after a stroke, minimizing brain damage, which underscores its preventive potential. Prevention, after all, is far simpler and more effective than cure. As the authors of a relevant review note:

> The therapeutic effects of lithium are indirectly related to its ability to activate a cascade of fail-safe pathways, designed to protect cells from a variety of threats such as posttraumatic brain injury and depletion of the triphosphate level. Lithium mimics the lowered level of the cellular environment in a way that makes it incredibly effective at protecting cells. These systems are designed to protect neuronal cells from conditions that are detrimental to their survival, such as low levels of magnesium.[320]

However, instead of seeking natural solutions for preventing vascular dementia, the focus, as with all chronic diseases, remains on new pharmaceutical targets and "magic bullets." This is largely because the market is enormous, as another review article highlights: "Vascular dementia (VaD) is a significant form of cognitive impairment with a vascular cause and is considered the second most common form of dementia after Alzheimer's disease (AD). VaD affects around 15–20 percent of dementia patients in North America and Europe and up to 30 percent in Asia and developing countries. Additionally, more than 60 percent of older AD patients suffer from a combination of AD and VaD, further increasing the number of VaD cases."[321] According to the authors, while vascular dementia is "considered a preventable and treatable form of cognitive impairment," the article aims solely for a pharmacological solution: "However, the pathomechanisms and therapeutic targets associated with VaD are still unexplored, and highly effective therapeutics are needed."

Neuroinflammation As Far As the Eye Can See?

A lifestyle largely alien to our species, exacerbated by an equally unnatural chronic lithium deficiency, leads to a life characterized by chronic, creeping neuroinflammation, and consequently, chronic exhaustion. Some may even grow accustomed to it, even if their mental immune system and ultimately their soul suffer. And even if neuropsychiatric developmental disorders don't emerge early, neurodegenerative diseases threaten in the medium term once the vicious cycle of neuroinflammation has begun. While no essential facet of a species-appropriate lifestyle is more important than another, lithium plays a truly unique role among vital substances. This is due to its involvement in all biological functions and its calming influence as a trace element, stemming from its stress-reducing, anti-neuroinflammatory, and neurogenesis-promoting functions—it impacts all possible bodily functions like no other trace element. The fact that even severe deviations from what would enable optimal health don't immediately kill us by no means implies that they wouldn't severely impair our lives. However—and I'm happy to reiterate—it's usually not too late to embark on a different life path, as the Roman philosopher and writer Seneca (4 BCE–65 CE) aptly put it:

No man was ever wise by chance.
It takes the whole of life to learn how to live.

CHAPTER 5

CYTOKINE STORM—ACUTELY LIFE-THREATENING LITHIUM DEFICIENCY

For they sow the wind, and they shall reap the whirlwind.
—Bible: Hosea 8:7, King James Version (KJV)

The Story of a Life-Threatening Inner Storm

A cytokine storm is an excessive, uncontrolled, and life-threatening release of pro-inflammatory cytokines. This phenomenon has captured wider public attention since at least COVID-19, with even mainstream media like *Der Spiegel* reporting on it quite early.[1] The first description comes from a 1993 article on the so-called graft-versus-host reaction.[2] A graft-versus-host reaction occurs when immune cells in donor tissue recognize the recipient's body tissue as foreign and trigger a massive immune response that can rapidly lead to death. In the early 2000s, the term "cytokine storm" appeared in reports as a cause of severe viral and bacterial infections, including in 2005 when describing the fatal progression of avian flu caused by the H5N1 influenza virus.[3] "Public interest in 'bird flu' also brought the term cytokine storm into the popular media," one reads in the review article "Into the Eye of the Cytokine Storm."[4] After the article's publication in 2012, an internet search for "cytokine storm" already yielded 323,000 hits. The fact that the cytokine storm is the main health problem of a coronavirus infection was already known from SARS-CoV (the first SARS-CoV wasn't yet numbered), as can be seen from a 2005 publication.[5] This was the precursor to the pathogen that causes COVID-19; in 2002, it also threatened to unleash a pandemic but ultimately didn't have the optimal conditions to do so. Since this was over two decades ago and made relatively few waves, most people likely only learned about cytokine storms from the media around late 2019, when SARS-CoV-2, genetically modified into a bioweapon, was unleashed on humanity.[6]

Just like meteorological storms, cytokine storms also have varying intensities. The article "The amount of cytokine-release defines different shades of SARS-CoV-2 infection," published at the end of 2020, differentiates as follows: "Depending on the amount of cytokines released as the result of the immunological activation induced by SARS-CoV2, three major clinical phenotypes can be identified: 'mild,' symbolized as a 'drizzle' of cytokines; severe as a 'storm'; and critical as a 'hurricane,'" explain immune specialists from the University of Padua in Italy (see Fig. 29).[7] The strength of the cytokine storm is clearly not directly related to the virus itself, as everyone in a pandemic wave is infected by the same virus. Rather, it's about how the individual immune system reacts to the infection. Pierre Miossec, an immunologist at the University of Lyon in France, also writes in his review article, "Understanding the cytokine storm during COVID-19: Contribution of preexisting chronic inflammation," that "the cytokine storm in COVID-19 results from inflammation, rather than from the virus itself." Preexisting chronic inflammation ("silent" inflammation, see Fig. 29) provides a "constant wind" and massively increases the risk of a severe storm developing: "In these high-risk patients, all it takes is another event that, again through synergy, will give this acute situation in a very short time,"[8] writes Miossec. If you compare Fig. 29 with Fig. 4 in Chapter 1, you can understand how the slight increase from left to right is based on a disregard of the Laws of Minimum and Maximum. As I showed in my book *Herd Health*, people with severe or fatal infections (such as pneumonia or other infections) or those experiencing a cytokine storm usually have a dysfunctional immune system from the outset, often due to a significant deficiency of one or more essential nutrients.[9] With such an overstimulated or imbalanced immune system, even a seemingly harmless infection is enough to trigger a storm. Instead of reacting with a gentle breeze—a temporary increase in pro-inflammatory cytokines followed by a calming increase in anti-inflammatory cytokines (see Fig. 29)—a cytokine storm initially rages in the lung tissue, which in extreme cases spreads to other organs like a "hurricane."

As previously mentioned, these connections had been known for decades before COVID-19. Therefore, from the very beginning, it was understood how life-threatening disease progressions could be prevented through natural means. For example, Professor Andrea Giustina, President of the European Society of Endocrinology (the science of hormones), and his colleague Anna Maria Formenti—both from the Vita-Salute San Raffaele University, Milan, Italy—recommended as early as February 28, 2020, in their *British Medical Journal* article, "Preventing a COVID-19 pandemic": that the population should be sufficiently supplied with vitamin D, and that fatal outcomes were most likely due to a deficiency.[10] The fact that the majority of infections run a completely harmless course attests to the virus's inherently low virulence (a measure of the severity of disease caused by a pathogenic agent). Only when a vitamin D deficiency leads to the overproduction and release of a wide range of pro-inflammatory cytokines does this result in severe, life-threatening, and fatal infection outcomes. However, this simple logic, which, if implemented, would have immediately ended the entire pandemic, was deliberately ignored. As more and more studies confirmed his assumptions, Giustina reiterated on December 11, 2020, to *Medscape:* "Patients with low vitamin D levels are at high risk of hospitalization for COVID-19 and developing severe and lethal disease. [. . .] This is likely due to the loss in the protective action of vitamin D on the immune system and against the SARS-CoV-2-induced cytokine storm."[11] A large meta-study then also found that at a natural vitamin D prohormone level of 125 nmol/l, the risk of experiencing a fatal cytokine storm approaches zero (see also Chapter 1, Additional Info: *Vitamin D Deficiency—A Deadly Example of Ignoring the Law of the Minimum*).[12] The ignorance displayed by the medical establishment toward the (unquestionable) essentiality of vitamin D is comparable to the ignorance toward lithium's essentiality—both factors are likely ignored and combated for the same reasons—with very similar impacts regarding cytokine storms.

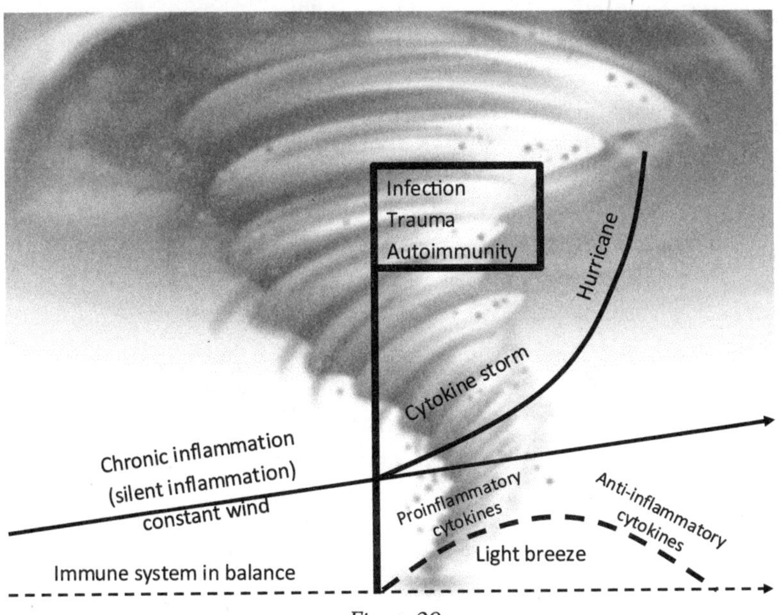

Figure 29

Lithium Protects Against Viral Cytokine Storms

Lithium must be considered a vital trace element not only for its considerable effect on mental health but also for its regulatory function in the body's immune system. This is why a deficiency not only in vitamin D but also in lithium enables the development of the so-called cytokine storm in the first place. Therefore, in my book *The*

Corona Syndrome (June 2021), I strongly recommended not only sufficient vitamin D supplementation for the population (a level of 125 nmol/l) but also addressing a lithium deficiency by taking 1 mg of lithium. In this way, the risk of a life-threatening dysregulation of the immune system in the event of a SARS-CoV-2 infection can be greatly reduced, if not eliminated; ultimately, this principle applies to any infection. I wrote at the time: "The multipotency of microdosed lithium did not go unnoticed during the COVID-19 pandemic, especially since the central mechanisms in both chronic and acute inflammation have long been known. The trace element inhibits several central molecular switching points whose activity contributes to the development of the cytokine storm. However, there is currently no drug that works as well as lithium, which is why it is time to recognize lithium as an essential trace element and thus enable unproblematic microdosed supplementation for all people."[13] As early as August 2020, when the population began to be terrified by horror scenarios of seriously ill, intubated patients apparently in their death throes, two things became clear: First, vitamin D prohormone administration can be considered a simple and effective solution, as was subsequently confirmed in further studies.[14] Second, the successful clinical application of lithium in six separate COVID-19 cases, also published in August 2020, should have been celebrated as a welcome solution, yet it was simply ignored.[15] Instead, Munich newspaper *Süddeutsche Zeitung*, for example, reported on November 12, 2020, that three more pharmaceutical "hopefuls" against COVID-19 had failed and quoted from an article in the *New England Journal of Medicine* (the same journal that two years later, in an editorial, deemed vitamin D unnecessary even if a deficiency existed; see Chapter 1) as follows: "However, we found no significant effect [of these new agents] on the risk of mechanical ventilation, exacerbation of symptoms, number of deaths, or duration of oxygenation."[16]

Saving lives with vitamin D prohormone or lithium would have been an option, especially since the authors of the lithium study fully and logically justified their approach and positive results: "Lithium has shown the capacity to: a) inhibit the replication of several types of viruses, some of which are similar to the SARS-CoV-2 virus, b) increase the immune response by reducing lymphopenia, and c) reduce inflammation by preventing or reducing the cytokine storm." The study's success and the clarity of its underlying scientific argumentation paved the way for a larger, placebo-controlled clinical trial, published in 2022.[17] The results of this lithium-COVID-19 study were remarkable: compared to COVID-19 patients in the control group who received no lithium, the duration of necessary hospital stays was halved (!) in the lithium group. The disease stabilized and improved so quickly under lithium that no patient had to be transferred to the intensive care unit. In contrast, two out of fifteen patients in the control group were, one of whom eventually died. The authors summarized their results: "SARS-CoV-2 infection induces exaggerated inflammation driven by components of innate immunity. We demonstrated that lithium was able to reduce the number of days of hospital and ICU admission as well as the risk of death. Lithium, through its immunomodulatory action, reduces inflammatory cytokine levels by preventing cytokine storms, thus reducing the severity of the infection and the risk of death." Regarding lithium's superior effect, the study's authors also formulated the statement quoted in the Essential Opening Remarks at the beginning of this book, that "the third element of the periodic table is perhaps the most important element in life." Even though the dosages of 40 mg of lithium (in the form of lithium carbonate) twice daily were significantly higher than what's needed for daily intake as an essential trace element due to the therapeutic emergency situation, this trace element's fundamental ability to quickly and decisively defuse life-threatening inflammatory reactions was demonstrated. However, it wasn't investigated whether lower amounts of lithium would also have been sufficient, nor was lithium orotate (see Chapter 3), which in my opinion is more suitable, used.

Antiviral Lithium. Lithium has an antiviral effect, and this was known long before the SARS-CoV-2 pandemic.[18] It has been established since the 1970s that it reduces the infectivity of a variety of DNA and RNA viruses.[19] Accordingly, lithium is also effective against herpes infections.[20] Particularly relevant for SARS-CoV-2: it was shown as early as 2007 that lithium specifically inhibits the growth of coronaviruses via the Lithiome (IMPase, GSK-3, etc.).[21] The importance of lithium was further confirmed in 2021 in a large retrospective study: "Our analysis of clinical data from over 300,000 patients in three major health systems demonstrates a 50 percent reduced risk of COVID-19 in patients taking lithium, a direct inhibitor of glycogen synthase kinase-3 (GSK-3)."[22] According to the authors, these results show "that GSK-3 is essential for phosphorylation of the SARS-CoV-2 nucleocapsid protein and that GSK-3 inhibition blocks SARS-CoV-2 infection in human lung epithelial cells." According to the authors, "These findings suggest an antiviral strategy for COVID-19 and new coronaviruses that may arise in the future." Although the causal relationship between the antiviral effect of a lithiome regulated by sufficient or essential amounts of lithium, highlighted in this publication, should have been treated as a sensation, this inexpensive, readily available (but not patentable and therefore not very lucrative) trace element did not receive the life-saving attention it deserved. Additionally, the aim was to genetically "vaccinate" humanity with spike mRNA. Successfully defusing the problem naturally was therefore not desired. In this respect, those involved in staging the COVID-19 pandemic also benefited financially and in terms of power politics from the non-recognition of lithium's essentiality and its ban as a dietary supplement.

One study showed that neurogenesis in human hippocampal progenitor cells is disrupted when exposed to serum samples from hospitalized COVID-19 patients with neurological symptoms.[23, 24] This is due to the massive increase in pro-inflammatory messengers in the blood. The vicious cycle of neuroinflammation triggered by the spike protein or its S1 subunit, as shown in Fig. 24 (Chapter 4), is also responsible for the brain fog phenomenon, according to compelling studies, and thus for the psychological effects of long COVID after infection and post-vac syndrome after spike mRNA injection.[25] Due to their common trigger, these problematic conditions of the mental immune system are collectively termed spikeopathy.[26] These long-term effects, manifesting as chronic sickness behavior—even long after an infection has been overcome—have several reasons. For example, a severe infection leads to a massive consumption of essential vital substances, which paralyzes the mental immune system just as much as persistent spike proteins. These can continue to cause neuroinflammation even after the viruses have been eliminated, with both causes synergistically damaging the brain. This is also a major problem with spike mRNA injections, as the viral genetic material has been genetically optimized for long-term spike production. This, in turn, could explain the success of the lithium intervention: "Finally, one data that reaffirms the improvement of lithium-treated patients is that when we examined patients 1 month after hospital discharge and long-term neurological effects were recorded, 40 percent of lithium-treated patients were observed to have symptoms, while 73 percent of the control group had some neurological symptoms."[27] Therefore, according to the authors, "lithium can also be studied for earlier use, such as at the time of diagnosis, in order to avoid hospitalization altogether, as well as in the treatment of 'Long Covid' syndromes."

A small number of patients "reported greater declines in fatigue and brain fog with the higher dose of 40-45 milligrams per day [in the form of lithium aspartate]" than 15 mg/day, as a study from the University at Buffalo found.[28] Due to the presumably lower efficacy of lithium aspartate compared to lithium orotate (see Chapter 3), I recommend low-dose lithium for spikeopathies, i.e., 5 to 10 mg/day in the form of lithium orotate monohydrate, which corresponds to a total weight of 128.6 to 257.1 mg. As always, this recommendation also applies within the framework of systemic therapy (see the preceding chapter). Ideally, synergistically activated autophagy through lithium and ketogenesis can support the elimination of the toxic S1 subunit, especially since the suppression of autophagy, or even the "hijacking" of its decisive signaling molecules, is part of viral pathogenesis. SARS-CoV-2 is also so successful partly because it disrupts or blocks this protective process of autophagy.[29] In fact, there is already convincing evidence that cleansing autophagy also helps to minimize the symptoms of an infection and eliminate viral components.[30] Active microautophagy, in particular, could support the degradation of the neurotoxic spike protein or its S1 subunit, thereby helping to complete the healing process.[31]

Lithium Protects Against Bacterial Cytokine Storms

Sepsis (colloquially known as blood poisoning) is a life-threatening complication of a bacterial infection where the body's immune system overreacts with a cytokine storm.[32] It is one of the most common causes of death worldwide.[33] In 2017, for instance, there were almost 50 million cases of sepsis globally (nearly half of which were children under five) with around 11 million deaths, equating to almost a fifth (!) of all worldwide fatalities.[34] Here again, it's typically not the bacterial infection itself that leads to death, but rather an excessive immune system reaction. The self-destruction caused by the massive release of pro-inflammatory cytokines (blood pressure drop, increased permeability of blood vessel lining leading to edema, impaired blood clotting, etc.) can rapidly lead to shock, multi-organ failure, and unfortunately, death, especially if not recognized and treated immediately.

An effective way to treat cytokine storms would be the preventive administration of low-dose lithium. The mechanism of action is easy to understand: bacteria are recognized by the body via specific cell wall components such as LPS (lipopolysaccharides). Responsible for this are danger sensors like TLR-4 (see Chapter 4, Fig. 24)—among the most studied PAMPs. Lithium intervenes protectively on several levels here. Firstly, the essential trace element preventively limits the number of TLR-4s, as discussed in Chapter 4[35] Furthermore, by inhibiting GSK-3, it reduces the risk of excessive cytokine release. In addition to the known inhibition of NF-κB activation, another anti-inflammatory mechanism has been discovered: "Another target of GSK-3β," the authors write, "is the transcription factor Signal Transducer and Activator of Transcription (STAT). It was found that inhibition of GSK-3β by lithium led to decreased activation of STAT, It was found that inhibition of GSK-3β by lithium led to decreased activation of STAT, which was associated with a marked decrease in the release of pro-inflammatory cytokines."[36] All these mechanisms together protect against "stormy" activation of the pro-inflammatory signaling cascade. If there's a risk of a septic cytokine storm developing—despite a good basic supply of all essential vital substances including lithium—higher doses of lithium should be used (in addition to standard antibacterial therapy and emergency care) to halt life-threatening cytokine production, as demonstrated by the results of the lithium-COVID-19 study discussed above. This would represent a new and promising therapeutic approach that could save many millions of lives worldwide annually, especially since it has already been successfully tested numerous times in animal models, comparable to the success of the human lithium-COVID-19 study mentioned earlier.[37] This isn't surprising, because although LPS and spike are different PAMPs, the pathomechanism from the perspective of our immune cells is the same. However, this possibility of saving human lives from a septic cytokine storm with higher doses of lithium (in this case, dose-dependent as a medication) has been obvious for many years. Is lithium therapy for sepsis, which is also inexpensive, perhaps being ignored for the same reasons as lithium therapy for viral respiratory infections, as we've seen with the example of COVID-19? Is it all about profit?

Lithium Protects Against Traumatic Cytokine Storms

Traumatic injuries are responsible for approximately 6 million deaths per year worldwide.[38] Most people believe that in such cases, death results from the injuries themselves. It may therefore surprise many to learn that numerous individuals actually die from the body's reaction to the enormous release of DAMPs. Doctors and researchers from various trauma centers at Peking University in China state in a review article published at the end of 2023:

> Severe trauma is an intractable problem in healthcare. Patients have a widespread immune system response that is complex and vital to survival. Excessive inflammatory response is the main cause of poor prognosis and poor therapeutic effect of medications in trauma patients. Cytokines are signaling proteins that play critical roles in the body's response to injuries, which could amplify or suppress immune responses. Studies have demonstrated that cytokines are closely related to the severity of injuries and prognosis of trauma patients and help present cytokine-based diagnosis and treatment plans for trauma patients.[39]

Here again, it is (supposedly) unmanageable cytokine storms that present physicians with a (supposedly insoluble) problem.

Postoperative Delirium. Postoperative delirium (POD) is an acute mental dysfunction primarily characterized by memory and consciousness disturbances.[40] POD can prolong hospital stays, for instance, by leading to the development of postoperative depression. It also increases patients' general mortality risk, as they are harder to mobilize. Approximately half of all high-risk surgeries result in POD. Neuroinflammation as a pathogenetic mechanism of POD can arise from various factors, but trauma-induced neuroinflammation from the surgical procedure or sometimes sepsis is crucial. A variety of inflammatory mediators like Interleukin IL-6, IL-1β, and TNF-α can cross the blood-brain barrier and cause damage to central neurons and synapses.[41] To prevent POD, low-dose lithium would be a suitable option even before surgery.

Despite intensive searching across all publicly accessible databases, I could not find any work that attempted the obvious solution: calming the self-destructive cytokine storm in trauma patients through the acute administration of lithium. Only with regard to severe head injuries have attempts been made. In the article "A New Avenue for Lithium: Intervention in Traumatic Brain Injury," published in 2014, the authors from the National Institute of Mental Health wrote: "Traumatic brain injury (TBI) is a leading cause of disability and death from trauma to central nervous system (CNS) tissues. For patients who survive the initial injury, TBI can lead to neurodegeneration as well as cognitive and motor deficits, and is even a risk factor for the future development of neurodegenerative disorders such as Alzheimer's disease. Preclinical studies of multiple neuropathological and neurodegenerative disorders have shown that lithium, which is primarily used to treat bipolar disorder, has considerable neuroprotective effects. Indeed, emerging evidence now suggests that lithium can also mitigate neurological deficits incurred from TBI."[42] Throughout the article, the authors' excitement about finally finding a solution is palpable: "In summary, recent preliminary data using lithium have demonstrated robust beneficial effects in experimental models of TBI. These include decreases in TBI-induced brain lesion, suppression of neuroinflammation, protection against blood-brain barrier disruption, normalization of behavioral deficits, and improvement of learning and memory, among others." The authors concluded that "additional clinical research is clearly warranted to determine its therapeutic attributes for combating TBI." However, this has not occurred. Here too, the simple, effective, and extremely cost-effective option of mitigating the acute and chronic consequences of severe trauma with lithium has not been pursued further.

Traumatic Birth. Intraventricular hemorrhages (IVH) are a common complication in infants born prematurely between the 23rd and 28th week of gestation. If these children survive, they face a risk of neurological developmental disorders, including impaired growth of the cerebral cortex. This is attributed to hyperactive GSK-3, which is why its inhibition reactivates neurogenesis and can thus improve the neurological development of premature babies with IVH.[43] As a natural GSK-3 inhibitor, essential lithium is a promising option.

Lithium Protects Against Autoimmune Cytokine Storms

Our body's immune system is tasked with recognizing and eliminating invading microorganisms. However, life-threatening "microorganisms" also include the body's own cells if they've mutated, exhibit abnormal behavior, and threaten the entire organism.

The high art of our immune defense lies in distinguishing between healthy body cells and mutated cancer cells—recognizing only the latter as problematic and destroying them. Yet, sometimes the immune system makes a (learning) mistake, identifying healthy body tissue as a foreign threat and attacking it. These often intense attacks against the body's own structures are also referred to as cytokine storms due to the excessive immune response and, as a 2021 article calls it, result from "imbalances between innate and acquired (adaptive) immunity."[44] The consequence of this immune system misdirection is an autoimmune disease. The exact clinical picture depends on which structures are mistakenly attacked. The causes are diverse. Overall, there's increasing evidence of numerous environmental factors that can be linked to the development of autoimmunity and autoimmune diseases. Our increasingly unnatural way of life explains the rising disease rates. After all, about 10 percent of humanity currently suffers from one or more autoimmune diseases. Type 1 diabetes is the best studied, with a steady increase in incidence of 3 to 4 percent per year since the 1970s.[45] "Most worrying are the major changes in our diets and their impact on the microbiome, exposure to xenobiotics [from the Greek *xénos* meaning 'foreign' and *bíos* meaning 'life'; i.e., chemical substances not naturally formed but synthesized by humans and foreign to the biological material cycle, such as dyes, pesticides, or microplastics] and chlorinated solvents, infections, personal lifestyle and the associated increase in obesity and lack of sleep, stress, air pollution," writes Frederick W. Miller, former head of the Environmental Autoimmunity Group of the US National Institute of Environmental Health. For incomprehensible reasons, he also speaks in this context of "the effects of climate change as possible factors contributing to these increases."[46] A downright absurd assertion, if only because autoimmune diseases are known to be very rare in countries of the global South.[47] What Miller notably omits in his extensive review article, however, is the influence of the increasingly extensive vaccination programs since the 1970s, which massively—and certainly much more directly—impact and impair the maturing immune system of infants and young children than the climate. At least, epidemiological studies here suggest a connection.[48] The ignorance required for such statements and omissions is highly astonishing. Is there an agenda behind this?

Multiple Sclerosis

Many of our nerve cells are surrounded by an insulating layer called the myelin sheath, which enables faster and more efficient information transmission. This fat-rich layer is formed from the myelin-containing cell membranes of glial cells. In multiple sclerosis (MS), immune cells attack this myelin sheath, damaging it and even inhibiting myelin production. In demyelinating diseases like MS, as well as in experimental autoimmune encephalomyelitis (EAE)—the animal counterpart to MS—GSK-3β activity is massively increased.[49] This activation occurs via the TLR4-dependent signaling pathway and generates a neuroinflammatory cytokine storm.[50] Dr. Karl Frei, a leading Swiss neuroscientist, recently shared with me that he and his team discovered decades ago that EAE mice, if molecularly stripped of the IL-6 gene, are resistant to developing EAE and therefore don't exhibit MS symptoms.[51] Thus, IL-6 appears to play a special role within the broad spectrum of pro-inflammatory messengers. This is also confirmed by the 2019 study "Interleukin-6 Disrupts Synaptic Plasticity and Impairs Compensation for Tissue Damage in Multiple Sclerosis."[52] Inhibiting IL-6 activation would therefore be of clinical relevance to prevent further cytokine storms or disease relapses.

Instead of eliminating the IL-6 gene, it's more practical for treating MS patients to simply inhibit the pro-inflammatory signaling chain with essential or low-dose amounts of lithium. The authors from the Department of Psychiatry and Behavioral Neurobiology at the University of Alabama write in their article "Lithium Prevents and Ameliorates Experimental Autoimmune Encephalomyelitis": "In relapsing/remitting EAE induced with proteolipid protein peptide 139–151, lithium administered after the first clinical episode maintained long-term (90 days after immunization) protection, and after lithium withdrawal the disease rapidly relapsed."[53] The relevance for MS patients is obvious. However, the authors don't view lithium as the solution to the MS problem, but only as the presumed target of lithium's action: "These results demonstrate that lithium suppresses EAE and identify GSK3 as a new target for inhibition that may be useful for therapeutic intervention of multiple sclerosis and other autoimmune and inflammatory diseases afflicting the CNS.' GSK-3 is identified as a target for new drugs or lithium mimetics, yet curiously, lithium itself isn't further discussed as a therapeutic option, even though a therapeutic intervention could be started immediately based on this finding. And this isn't an isolated case—this mindset permeates medical literature like an unwritten law. In another rat model, treatment with lithium significantly delayed the onset of EAE; furthermore, it mitigated the disease's severity by inhibiting pro-inflammatory TNF-α and inactivating GSK-3β.[54] But here too, lithium wasn't proposed as a therapeutic option, only GSK-3 as a pharmacological target—despite knowing that lithium acts on the entire Lithiome, meaning it has far more complex effects on central nervous autoimmune diseases than an isolated pharmacological inhibition of GSK-3.[55] I consider it highly problematic to reduce lithium's health-promoting effect solely to GSK-3 inhibition—even if this is intended to serve the pharmaceutical interest of target identification. After all, in the case of Alzheimer's, the disease hypothesis reduced purely to β-amyloid has already failed. An animal study published in 2023 even showed that lithium improves the differentiation of stem cells into oligodendrocytes and thus promotes remyelination in MS—meaning it not only slows down the disease but even treats it or initiates a healing process.[56] A review article published in September 2024 only concludes: "Thus, exaggeration of GSK-3β is linked with MS neuropathology, and GSK-3β inhibitors may be effective in the management of MS."[57] But why should MS patients wait for artificial GSK-3β inhibitors—xenobiotics that are even suspected of triggering autoimmune diseases—when natural lithium is already available today, and the entire Lithiome can be leveraged to stop the disease process? However, as repeatedly mentioned, this should always be low-dose as part of systemic therapy or relapse prevention.

Anything that reduces neuroinflammation should be promoted, and anything that promotes it should be reduced, as extensively discussed in the previous chapter and Chapter 1. For example, the vitamin D hormone also lowers IL-6 levels.[58] High-dose supplementation (Keyword: Coimbra Protocol) can be helpful if vitamin D resistance is present.[59] From a systems biology perspective, however, even better results can be expected if all deficiencies are corrected, as discussed in Chapter 4, to enable therapeutic and relapse-preventing effects even with moderate doses. This, of course, applies to all autoimmune diseases.

Autoimmune Depression. "A high prevalence of depression and anxiety, and a higher rate of suicidal ideation were identified in MS patients compared to the general population. The presence of depressive symptoms appeared to have a direct influence on the risk of suicide." This is the conclusion of a clinical study from 2018.[60] "Neuroinflammation drives anxiety and depression in relapsing-remitting multiple sclerosis" is the unsurprising title of another psychiatric study from 2017.[61] Within a systemic therapy framework, lithium, due to its complex effect across the entire Lithiome, would not only generally inhibit neuroinflammation but also reduce the propensity for depression in MS patients by reactivating adult hippocampal neurogenesis. Indeed, a clinical study on MS patients reported that depression significantly improved with low-dose lithium treatment compared to the untreated group.[62] A meta-study examining eighteen low-dose neuropsychiatric trials found that "Low-dose lithium was not reported to be associated with a greater risk of AEs or SAEs compared to placebo."[63]

Type 1 Diabetes Mellitus

When the insulin-producing β-cells of the pancreas are destroyed, Type 1 diabetes develops. Only with the complete elimination of these cells—the target—is the immunological attack concluded, which also manifests as a cytokine storm (biomarkers include high IL-1β levels in the blood).[64] In Type 1 diabetes, affected individuals are lifelong dependent on medicinal insulin. Since an autoimmune cytokine storm very likely occurs independently of the immune system's cellular target, I suspect that all findings from MS research can largely be applied to other autoimmune diseases. For instance, pharmacological blockade of TLR-4 in the NOD mouse model (standing for nonobese diabetic) prevents the spontaneous development of Type 1 diabetes.[65] Since lithium downregulates TLR-4 and can also inhibit the TLR-4 signaling cascade at the level of GSK-3 (see Chapter 4), one could reasonably assume that an essential amount of lithium reduces the risk of disease, while a low-dose amount at least delays the disease process—always within the systemic approach presented in Chapter 4. An animal study from 2024, for example, showed that mice with Type 1 diabetes benefited from a combination of lithium and physical activity.[66]

Rheumatoid Arthritis

Rheumatoid arthritis (RA) is an autoimmune disease characterized by chronic inflammatory joint destruction. The enormous pro-inflammatory cytokine production present in this case is also referred to as a cytokine storm. A significant increase in pro-inflammatory cytokines (TNF-α, IL-6, IL-1β) is typical in RA patients compared to healthy controls, also serving as biomarkers for disease prognosis and progression.[67] RA treatment typically involves taking medications to suppress the immune system, such as methotrexate, leflunomide, and hydroxychloroquine, which have many side effects. However, as with all autoimmune diseases, low-dose lithium treatment as part of a systemic therapeutic approach is also beneficial here, especially since monotherapy with lithium has already been shown in arthritis models to prevent IL-1β-induced cartilage degradation and loss of mechanical properties.[68] Microdosed lithium also protects against the destruction of pancreatic islets and renal dysfunction (see Chapter 6) in Type 1 diabetes, which is triggered by a cytotoxin (streptozotocin) that mimics the autoimmune disease in animal models.[69]

Systemic Lupus Erythematosus

Systemic lupus erythematosus (SLE) is another autoimmune disease driven by cytokine storms, where the immune system indiscriminately attacks healthy tissue. Symptoms of an SLE-related cytokine storm usually begin in early adulthood and range from fever and skin rash to organ failure. In particular, the kidneys can become clogged by immune complexes and lose their filtering function. Even before the blood-brain barrier is breached, NZB/W mice, which spontaneously develop an autoimmune syndrome remarkably similar to human SLE, also exhibit corresponding neuropsychiatric behavioral deficits, explained by impaired hippocampal neurogenesis.[70] IL-6 and IL-18 (see Fig. 24, Chapter 4) directly activate apoptosis of adult hippocampal neurons. Treatment for cytokine storms in SLE, as in almost all autoimmune diseases, typically involves administering immune-suppressing drugs like corticosteroids, methotrexate, and hydroxychloroquine. Here too—and this will be the final example from this complex of autoimmune diseases—lithium chloride was able to drastically improve the survival chances of NZB/W lupus mice.[71]

Lithium Protection Undesirable?

Those who disregard the basic laws of nature sow the wind—a pathological development of our modern culture. In extreme cases, the central pathological process of silent inflammation, which underlies almost all so-called diseases of civilization, escalates into a cytokine storm. This is the culmination of chronic and usually unrecognized inflammation, when it occurs in fast motion and then suddenly is no longer silent, but devastatingly destructive. While lithium isn't the only vital substance deficiency affecting many people's health, its intake in essential dosage not only reduces the risk of neuroinflammation, as detailed in the previous chapter, but should also, for the same reasons, protect against a cytokine storm. This is especially true within a species-appropriate lifestyle where no other vital substance deficiencies should exist. If a traumatic, bacterial (sepsis), or autoimmune cytokine storm is imminent (as illustrated in the four examples given), low-dose lithium could preventively avert the danger, while slightly higher doses can calm a storm already underway.

However, the nonrecognition of lithium as an essential trace element—much like the propaganda against vitamin D—exemplifies that health authorities, who are obligated to educate the public and tasked with developing medical guidelines for health crisis scenarios, are not primarily concerned with health. They seemingly operate mainly to generate and preserve markets for the pharmaceutical industry. Against the background of the studies presented in this chapter, it becomes unmistakably clear how long it has been ignored that seriously ill people could be saved by simple means. This knowledge is simply not utilized because lithium is apparently not desired as either a preventive or therapeutic agent—except for the treatment of bipolar disorder (presumably because the wheel of time cannot be turned back). Thus, there are practically no clinical studies—with the exception of the very successful lithium intervention in COVID-19 patients mentioned above—that have even attempted to replicate the animal experimental successes in humans facing death from a cytokine storm. This is despite a century of experience with lithium and how it could be therapeutically used in such extremely threatening situations for the benefit of patients.

CHAPTER 6

LITHIUM FOR LIFE

It is the mind that builds the body.
—Friedrich Schiller (1759–1805)

Can the Clock of Life Only Be Wound Once?

The "Evolution of the Grandmother" is not only based on the intergenerational transmission of knowledge. Besides their life experience, which increased their grandchildren's survival probability (demonstrable at least until the late nineteenth century; see Chapter 1), their active assistance with daily chores was also a crucial pillar of the extended family. Our genetic potential to remain both mentally and physically fit into old age is, therefore, closely interconnected. These potentials developed in the presence of lithium—which is why this trace element is indispensable for maintaining not only mental but also physical fitness. As we already observed in Chapter 2 (Argument 5), lithium proves to be an "antiaging vital substance"; after all, population groups that consume slightly larger amounts of lithium through drinking water exhibit a lower overall mortality rate.[1]

In their 2016 review article "Chronic Overeating and Dysregulation of GSK3 in Disease," two Canadian researchers hypothesized that nature chose GSK-3 as the guardian of our lifespan. The reason: while we aren't viable without GSK-3, as animal gene knockouts show, excessive GSK-3 activity leads to earlier death.[2] "Despite the complexity of cellular signaling pathways and their interactions," the two authors are certain on one point based on their comprehensive analyses: "under all harmful conditions, there is a strong activity of GSK3." As demonstrated so far, at least to some extent, using popular examples, there's very likely no life-shortening, chronic disease not (directly or indirectly) caused by a dysregulated Lithiome—due to lithium deficiency. An indirect connection can sometimes exist insofar as our physical health is influenced by our life choices. The quality of such decisions depends on the performance of our (lithium-dependent) mental immune system. The same also applies to our parents and grandparents, as long as they make vital decisions for us: For a long time, a decisive role is played by whether we grow up in a loving family, whether we are nourished without a lack of vital nutrients, or how many vaccinations we have to endure—to name just a few examples. While the potential for a healthy life is genetically evolved and physically inherent in us, whether we fully realize it depends on family and, not least, cultural conditions. In this sense, the failure to recognize the essential nature of a vital trace element unfortunately forms the basis for children today growing up in a culture detrimental to their health.

In my opinion, attempting to alter our genetic material to adapt it to a profit-oriented culture and increasingly unnatural living conditions is impossible from a systems biology perspective alone (even if the pharmaceutical industry dreams of genetically optimizing humans).[3] However, it would always be possible to re-align our way of life and cultural development with our genetic needs. Then, many people whose biological clocks have long ticked faster than necessary could even rewind their biological time a little, gaining several healthy years of life. While lithium is no miracle cure, it plays a central role within the structure of all vital functions, as life evolved around this trace element. The life-prolonging effect observed after correcting a lithium deficiency in both humans and animal models has already been discussed in detail in Chapter 2 (Argument 5).[4] The pharmaceutical industry sees much hope in this, as the authors of a corresponding "fly study" state: "The extension of life expectancy by pharmacological means is becoming increasingly urgent from both a health and an economic point of view."[5] They point out that lithium is "a drug approved for human use," and that in low doses in fruit flies it "promotes longevity and healthspan [. . .] when administered throughout adulthood or later in life." Nevertheless, lithium here is considered only a means to gain knowledge; it serves solely to identify the targets that could then be used to develop pharmaceutical lithium mimetics: "The life-prolonging mechanism involves inhibition of glycogen synthase kinase-3 (GSK-3) and the activation of the transcription factor Nuclear Erythroid-2-related Factor (NRF-2)." Putting aside this pharmaceutical-driven perspective, these studies provide valuable evidence as to why lithium in essential amounts is indispensable for a long and healthy life. For example, this study showed that fruit flies live longer with low lithium doses than under lithium deficiency, but that lifespan is shortened at extremely high lithium doses—which experimentally confirms the results from Fig. 22 in Chapter 3 on the dose-response relationship in humans. Thus, lithium's life-prolonging effect is not achieved through high-dose application (as in the treatment of bipolar disorder), but only in essential (low-dose) quantities or its essential breadth, as also discussed in Chapter 3.

The Clock of Life Ticks in Telomeres

At the cellular level, the clock of life ticks, among other ways, through the constant shortening of telomeres. Their role is to keep our genetic material intact during cell division.[6] A successive degradation of telomeres with each DNA replication (which must precede every cell division) ultimately leads to cellular dysfunction, blocks further cell divisions, and thereby impairs tissue regeneration—in short, we age. How quickly we age, or how rapidly telomere length decreases, depends on a large number of influencing factors that one might not have even considered (without knowing the Methuselah Formula). Telomere- and lifespan-shortening influences include "stress factors in childhood, increased depressive episodes, and constant anger." Conversely, we age more slowly if our telomeres are "lengthened through physical activity, good nutrition, mindfulness/meditation, and positive and altruistic life goals." This is how a 2017 review article summarizes the impact of our lifestyle on our biological clock, providing a genetic example for the Methuselah Formula (Chapter 4).[7] And good nutrition also includes an adequate intake of essential lithium. For example, there's a causal link between oxidative stress—a high production rate of cell-damaging oxygen radicals—and accelerated telomere degradation in human tissue.[8] Lithium counteracts this because, in addition to its anti-inflammatory effect, it has also been shown to have an antioxidant function by inhibiting the production of free radicals while simultaneously increasing the activity of endogenous antioxidant systems like glutathione peroxidase.[9]

The telomere-lengthening function of lithium is therefore essential not only for preserving our mental health, as we've already seen in Chapter 2 (Argument 2), but also for our physical well-being. In the 2018 research paper "The Polygenic Nature of Telomere Length and the Anti-Aging Properties of Lithium," telomere length was identified as a promising biomarker for age-related diseases. Concurrently, it was emphasized that their complex regulation presents a promising target for antiaging drugs—and lithium's potential clearly emerged here.[10] Through the functions of the Lithiome, it promotes telomere lengthening, thereby exerting a significant antiaging effect. Combined with other species-appropriate measures, this health-promoting effect can be further enhanced. Another fly study by the same research group showed that the pharmacological inhibition of another signaling protein, mTOR (mammalian or mechanistic Target of Rapamycin), by the immunosuppressive rapamycin, synergistically works with lithium to prolong life.[11] However, in addition to lithium and rapamycin, the cancer drug trametinib—another medication with numerous side effects—was used to mask rapamycin's side effects, as pharmacist and emeritus professor of pharmaceutical biology Theo Dingermann wrote for *Pharmazeutische Zeitung*: "It partially cancels out the side effects caused by rapamycin, which manifest themselves as a derailment of the metabolism."[12] Dingermann also explained the significant media attention this fly study garnered: "This approach also seems possible for humans."

Although mTOR was originally discovered through the action of rapamycin and named as its mechanistic target, there's no need to ingest drugs with many side effects if you want to extend your life in a truly healthy way via mTOR inhibition. As has long been known, longer daily eating breaks are sufficient to regulate mTOR in a healthy manner (see Chapter 4).[13] Combining this simple and completely natural measure with a little physical activity leads to a massive inhibition of mTOR.[14] Intermittent calorie reduction is more effective than rapamycin, as the fruit fly study impressively demonstrated, and explains the side-effect-free, life-prolonging effect of intermittent fasting. There's also an evolutionary biological explanation for this, based on mTOR's function in energy metabolism regulation. During fasting, the inhibition of this cellular processor, among other things, stimulates autophagy to gain energy through healthy "self-digestion." However, initially, only aged cell organelles and dysfunctional proteins are broken down. In the spirit of evolutionary optimization—"use every crisis as an opportunity"—this makes the organism more efficient and rejuvenates it. It's not only telomere length that determines a cell's biological age, but also the intracellularly accumulated "waste" that impairs its function. As shown in an animal research paper published in 2019, "Intermittent Fasting Increases Neurogenesis in the Adult Hippocampus," as a result, the mental immune system is rejuvenated, and people can retain their youthful curiosity.[15] Micro-amounts of lithium slow down the aging of brain cells.[16] Since it also prevents neuroinflammation and neurodegeneration, researchers conclude "that very low doses of Li_2CO_3 [lithium carbonate] may play an important role in neuroprotection by reducing neuronal loss and neuroinflammation in the elderly."[17]

The synergistic combination of essential lithium, physical activity, and intermittent fasting allows us to age biologically more slowly, rejuvenating our cells (e.g., by activating telomerase and autophagy) and our organs (e.g., by regeneratively replacing old cells with new ones), including the mental immune system through adult hippocampal neurogenesis. Therefore, we don't need pharmaceuticals to fully exploit our natural life potential and stay healthy into old age—on the contrary, their xenobiotic products are unnatural and anything but health-promoting. These cellular and organic effects apply to all body systems, as we'll see below.

Lithium Protects the Cardiovascular System

Maintaining a healthy vascular system is vital, not just for heart health, but for overall well-being. Arteriosclerosis, for instance, contributes to the onset and progression of numerous organ diseases by reducing oxygen supply and disrupting the transport of metabolic products. Among all its consequences, coronary heart disease remains a leading cause of death in countries of the global North, such as the USA.[18] Chronically elevated blood pressure, whether a cause or consequence of arteriosclerosis, leads to a sustained burden and eventual overload of the heart muscle. This increased energy demand, coupled with arteriosclerosis-induced reductions in energy supply via the coronary arteries, risks fatal cardiac arrhythmias (ventricular fibrillation) or a life-threatening heart attack. The risk of stroke due to acute ischemia (embolism, thrombosis, or vascular rupture/hemorrhage) and the risk of vascular dementia from chronic circulatory disorders are also significantly elevated. As early as the late 1960s and early

1970s, epidemiological studies analyzing data from over a hundred American cities revealed an inverse relationship between lithium in drinking water and the likelihood of developing and dying from atherosclerotic heart disease.[19] The results of even the first of these studies were apparently conclusive enough for the WHO to mention them in its 2016 compendium, *Trace Elements in Human Nutrition and Health*. This is notable, as the WHO, despite all the evidence of lithium's essentiality in animals, remains unwilling to acknowledge human need for this trace element.[20] This finding was further confirmed by a 2018 study involving over 4.2 million patient records: a lithium concentration in drinking water of less than 40 μg/l, compared to over 40 μg/l, correlated with a 16 percent increased risk of heart attack.[21] Even though the study employed some statistical maneuvers to try and obscure this health benefit, this inverse correlation, clearly evident in the raw data, is indeed causal.

For example, GSK-3 regulates blood pressure within the central nervous system. Inhibiting GSK-3 activity lowers blood pressure, thereby counteracting the development of arteriosclerosis.[22] However, it has been shown that lithium only lowers blood pressure at low concentrations, but can increase it at extremely high concentrations (as used in the treatment of bipolar disorder).[23] This phenomenon of a dose-dependent, opposing effect, which we've also seen with Alzheimer's dementia (see Chapter 4), is not atypical for this trace element—we'll revisit this in a comprehensive overview at the end of this chapter. Accordingly, low-dose lithium supplementation in animal experiments leads to a physiological strengthening of the heart muscle's performance,[24] while high doses of lithium in patients treated for bipolar disorder can have negative effects on cardiac function, ranging from arrhythmias to decreased cardiac output, depending on lithium levels. These range from cardiac arrhythmias to reduced cardiac output.[25] The review article "Beyond Its Psychiatric Use: The Benefits of Low-dose Lithium Supplementation," authored by a clinical expert group from Brock University, St. Catharines, Ontario, Canada, in 2023, therefore concludes that low-dose lithium strengthens the heart muscle through various synergistic mechanisms, enhancing its adaptability and promoting vascular health.[26]

Lithium thus acts preventively, but also therapeutically, after a brain or heart attack, and consequent tissue destruction. This is because DAMP-activated inflammation plays a crucial role in *post*-infarction dysfunction. Therefore, inhibiting GSK-3β not only reduces heart damage from an infarction but also improves the regeneration of cardiac function after the injury.[27] In the case of a stroke, addressing lithium deficiency also ensures better regeneration. An animal study found that after a stroke, lithium reduces oxidative stress, promotes blood flow, and thereby decreases ischemic (blood-deficiency-related) damage. Due to its antiapoptotic function, it prevents nerve cell death and maintains the integrity of the blood-brain barrier. This neuroprotective effect was also observed in a dose-dependent manner.[28] Another animal study showed that lithium reduces infarct volume, improves brain function, and alleviates associated cognitive and depressive impairments. In summary, according to the authors, this study "provides the molecular mechanisms of lithium neuroprotection during cerebral ischemia and thus the theoretical background for expanding the clinical use of lithium for the treatment of ischemic stroke."[29] Further systematic review articles confirm that, based on this evidence, it can be assumed that stroke patients should benefit from lithium administration.[30] However, there is only one randomized clinical study from 2014 where, among other things, improved hand motor function was demonstrated in the lithium group thirty days after a stroke, with two doses of 300 mg lithium carbonate (a total of about 120 mg lithium per day).[31]

Even if the clinical data remains sparse despite these positive effects and clear animal experimental evidence, I would request lithium be administered by treating physicians in the event of a stroke or heart attack—in addition to standard therapy. Merely because there's seemingly no interest in clinical studies, I would want to minimize the damage from an infarction using this neuroprotective trace element. Here too, a systemic approach certainly offers the best chances for the best possible recovery. For example, animal experiments have shown that the intake of aquatic omega-3 fatty acids also protects brain tissue.[32] These omega-3 fatty acids also have a preventive effect, as a large meta-study found.[33]

Lithium Promotes Kidney Function

According to a 2013 epidemiological study, approximately 700 million people worldwide suffer from kidney problems; that's nearly 10 percent of the global population, and the true figure is likely even higher.[34] One indicator of impaired kidney function is the excretion of proteins in urine, known as albuminuria (albumin, at about 60 percent, is the most common protein in blood). This points to impaired filtration, which occurs in the renal corpuscles (glomeruli). Here, specialized cells called podocytes form narrow slits; in healthy kidneys, primary urine, free of proteins, is filtered through them. GSK-3 is essential for the development and function of podocytes. Whether in mice or fruit flies, a complete absence of GSK-3 leads to massive albuminuria and neonatal death, as a "podocyte study," as I'll refer to it, discovered.[35] Thus, a complete genetic removal of GSK-3 function is extremely detrimental to the kidneys. Unfortunately, high doses of lithium also cause long-term kidney damage in some patients: "The most common [damage] is so-called diabetes insipidus," write the authors of the podocyte study, "but the most serious is terminal renal failure, which occurs six to eight times more frequently in these patients, especially if they have taken the drug over a long period of time." The scientists hypothesize "that in some patients receiving lithium, GSK3α/β activity in their podocytes is excessively suppressed, leading to glomerular and renal damage." Interestingly, low levels of lithium have precisely the opposite effect—they protect the kidneys. This is summarized in the 2016 review article "Lithium in the Kidney: Friend and Foe?": "Several recently conducted animal studies have shown that short-term administration of low doses of lithium prevents various forms of experimental acute renal failure."[36] According to the authors, the finely tuned regulation of GSK-3 plays the crucial role in this seemingly paradoxical effect of lithium. They argue that "new discoveries regarding the protective effect of lithium against AKI in rodents call for follow-up studies in humans." They also "suggest that long-term therapy with low lithium concentrations could be beneficial in CKD [chronic kidney disease]," adding: "It will be very interesting to find out in future studies whether this also applies to humans and whether long-term treatment with low levels of lithium is also beneficial in the prevention of chronic kidney disease."

The promising results from animal experiments have been confirmed. For example, a preliminary study on the effect of small amounts of lithium on the human kidney, published in 2023, revealed that low amounts of lithium even protect transplanted kidneys, which are extremely vulnerable organs due to the risk of rejection.[37] For every doubling of lithium intake via drinking water and diet (precisely measured by lithium excretion), the risk of transplant failure was reduced by 46 percent, the risk of kidney function deterioration by 27 percent, and even overall mortality by 36 percent. This health-promoting effect can be situated on the ascending limb of the sigmoid dose-response curve in Chapter 3 (Fig. 22), which begins at an intake of about 10 μg lithium per day. "It should be noted, however," the authors write, to explain the discrepancy between their results and the fear of kidney damage from lithium, "that the daily lithium intake in patients with bipolar disorder is in the order of 1000 mg, which is more than 40,000 times higher than the average dietary lithium intake of the kidney recipients in the present study (21 μg)." Lithium in essential concentrations regulates and protects kidney function. Conversely, this organ is simultaneously susceptible to dysregulation and dysfunction if the trace element is supplied in unnatural amounts as part of medical treatment. This understanding—that essential quantities of lithium protect the kidneys—should now be implemented in clinical practice, especially since there is no healthy alternative for almost a billion people with kidney problems.

Lithium Activates Energy Metabolism

"The prevalence of obesity has increased worldwide in the past ~50 years, reaching pandemic levels," writes Matthias Blüher, Professor of Clinical Obesity Research at the University of Leipzig, in a 2019 review article.[38] According to Blüher, "Obesity represents a major health challenge because it substantially increases the risk of diseases such as type 2 diabetes mellitus, fatty liver disease, hypertension, myocardial infarction, stroke, dementia, osteoarthritis, obstructive sleep apnoea and several cancers, thereby contributing to a decline in both quality of life and life expectancy." Obesity is defined as a body mass index (BMI) of over 30 kg/m^2 and is caused by an energy imbalance, specifically when the amount of calories consumed far exceeds the amount of calories expended.

Many patients with bipolar disorder who are treated with high doses of lithium tend to gain a significant amount of weight.[39] Conversely, an epidemiological drinking water study conducted in Texas showed that not only does Alzheimer's disease develop significantly less frequently when traces of lithium are present in tap water, but so too do obesity and type 2 diabetes. The authors noted: "Traces of lithium in water are negatively associated with the risk of dying from Alzheimer's disease as well as obesity and Type 2 diabetes, which are important risk factors for Alzheimer's disease."[40]

Studies on animal models have shown that this health-promoting effect can indeed be causal. Low-dose lithium supplementation was able to attenuate diet-induced obesity compared to the control group.[41] In another animal study, low-dose lithium was shown to prevent weight gain from overeating about as effectively as physical activity.[42] Experimentally, it was found that lithium stimulates adaptive thermogenic mechanisms, thereby ensuring that excess energy isn't stored as fatty acids but is released into the environment with heat production.[43] Thus, lithium would aid in weight loss or at least prevent weight gain. This would also explain why individuals who consume essential amounts of lithium have reported increased sweating; perhaps thermogenesis is particularly stimulated in them.

These animal models demonstrated similarly positive effects concerning blood glucose regulation. As already discussed in Chapter 2, many lithium targets regulate blood glucose, foremost among them GSK-3, which owes its name to this very fact. Insulin activates Akt (as does lithium), which inactivates GSK-3 (also like lithium), collectively activating glycogen synthesis and thereby lowering blood sugar. Lithium thus acts like insulin—an effect derivable from the trace element's influence on improving glucose homeostasis. Animal experiments revealed that low-dose lithium, within the context of a high-calorie diet, helps reduce blood glucose more effectively compared to control animals.[44] Again, remarkably, the authors of the previously mentioned review article "Beyond its Psychiatric Use," "this effect of lithium treatment was similar to the compensatory effect of physical activity on a high-fat diet, suggesting that lithium may have at least some degree of similar metabolic health benefits as regular physical activity."[45] Of course, it would be even better to combine the two: "Therefore, we can only speculate that consistent treatment over a period of time in animal models may elicit similar effects as those shown above and that perhaps combining exercise and insulin treatment with low-dose Li may have synergistic effects in combatting obesity and obesity-induced insulin resistance and hyperglycemia."

Lithium Keeps the Musculoskeletal System Fit

Movement is essential for life, if only because for the longest period of human history, it was a prerequisite for our ancestors to find sustenance. When we're physically active, our bodies release hormones that penetrate the brain, triggering the release of neurotransmitters. This entire process strengthens the mental immune system and reduces the risk of excessive neuroinflammation both directly (anti-inflammatory effect) and indirectly (by strengthening psychological resilience). The loss of bone and muscle mass in advanced age, often considered normal in modern societies, is, in fact, only seemingly age-related.

A lack of physical activity contributes significantly, as does an often lifelong vitamin D deficiency. It's crucial for this to become common knowledge—even if, according to new medical guidelines, vitamin D deficiency symptoms should practically only be considered once a patient is hospitalized with an osteoporotic fracture. Taking into account the multilayered effects via the Lithiome, which have been discussed, it's not surprising that lithium in essential quantities also contributes to the maintenance of the musculoskeletal system into old age—in the spirit of the Evolution of the Grandmother.

Patients with bipolar disorder have a lower risk of osteoporosis, but only when treated with lithium.[46] The situation here is similar to Alzheimer's: the increased risk of disease is reduced by lithium—though significantly lower, side-effect-free dosages would also be sufficient. The fact that lithium (among other aspects) links these two clinical pictures hasn't escaped the authors of the review article "From the Mind to the Spine: The Intersecting World of Alzheimer's and Osteoporosis." This again demonstrates how the body and mind are influenced in their function via the same fundamental signaling pathways.[47] For example, hyperactive GSK-3β impairs bone formation in animal models, but even low-dose lithium supplementation is enough to inhibit GSK-3β and thereby promote bone formation.[48] This finding could also be applied in bone surgery, where, for example, calcium phosphate cement (CPZ) is used, which itself stimulates bone formation. Animal experiments have shown that bone healing is even more efficient when some lithium is added to the CPZ.[49] Furthermore, in a genetic model for osteoporosis, using so-called Lrp5-knockout mice, it was also demonstrated that low-dose lithium intake is sufficient to stimulate genetically impaired bone formation, increase bone mineral density, and reduce increased skeletal fragility.[50] In a menopause model of osteoporosis (estrogen deficiency is implicated in accelerated osteoporosis), a two-week treatment of rodents with low-dose lithium improved fracture healing.[51] Overall, lithium's contribution to osteogenesis (bone formation) is based on the natural inhibition of GSK-3, which prevents or slows the progression of osteoporotic bone loss and even improves the healing of osteoporotic fractures.

Sarcopenia (from Greek sarx for "flesh" and penia for "deficiency") refers to the decrease in muscle mass and muscle strength, and the associated functional limitations in older adults. It's often caused by a combination of reduced physical activity, malnutrition (e.g., low protein intake), hormonal changes, systemic inflammation, and loss of motor neurons due to neuromuscular diseases.[52] GSK-3, which is upregulated during infections, for example, facilitates energy production from muscle mass by inhibiting the synthesis of muscle proteins and promoting their breakdown. It's therefore considered a pharmacological target for preventing muscle breakdown.[53] However, the problem is a constantly elevated GSK-3 activity, as occurs in the context of silent inflammation due to an unnatural lifestyle. The supposed age-related muscle loss is thus another facet of the unnatural "normality" we face in many areas of life today—the lack of essential lithium contributes significantly to this.

As discussed in Chapter 4, it has been shown in animal models that lithium and low-intensity endurance training synergistically increase the production and release of the neuroprotective BDNF in the hippocampus—this can prevent neurodegenerative diseases such as Alzheimer's dementia.[54] That the same combination of exercise and lithium not only protects the brain but also strengthens the muscles more effectively than either measure on its own could be demonstrated in humans for the first time.[55]

Essential Lithium Protects Against Cancer

An intact immune system serves as the best and most highly effective natural defense against life-threatening cancers. Approximately 3.8 million new body cells are created every second, totaling about 330 billion cells daily.[56] It is entirely natural for copying errors to occur occasionally (without such "errors" in germ cell production, there would be no evolution). However, it's equally natural for such cells to be recognized and destroyed by our immune system if genetic programming errors have caused them to become malignant. For example, a report on findings by an Australian research group titled "The immune system spontaneously kills blood cancer cell" reveals that "T cells of the immune system carry out regular checks to find cancerous and pre-cancerous B cells" and eliminate them.[57] This means cancer cells are formed daily but also eliminated daily—provided the immune system is healthy.

Crucially, it's not just the physical immune system that protects us from cancer; our mental immune system plays a role too. Every day, we make choices about what we eat (e.g., organic or not), whether we inhale toxins (e.g., smoking or not), and generally how we structure our lives (stressful or relaxed). A healthy mental immune system promotes eustress, whereas distress, through high cortisol release, can paralyze the physical immune system, thereby hindering immune surveillance. Since we now know that both immune systems require lithium to function optimally, it's immediately clear that our likelihood of developing cancer, and if so, our chances of survival, also depend on this. These fundamental considerations are always based on the basic laws of life (Chapter 1) and a systemic approach (Chapter 4), particularly in cancer prevention and therapy. The fact that both immune systems and their interaction are essentially dependent on lithium explains the title of a review article published in February 2023: "Lithium: A

Promising Anti-Cancer Agent." In this article, the authors summarized the state of research as follows: "Lithium formulations [. . .] induce apoptosis, autophagy, and inhibition of tumor growth and also participate in the regulation of tumor proliferation, tumor invasion, and metastasis and cell cycle arrest. Moreover, lithium is synergistic with standard cancer therapies, enhancing their anti-tumor effects. In addition, lithium has a neuroprotective role in cancer patients, by improving their quality of life."[58]

Despite the unnaturally high intake of lithium as part of standard treatment for patients with bipolar disorder, this still reduces cancer risk by an average of about 27 percent compared to a control group treated with a different medication.[59] The authors attribute this to the fact that: "Lithium inhibition of GSK3β suppresses cadherin-11, which is involved in cell–cell adhesion, cancer cell invasion, and metastasis of cancer." In a previously cited 2018 study involving over 4.2 million patient records, researchers, analyzing raw data, found that a lithium concentration in tap water of less than 40 μg/l, compared to over 40 μg/l, correlated with an approximately 26 percent increased risk of developing prostate cancer, for example.[60] This suggests that lithium could protect against cancer even in very small quantities. This makes sense, considering that all bodily functions depend on homeostatic regulation by lithium and, consequently, by GSK-3. This is particularly true for both immune systems. However, optimal homeostatic regulation can only occur with physiological lithium concentrations. Therefore, it's highly probable that simply addressing a lithium deficiency is enough to achieve the best possible protective effect against cancer development. Indeed, dysregulated (or lithium-uninhibited) GSK-3β is "also participates in tumor cell survival, evasion of apoptosis, proliferation and invasion, as well as sustaining cancer stemness and inducing therapy resistance," according to a 2020 review article.[61] For this reason, it is one of the most studied targets. According to the authors' research, "A therapeutic effect from GSK3β inhibition has been demonstrated in 25 different cancer types." The aforementioned 2023 review article lists various cancer types where lithium's effect as a natural GSK-3 inhibitor was specifically investigated: "The anti-cancer effect of lithium has been reported for colorectal cancer, neuroblastoma cells, ovarian cancer cells, medullary thyroid TT cancer cells, esophageal cancer, medulloblastoma, and glioblastoma multiforme, hematological tumors (multiple myeloma), breast cancer cells, MTC, prostate cancer cells, pancreatic ductal adenocarcinoma cells, corneal endothelial cells, and colon cancer cells."[62] However, they also report "opposite effects of lithium . . . for neuroblastoma, hepatoblastoma cells, and malignant glioma cell lines." It's important to note that these are cell lines, making it unclear to what extent these and the unnatural situation (cell culture) reflect conditions in a patient. Unfortunately, this limitation applies to almost all such studies where lithium's effect is investigated as a single pharmacological substance, rather than in conjunction with other measures that support immune system function. From a systemic perspective, a cancer therapy that relies on the body's natural immune system's defense against cancer cannot be monotherapeutic if it aims to be as effective as possible. As discussed in detail in Chapter 4, all aspects of a species-appropriate lifestyle should always be considered and included. Only then can lithium exert its natural effect at physiological dosages.

The whole is usually greater than the sum of its parts, which is why a systemic approach is fundamental in cancer prevention and, by extension, in therapy. For example, the 2020 DO-HEALTH study demonstrated that a combination of vitamin D3 (at only 2,000 IU per day, which is actually too little to achieve an optimal vitamin D prohormone level of about 125 nmol/l), 1 g of aquatic omega-3 fatty acids per day (also too little; 2 g/day is at least necessary for an optimal omega-3 index of 11), and a simple home exercise program (some strength endurance three times a week, so not much either) can reduce the cancer risk in healthy individuals with an average age of about seventy-five years by up to 61 percent.[63] If the respective dosages were chosen closer to the optimum and—in addition to everything we discussed in Chapter 4 within a systemic approach to species-appropriate medicine—essential lithium were added to this healthy mix, the residual cancer risk would likely be minimal. A risk of zero is principally impossible (in nature, chance governs, so it's always about probabilities that can be influenced, never certainties)—but the chances are good of achieving a value very close to the desired zero.

However, even if one opts for a conventional medical approach following current standards, "there is increasing evidence that GSK3β inhibition protects normal cells and tissues from the harmful effects associated with conventional cancer therapies," reads another review article on the available data.[64] A 2012 article specifically addresses lithium's importance for the mental immune system during conventional cancer therapy:

> Neurocognitive impairment is being increasingly recognized as an important issue in patients with cancer who develop cognitive difficulties either as part of direct or indirect involvement of the nervous system or as a consequence of either chemotherapy-related or radiotherapy-related complications. Brain radiotherapy in particular can lead to significant cognitive defects. Neurocognitive decline adversely affects quality of life, meaningful employment, and even simple daily activities. Neuroprotection may be a viable and realistic goal in preventing neurocognitive sequelae in these patients, especially in the setting of cranial irradiation.[65]

Lithium—Simply Healthy

Many people, unfortunately including many doctors, fear lithium due to the known side effect profile of high-dose therapy. Paradoxically, the biological effect of lithium in essential or low-dose amounts is often the exact opposite of what lithium does at extremely high doses. This could be because the very biological functions most dependent on natural lithium regulation are impaired by high dosages. In the following table, I've listed the two extremes for some systems:

System / Biological Function	**Essential to Low-Dose**	**High-Dose Treatment**
Alzheimer's Prevention	Good effect (compared to deficiency)	Increased risk with bipolar disorder, reduced to "normal"
Alzheimer's Therapy	Good effect compared to all medicines	The higher, the worse
Blood Pressure	Improved (compared to deficiency)	Slightly worse
Cardiac Output	Improved (compared to deficiency)	Slightly worse than "normal"
Heart Rhythm	Improved (compared to deficiency)	Slightly increased risk of cardiac arrhythmia
Kidney Function	Improved (compared to deficiency)	Renal dysfunction, risk of kidney failure
Blood Sugar	Improved (compared to deficiency)	Blood sugar lowering
Fat Metabolism	Improved (compared to deficiency)	Poorer blood values
Obesity	Good effect (compared to deficiency)	Increased risk of weight gain
Bones	Protects against degradation, helps to build up	Unclear
Musculature	Protects against degradation, helps to build up	Unclear
Cancer Prevention	Protects strongly	Protects somewhat

Overall, lithium in essential to low-dose concentrations exhibits consistently positive effects, whereas high-dose treatment presents a very mixed picture. It's therefore not surprising that low doses of lithium are associated with a long, healthy life, although an observational study in patients with bipolar disorder treated with high-dose lithium failed to demonstrate this effect.[66] However, it's not the aim of this book to defend the high-dose administration of lithium. As indicated in Chapter 4, even for bipolar disorder, a smaller amount of lithium within a systemic therapy could be sufficient, with fewer or no side effects, especially since monotherapy only truly helps a fraction of patients. My aim with this book is to show that an appropriate lifestyle enables us to live the longest and healthiest life possible—and essential lithium is an indispensable building block for this. However, a lot of educational work is still needed, especially as the ban on lithium as a dietary supplement is an indication that the medical establishment has no interest in applying the existing knowledge of its numerous benefits. But as Arthur Schopenhauer (1788–1860) once declared:

> ***All truth passes through three stages.***
> ***First, it is ridiculed.***
> ***Second, it is violently opposed.***
> ***Third, it is accepted as being self-evident.***

CHAPTER 7

ESSENTIAL LITHIUM—KEY AND SYMBOL FOR THE RENAISSANCE OF HUMANE MEDICINE

Truth is born of the times,
not of authority.[1]
—Bertolt Brecht (1898–1956)

Lithium Ban—Unmasking a Misanthropic Industry

The proof is established: lithium is an essential trace element. As summarized in Fig.30, a seamless causal chain exists, linking lithium (or a deficiency thereof) to its molecular targets (the lithiome), their function or dysfunction under lithium deficiency, leading to mental immunodeficiency syndrome, and finally to negative societal consequences. A social immunodeficiency, in addition to the problems depicted in the graphic that complicate communal living, makes society more susceptible overall to propagandistic manipulation, which is detrimental to genuine democracy. After all, a democracy requires a society where at least a majority of citizens possess a functioning mental immune system, which, given the presented evidence, is problematic in the context of essential lithium deficiency. Although not all aspects of this essential trace element's mechanism of action are yet fully understood, the existing factual basis is sufficient to inform updated intake recommendations. The vital necessity of lithium in small quantities has been ignored for far too long. Evidence and indications from the fields of epidemiology, biochemistry, evolutionary biology, genetics, molecular genetics, as well as from animal experiments, clinical observations, and simple combinatorial logic, leave no serious doubt. Scientific interest in delving deeper into these connections still offers ample opportunity for decades of research and countless further studies. However, this should not distract us from the fact that sufficient evidence already exists today to recognize the fundamental intake recommendations for lithium as an essential trace element and to ensure a cost-effective supply for the world's population.

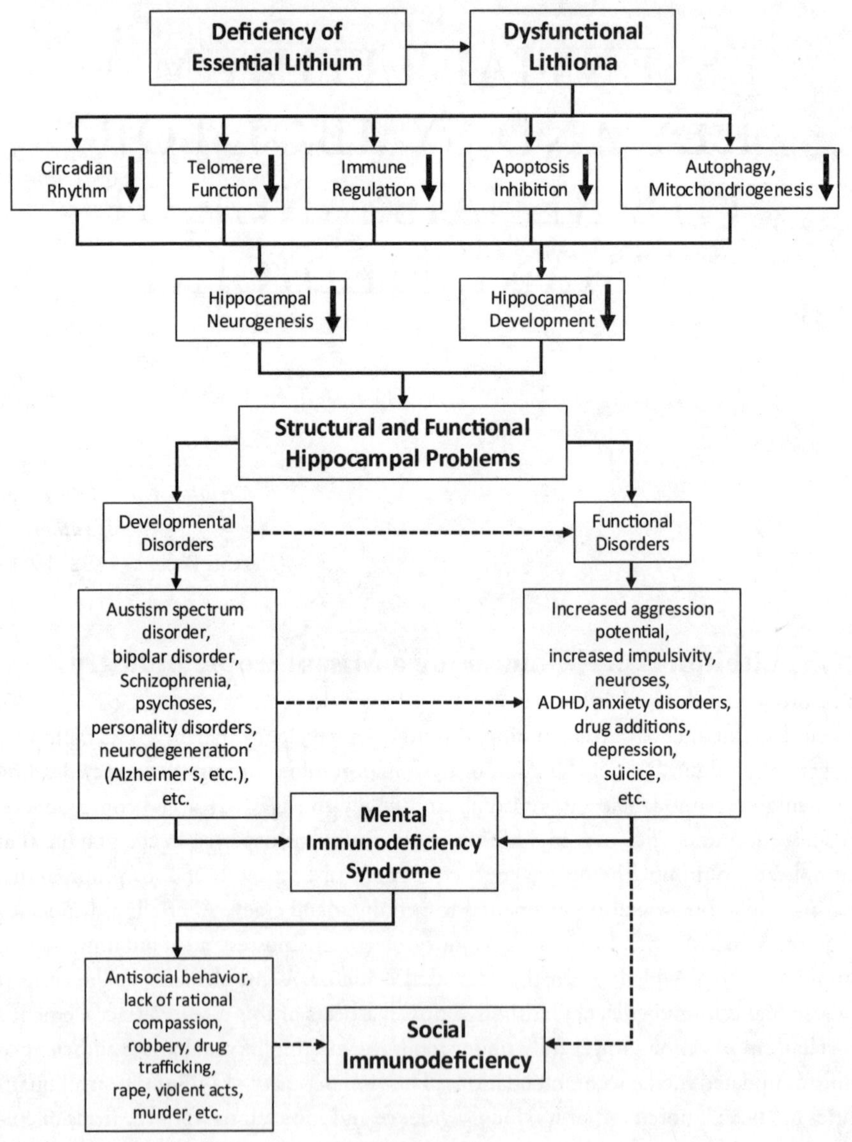

Figure 30

The repeatedly cited toxicity of high doses can be logically refuted as a counterargument to the essentiality of small amounts of lithium. However, while some scientists continuously dwell on these points, endlessly circling and denying the general evidence, the prevailing lithium deficiency impairs the mental development and health of a large proportion of humanity. This will not change as long as we continue to overemphasize unresolved questions and await specific study results, instead of recognizing, based on already accumulated evidence, that taking lithium in essential dosages is completely risk-free (because it's natural!). Furthermore, convincing evidence exists for the far-reaching benefits of its intake and its systemic indispensability. Indeed, there is no civilization disease not directly caused by a lithium deficiency or at least indirectly promoted by it. However, as the development and performance of the human brain depend to a particular extent on lithium-controlled functions, our "humanity" suffers particularly from a lack of this trace element.

Since I coincidentally stumbled upon the normalization of increased Alzheimer's risk in bipolar disorder through high-dose lithium in 2012, it quickly became clear to me in my molecular biology research that lithium must be an essential trace element. I presented this finding in 2016 in my systems biology article on the "Unified Theory of Alzheimer's Disease" (UTAD), where I showed that lithium stands on par with essential trace elements like selenium and zinc. Since lithium was not only unrecognized as an essential trace element but also banned as a food supplement in the EU, and thus in Germany, I could only refer in my article to drinking water sources with naturally high lithium content. Today, in spring 2025, approximately ten years after the UTAD article was written, the evidence for lithium's essentiality has become even stronger. Particularly in recent years, scientific understanding has advanced to such an extent that, in my opinion, previous objections to recognizing lithium's essentiality have been completely invalidated. Its essential function for brain development and the lifelong maintenance of mental health is beyond question. But I was far from the first to suggest that lithium is an essential trace element. As early as 1949, John Cade expressed the specific suspicion that lithium plays an essential role in the development of mental health: "The effect on patients with pure psychotic excitement—that is, true manic attacks—is so specific that it inevitably leads to speculation as to the possible aetiological [causative] significance of a deficiency in the body of lithium ions in the genesis of this disorder."[2] And Cade consistently concluded "that lithium may well be an essential trace element." Unfortunately for all of humanity, however, he reached this realization in connection with his work that established the medicinal use of high-dose lithium in bipolar disorder, which, perversely, has prevented its recognition as an essential trace element to this day (as of spring 2025), as we will see.

Gerhard Schrauzer was likely one of the first scientists who, as early as 2002, did not merely suspect but was very certain that lithium is essential.[3] Yet, his extensive work and that of many other doctors and scientists in epidemiological, clinical, biochemical, physiological, and experimental research—all confirming Cade's assumption and Schrauzer's certainty—fell on deaf ears within the medical establishment. They stubbornly continued to claim it wasn't essential, astonishingly without seriously engaging with the evidence. We see this in the way, for instance, David A. Hart, a professor at the Faculty of Kinesiology at the University of Calgary, Canada, simply states in a July 2024 article: "Lithium ions have been excluded from inclusion as an essential element during the evolution of increasingly complex biological systems."Or: "While a common element, it has not been found to be an essential element in biological processes, ranging from single cell organisms to *Homo sapiens.*" And further: "Although lithium ions are widely distributed in the environment, they are not essential co-factors in biological systems."[4] However, based on the evidence compiled in this book, it becomes clear: those who brazenly claim the opposite of what is demonstrably true inevitably find themselves in significant need of explanation when faced with simple inquiries. And it's almost amusing to read the lengths to which Hart goes to justify that what cannot be (lithium is essential) must not be (lithium is the cost-effective natural substitute for profitable drugs). The scientific data is well-known, as an expert he's necessarily aware of it and knows it contradicts his statements. Thus, he can't completely ignore central findings when he writes: "Although lithium ions are not absolutely necessary in biological systems, they appear to exert a biological regulatory function in some systems." And that's not all; he even concedes that lithium is "capable of modulating a variety of biological processes and 'correcting' deviations from normal activity, as lithium deficiency can have biological effects [. . .] and lead to adverse health effects." If a lack of lithium has been proven to lead to health problems, this speaks to it being an essential trace element. Hart must furthermore confirm that "low-dose lithium may have excellent potential, alone or in combination with other interventions to prevent or alleviate aging-associated conditions and disease progression." And he wants all this to be taken into account while still classifying lithium as nonessential—this undeniably begs the question: What is the real motivation behind his refusal to adequately interpret the obvious correlations?

Recognizing the essentiality of lithium would fundamentally change medicine, as will become even clearer as we proceed. It seems some are trying to prevent this dam from breaking. This is an ultimately impossible task for all who seriously attempt to contradict nature. To avoid the judgment of essentiality, Hart clumsily instrumentalizes the concept of "Hormesis" (from Greek *hormáein*, "to stimulate") as an explanation for lithium's effect as a trace element. Hormesis describes a biological mechanism where low exposure has no direct positive effect but triggers adaptive counter-reactions in the organism that are ultimately beneficial. Hart writes:

> The effect of lithium salts on cells in vitro [in "glass" or here, in Petri dishes] exhibits a property known as "Hormesis" [. . .] in which low doses can have positive effects and higher doses can have negative effects. This property of hormesis by lithium can also be observed *in vivo* [in organisms], where high doses

> (hundreds of mg/day) are required to treat bipolar disorder, while very low doses (<1–10 mg/day) may be sufficient to act as a microregulator of biological processes and/or prevent dysregulation.

He cites the review article "Lithium and Hormesis: Enhancement of Adaptive Responses and Biological Performance via Hormetic Mechanisms." In it, the authors point out, "These findings suggest that lithium may have critically valuable systemic effects with respect to those therapeutically treated with lithium as well as for exposures [meaning an intake level]that may be achieved via dietary intervention[such as lithium-rich drinking water, etc.]."[5] Drawing on the principle of Hormesis, an attempt is also made here to explain why lithium—misleadingly interpreted as a "stressor" in dietary quantities—coincidentally "triggers" or "stimulates" something positive, even though it's not essential in itself. This includes, among other things, a life-prolonging effect, as a 2016 article already admits in its title: "Lithium Promotes Longevity Through GSK3/NRF2-Dependent Hormesis."[6] According to this interpretation, people don't live longer because lithium is essential, but because lithium triggers something that coincidentally promotes health and longevity. The nonessential yet beneficial effect of lithium, according to the Hormesis explanation, would then fall into a category that also includes ice bathing and its positive effects after brief exposure. While bathing in water temperatures near 0°C for a few minutes can indeed bring health benefits—provided one doesn't suffer a heart attack—100 minutes in ice water is usually fatal.[7] However, for lithium to exert its life-prolonging effects, it must be continuously present in low concentrations, which contradicts the fundamental idea of Hormesis as a "nudge." This helpless explanation of positive effects, designed to avoid accepting essentiality, is thereby fully rendered absurd.

Nevertheless, let's consider what this absurd assumption would imply, for instance, regarding the animal experimental findings we discussed in Chapter 2, Argument 5: If lithium were understood as a nonessential stressor, a life without this stressor should be readily possible and natural. However, this would mean that rats and goats, under natural conditions that might well lack lithium (since it's purportedly nonessential), would produce little to no offspring, and their underweight newborns would have very low survival rates. Only when the allegedly harmful lithium "nudges" certain systems would the tide turn, according to the hormesis hypothesis, with litter sizes increasing and offspring surviving more often. Nonsensically, this stimulus would even need to be permanent; this would certainly no longer be a "nudge," but—to use the earlier example—comparable to constant ice bathing. The explanation via "Hormesis" is therefore a completely contradictory and, t apparently desperate attempt to both deny lithium's essentiality and explain the undeniable health-promoting effects of small quantities. The authors of the longevity study cited earlier should also have recognized this logical flaw. After all, they state that lithium's life-prolonging effect is closely linked to its inhibitory effect on GSK-3—and this inhibition must be permanent to exert the desired effect.

However, in their article, the authors dream of new drugs that could achieve precisely this lasting effect, and the presence of essential lithium would only interfere. They write: "The discovery of GSK-3 as a therapeutic target for aging will likely lead to more effective treatments that can modulate mammalian aging and further improve health in later life."[8] From all this, we can conclude: The pharmaceutical industry is seemingly willing to go to extreme lengths just to develop a patentable compound that, even without side effects, might not be as effective as essential lithium—their billions spent on failed developments and clinical trials are well documented. Recognizing the essentiality of lithium would dash these hopes—an obvious reason why some are trying to prevent this against all evidence and logic.

The review article by Professor Hart, as well as all other papers explaining the lithium effect through hormesis, should therefore not be viewed as isolated mistakes or misguided ventures. Instead, they represent the prevailing doctrine. Their argumentation, which is logically unsound on its own, ultimately managed to pass the peer-review process—the standard procedure for quality control prior to publication in scientific journals—without contradiction. This means that during the review of their work by independent scientists from the same field (the "peers"), no one noticed what, upon closer inspection, would strike even a layperson. The journal's publisher, the article's editor, and indeed the independent scientists seemingly failed to notice the argumentative contradiction. The hormesis thesis not only aims to deny lithium its essentiality—despite its health-promoting properties and the fact that an absolute deficiency runs counter to the evolutionary imperative "be fruitful and multiply," a fundamental law of living nature. Instead, it attempts to create the impression that high GSK-3 activity—though harmful—should be considered natural. This happens when one fails to recognize that lithium is necessary in certain amounts for proper regulation. A 2011 review published in the *International Journal of Alzheimer's Disease* on the fine regulation of GSK-3 activity actually states, as a logical consequence of this completely misguided hormesis argument: "The protein kinase GSK-3 is constitutively active in quiescent cells in the absence of growth factor signaling."[9] The authors use the word "unusually" to describe the permanent and exceptionally active state in non-stimulated or activated cells—which, as we now know, is solely due to a lack of essential lithium. In a 2015 review article, another research group also points out that GSK-3 possesses the unconventional property of being constitutively active within the large protein family of kinases (518 different kinases are encoded in the human genome):[10] "The protein kinase GSK-3 is constitutively active in quiescent cells in the absence of growth factor signaling."[11] Apparently, this shouldn't attract any attention, because what's truly "unusual" is that lithium isn't considered essential for GSK-3 regulation—and its absence is therefore deemed normal, even if it has "peculiar" or "unusual" consequences.

The question must arise as to why evolution would have produced such a life-threatening protein as GSK-3 (see Fig. 31). I, too, pondered this question until it first became clear to me that a nowadays commonplace, yet entirely unnatural, lithium deficiency is responsible for this unnatural (though unfortunately also considered normal today) and excessively high activity of GSK-3. This contradicts the concept of the Evolution of the Grandmother. In its constitutively active state, GSK-3 is implicated as at least partly responsible for virtually all diseases of civilization.[12] In its acutely hyperactive state (with additional activation by external stressors, i.e., PAMPs or DAMPs), GSK-3 is even significantly involved in life-threatening cytokine storms. Thus, by claiming that GSK-3 is naturally "constitutively active" and lithium is only "hormetically active," a clever attempt is made to deny the trace element its essentiality or to nip the question in the bud.

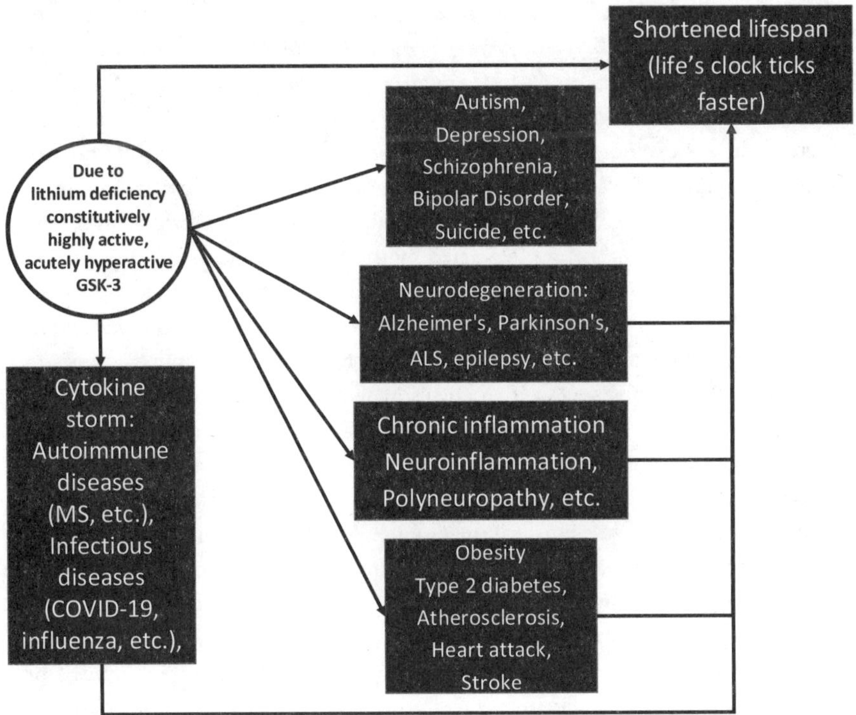

Figure 31

On October 16, 2024, just two months after Hart's attempt to cleverly disguise the natural role of lithium as an essential trace element with the term "hormesis," the German Consumer Centre provided information about the "effect of this dietary supplement" and "why it's so difficult and expensive to obtain in Germany."[13] Due to the content and timing, it's possible—and this also applies to Hart's articles—that they were inspired by my media engagement from around early 2024, where I've been educating people about lithium's essentiality and reaching an increasing global audience.[14] Perhaps they aimed to counteract the sudden surge of public interest. As the Consumer Centre, operating under a state mandate to protect consumers, writes: "Food supplements containing lithium may not be sold in Germany or in any other member state of the European Union. Annex II of the Food Supplements Directive lists all permitted vitamin and mineral compounds. [[15]]. The ultratrace element lithium is not one of them, so its use in food supplements is prohibited." In essence, they are relying on an existing classification, ignoring new evidence, and on this basis, prohibiting its availability as an essential dietary supplement. Interestingly, this ban rests solely with the European Commission, as revealed by a legal opinion prepared upon request by a law firm specializing in such matters.

Statement on the Legal Situation of Lithium as a Food Supplement in Europe by the Law Firm Dr. Brigitte Röhrig[16] (April 29, 2024): *It can be assumed that the sale of food supplements (NEM) containing lithium is prohibited throughout the EU. Extensive research might be needed to determine whether individual EU member states have issued special regulations.* However, their compatibility with EU law would be extremely questionable due to the explicit provision in Art. 3 of the Food Supplements Directive. The requirements for lawfully placing a food supplement on the market are regulated by the Food Supplements Directive, in addition to general food law provisions, particularly Regulation No. 178/2002/EC. NEMs are defined as "'Food supplements' means foodstuffs the purpose of which is to supplement the normal diet and which are concentrated sources of nutrients or other substances with a nutritional or physiological effect, alone or in combination, marketed in dose form[. . .]" (Art. 2 lit. a). Vitamins and minerals are "nutrients" (Art. 2 lit. b). In the case of lithium, the definition therefore covers a product intended to supplement the normal diet with lithium and containing a concentrate of lithium/lithium orotate. According to Art. 3 of the Food Supplements Directive, only NEMs that comply with the Directive's provisions may be placed on the market in the EU. The catalog of permitted vitamins and minerals and their forms (salts) for the manufacture of NEMs is listed in Annex I, Art. 4 para. 1. Lithium/lithium orotate are currently not listed in Annex I or Annex II, making it generally impermissible to place lithium-containing food supplements on the market in the EU. For lithium/lithium orotate to be included in Annex I of the Directive, Art. 4 para. 5 of Food Supplements Directive 2002/46/EC requires the adoption of an implementing regulation by the EU Commission, generally on its own initiative; the right of initiative for legislation in the EU lies almost exclusively with the EU Commission. The European Parliament has only very limited rights of initiative. This implementing regulation is adopted by the EU Commission after consulting the European Food Safety Authority (EFSA), specifically its Scientific Committee, Art. 14 of the NFD. The opinion of the EFSA Committee is not binding on the EU Commission.

To understand the scale of the absurdity of this lithium ban, a brief comparison with the trace element aluminum—which is not essential for humans, but rather toxic—is useful. In a 2002 review, researchers examined the effects of four trace elements (silicon, aluminum, arsenic, and lithium) on human health; they stated that all available research results at the time were compiled. They concluded that silicon and lithium play a protective role in the human diet, while aluminum and arsenic have a particularly toxic effect.[17] In contrast to brain-healthy lithium, brain-toxic aluminum is regulated in terms of food intake but is far from being banned as it should be. On the contrary, dietary intake occurs through a long list of still permitted food additives, identified by twenty-eight different E-numbers. The European Union (EU) determines which additives are allowed. Many sweets, for example, contain this brain-toxic light metal as an additive, including aluminum lacquers used for coloring. Hans-Ullrich Grimm's book, *Die Ernährungsfalle* (*The Food Trap*), draws attention to several additives that contain clearly declared aluminum: "In addition to the [previously shown] aluminum coloring lacquers, there are several other aluminum-containing additives: from pure aluminum (E 173) to aluminum sulfates (E 550 to E 553) to aluminum silicate (E 559). They are used for industrially packaged egg whites and for candied, crystallized, or glazed fruit and vegetables, also as a release agent for sauce powders and packet soups. They also ensure that packaged cheese slices do not stick together."[18] Other aluminum silicates (E 554, E 555, E 556) are also permitted as release agents.[19] But even additives that superficially appear to contain other substances may contain aluminum, including curcumin (E 100), carmine (E 120), amaranth (E 123), or anthocyanins (E 163) —as well as E 104, E 110, E 122, E 124, E 127, E 129, E 131, E 132, E 133, E 141, E 142, E 151, E 155, and E 180. These are also sometimes found in sweets and convenience foods.

The daily intake of 10 mg of aluminum permitted by the EU for a 70 kg adult is not considered seriously problematic—according to the Bavarian State Office for Health and Food Safety. This decision rests on an opinion from the European Food Safety Authority (EFSA), which advises the EU Commission: "Based on the available data, EFSA [in] 2008 derived a weekly tolerable intake of 1 mg aluminum/kg body weight ('Tolerable Weekly Intake' TWI) analogous to the World Health Organization (WHO) precautionary value from 2006."[20] The EFSA defines the "Tolerable Weekly Intake" (TWI) as: "The maximum intake of substances in food, such as nutrients or contaminants, that can be consumed weekly over a lifetime without risking adverse health effects."[21] According to the EFSA, this amount of aluminum in the form of industrial additives "can be ingested weekly over a lifetime without posing a health risk to humans, based on current knowledge." This is actually an impossible conclusion given the clearly proven brain toxicity and the fact that it accumulates in the brain—with dramatic consequences, particularly for the mental immune system.[22] The aforementioned "WHO precautionary principle" from 2006 changed in 2011; it became somewhat more generous, as stated there: "The Committee established a provisional tolerable weekly intake (PTWI) of 2 mg/kg body weight[. . .]. The PTWI applies to all aluminium compounds in food, including food additives. The previous PTWI of 1 mg/kg body weight was withdrawn. [. . .] Estimates of dietary exposure of children to aluminium-containing food additives, including high dietary exposures[. . .], can exceed the PTWI by up to 2-fold[!]."[23]

While the EFSA and WHO apparently have no problem setting a tolerable daily intake for toxic substances like aluminum, allowing the food industry (as well as the pharmaceutical industry for vaccines, antacids for heartburn, etc.) to liberally use aluminum with a de facto RDA, the EFSA apparently has great difficulty assessing lithium as a food supplement. The request of an applicant who sought to classify naturally enriched baker's yeast as a safe food supplement was rejected with the words: "*It is not possible to assess the safety of lithium-enriched yeast as a food source of lithium.*"[24] According to the applicant, the added amounts of lithium "depend on the product and are generally in the range of 0.5 to 2 mg/day" for the intended intake—almost precisely in the range of the provisional RDA for essential lithium proposed here in Chapter 3. Reason for rejection by EFSA: "Inability to assess the safety of lithium-enriched yeast added for nutritional purposes as a source of lithium in food supplements and the bioavailability of lithium from this source, based on the supporting dossier." Additionally, the Panel stated that, "No safety data or appropriate supporting references have been provided to substantiate the assumption of the safety of lithium-enriched yeast." The premise of this reasoning is that the lithium or yeast may have undergone toxic changes. This is absurd, especially since Gerhard Schrauzer had already demonstrated the mood-enhancing effect of 0.4 mg lithium in a placebo-controlled clinical study on former drug addicts using lithium-containing yeast.[25] No side effects from "yeast that has become toxic" were reported. There are two further sticking points in EFSA's assessment: "The Panel notes that neither EFSA's Scientific Panel on Dietetic Products, Nutrition and Allergies (NDA) nor other authorities have established reference values for the daily intake of lithium. There is no evidence that lithium is essential for humans." This statement seems designed to doom any further attempt to get a lithium-containing food supplement approved. An EFSA review can quickly become quite expensive "as extensive research is required to ensure consumer safety," as two Polish environmental physicians wrote in a review article on lithium as a micronutrient.[26]

The ideal way to make lithium a recognized staple food is therefore to evidence its essentiality. But this will be a rocky road if authorities, whose governmental mandate is to protect our health, allow industry to use even brain-toxic aluminum as a dietary supplement while simultaneously prohibiting people from taking brain-healthy lithium as a dietary supplement (with the USA as one of the few exceptions). This raises important questions. To understand who we're dealing with in our next steps on the high road to recognizing lithium's essentiality, we need to explore these questions:

- Is the recognition of lithium only being held back by the pharmaceutical industry's profit motive, or are there other interest groups at work?
- Who else could have an interest in the fact that more and more people are suffering from mental immunodeficiency syndrome—which, as we now know, encompasses a broad spectrum of mental performance deficits and mental illnesses—due to a lack of this essential trace element?
- Who benefits if we lose the ability to react calmly in conflict situations—both in our personal environment and on the global political stage—to "think peace" and even prefer war to diplomacy as a result?
- Who benefits from a society that is more easily ruled by fear due to a lack of resilience and in which the minds and psyches of its members can be manipulated?
- And last but not least: Who presumes to stubbornly deny the essentiality of an element that has demonstrated all the requirements of essential trace elements in numerous studies and consistently produces astonishing, confirming successes in countless practical applications, and even prohibits it as a dietary supplement?

To find answers to these questions, we need to look back into the history of medicine. Specifically, we'll now turn our attention to the beginning of the twentieth century, when some extraordinarily wealthy individuals realized how much richer they could become from common diseases if they gained control of the global healthcare system.

The Miscarriage of the Modern Disease Industry

In 1901, John D. Rockefeller Sr. (1839–1937), the world's wealthiest man thanks to his oil monopoly (Standard Oil), founded the Rockefeller Institute for Medical Research (now Rockefeller University). Among other sources, he financed it with funds from the "philanthropic" Rockefeller Foundation, which he established in 1913. His hidden agenda was to shape the future of medicine to be lucrative, aligning with his own interests. Rockefeller appointed the American pathologist Simon Flexner (1863–1946) as the first director of the Rockefeller Institute for Medical Research. Simon's brother, Abraham Flexner (1866–1959), an educator and science organizer, was commissioned in 1911 by the Carnegie Foundation (founded by steel magnate Andrew Carnegie, 1835–1919) to write a report on the future of medical education in America. This report became historically known as the "Flexner Report."

Inhumane Eugenics. To grasp the inhumane mindset of Rockefeller and his collaborators, the Abraham Lincoln Foundation, a German suborganization of the Rockefeller Foundation at the time, is noteworthy.[27] Between 1932 and 1935, it financed, among other projects, the notorious twin research of Otmar Freiherr von Verschuer (1896–1969), a racial researcher, eugenicist, and director at the Kaiser Wilhelm Institute for Anthropology, Human Heredity, and Eugenics in Berlin. His work laid a crucial foundation for National Socialist racial doctrine, with his institute providing no less than its "scientific legitimization."[28] The Kaiser Wilhelm Institute later became the Max Planck Society.[29] To advance their vision of public health, the Rockefeller Foundation also collaborated closely with the Carnegie Foundation from the outset. These two foundations were also the primary and most significant investors in the U.S. Eugenics Record Office, a research institute that collected biological and social information on the American population and served as a hub for eugenics and human heredity research.[30] American educator and eugenicist Harry Hamilton Laughlin (1880–1943) directed the Eugenics Record Office from its inception in 1910 until its closure in 1939.[31] He was one of the most influential figures in American eugenics policy, particularly concerning legislation for the forced sterilization of "defectives," public education on eugenic health issues, and the widespread dissemination of eugenic ideas.

The goal of the Flexner Report was to initiate a comprehensive overhaul of the American health-care system, aiming for a restructuring and centralization of medical institutions. What's particularly interesting in light of today's European health-care system is this passage: "It is the purpose of the Foundation to proceed at once with a similar study of medical education in Great Britain, Germany, and France, in order that those who are charged with the reconstruction of medical education in America may profit by the experience of other countries."[32] In other words: Europe became the medical testing ground for the new US health-care system. According to the analyses of award-winning investigative reporter James Corbett, a significant part of this restructuring toward a profitable, industry-controlled future of medicine focused on one crucial process: natural healing methods and medical treatments "focused on un-patentable, uncontrollable natural remedies and cures was now dismissed as quackery." The objective: only allopathic (from ancient Greek *állos*, for "other" or "different in nature," effectively: 'unnatural,' and *páthos*, German "suffering") medicines—that is, therapies based on foreign chemicals—were to be promoted and propagated from then on. Thanks to the Rockefeller network's already extensive control over the medical education system, they possessed the necessary financial, media, and political means to successfully implement this agenda.[33] Flexner, among other things, recommended the closure of schools that taught naturopathy and alternative approaches. Simultaneously, the Rockefeller and Carnegie Foundations generously offered funding to medical schools in exchange for influence over their curricula. The oiligarchy birthed entire medical industries from their own research centers and then sold their own products from their own petrochemical companies as the "cure."

Inhumane Research. According to James Corbett's historical research, it was Frank S. Howard (1890–1964), a senior executive at Rockefeller's Standard Oil in New Jersey, who convinced automobile magnates Alfred Sloan (1875–1966) and Charles Kettering (1876–1958) to donate their fortunes to the cancer center that would later bear their names. Howard appointed pathologist Cornelius P. Rhoads (1898–1959) as research director at the Sloan-Kettering Institute (SKI). Under Rhoads's leadership, Corbett continues, "nearly the entire program and staff of the Chemical Warfare Service were reformed into the SKI drug development program, where they worked on converting mustard gas into chemotherapy." This was based on the poison and mustard gas research Rhoads had conducted for the U.S. Army during World War I. A further indication of the inhumane nature of this developing industry: in his role as a researcher funded by the Rockefeller Foundation, Rhoads, by his own admission, intentionally infected several patients from Puerto Rico with cancer cells in 1931 to advance his program. In a letter, Rhoads stated he saw this as an opportunity to make his personal contribution to the extermination of the Puerto Ricans, the "most degenerate and thievish race of men ever inhabiting this sphere."[34] Despite these written statements being public knowledge (even the Vatican received a copy of his letter), they did not harm his career. On the contrary: it was not until decades after his death that the annual research prize awarded by the American Association for Cancer Research and named after him was renamed. This, however, was not an isolated incident.[35]

Many books have been written about the concerted efforts of extremely wealthy "philanthropists" and their institutions to globalize health care, with the aim of expanding their global power. Unfortunately, these "philanthropists" have been, and continue to be, very successful: they dominate medical education, state health-care systems, and, not least, the leading media, which consistently portray them in a positive light.

The Rockefeller Story—Copy and Paste. Rockefeller's "success story" found its imitators. Bill Gates, through the Bill & Melinda Gates Foundation, appears to be pursuing very similar ambitions and operating along comparable lines.[36] Vaccination is seemingly being promoted as a means to the same end of global control and manipulation. According to the WHO, the Rockefeller Foundation also took the lead here, as stated on the WHO's homepage: "More recently, and throughout Covid-19, the collaboration has focused on maintaining essential health services, expanding virus testing capacity, and strengthening genomic surveillance."[37] The WHO and the Rockefeller Foundation helped pave the way for the pandemic. Meanwhile, the Bill & Melinda Gates Foundation greatly benefited from the mRNA injection program and the genetic modification of much of the world's population during the staged SARS-CoV-2 pandemic. This was made possible, in particular, by suppressing compatible alternatives, including a specific campaign against the natural health approach of addressing the causal vitamin D deficiency, as discussed, for example, in Chapter 1, Additional Information: *Vitamin D Deficiency—Deadly Example of Ignoring the Law of the Minimum.*

To realize these global plans, the Rockefeller Foundation, through the League of Nations Health Organization (which was transformed into the WHO), became one of the midwives, if not one of the parents, of the WHO.[38] The Foundation participated as an observer at the first International Health Conference in June 1946, where the WHO constitution was signed. With the help of this first globally active UN special organization (the United Nations, also founded with the help of the Rockefeller Foundation), Rockefeller and his "philanthropic" partners were able to exert effective influence on the health policy of almost all nations.[39] In the following years and decades of close cooperation between the UN, WHO, and the Rockefeller and also the Bill & Melinda Gates Foundation, the belief in the "pharmaceutical narrative" was globalized.[40] Since then, this narrative has not only been instilled in the curricula of budding doctors but has also been culturally implanted in the minds of almost all of humanity.

The pharmaceutical narrative consists of three closely linked doctrines. From the perspective of the pharmaceutical industry and conventional medicine, they are regarded as axioms—fundamental assumptions that do not need to be proven and must not be questioned. Actual proof would not be possible here anyway, as they contradict nature:

1. Illness is a natural process inherent in our genes that becomes symptomatic with increasing age: Humans are a design flaw.
   Example: According to a 2012 review article, age is "the main risk factor for the prevalent diseases of developed countries: cancer, cardiovascular disease

and neurodegeneration"—and these may even be part of our (faulty) genetic program.[41] Accordingly, for example, pharmaceutical consultant Dennis Selkoe, in a 2015 *Süddeutsche Zeitung* article propagating an Alzheimer's vaccine, responded to the question of what to do to prevent dementia by stating there was only one rule to follow: "Choose the right parents and die early."[42]

2. Our lifestyle therefore has virtually no influence on our health. Supplementing vital substances (even if we have a deficiency) is also a waste of money or even dangerous to our health.
   Example: According to a 2022 *New England Journal of Medicine* editorial, which we discussed in Chapter 1, dietary supplementation is useless even at low 25-hydroxyvitamin D (25(OH)D) levels. Doctors should neither measure blood levels nor recommend supplementation. The new international guidelines for doctors published in 2024 largely reflect this absurd opinion. Despite countless studies demonstrating the benefits of a vitamin D prohormone level of around 125 nmol/l (see, e.g., Chapter 1, Additional Info: *Vitamin D Deficiency—Deadly Example of Ignoring the Law of the Minimum)*, it is claimed: "No clinical trial evidence was found to support routine screening for 25(OH)D in the general population, nor in those with obesity or dark complexion, and there was no clear evidence defining the optimal target level of 25(OH)D required for disease prevention in the populations considered; thus, the panel suggests against routine 25(OH)D testing in all populations considered."[43]
3. Which disease inevitably breaks out one day (cancer, Alzheimer's, heart attack, etc.) and ultimately ends life is pure fate or subject to the roulette of life. As a result, only pharmaceutical research with its drugs can save us from suffering and premature death (with the transhumanist idea, even death is ultimately eliminated).
   Example: "Personalized medicine," reads a 2024 review article by an employee of the pharmaceutical company Boehringer Ingelheim, "is a groundbreaking field in which tailor-made therapeutic solutions can be offered that are tailored to the genome of each individual patient."[44] There is talk of big data, AI, nanotechnology, and gene therapy, but tellingly, not a word is said about the actual cause of over 90 percent of all chronic diseases today: a lifestyle that is no longer appropriate for the species. Instead, in the future, personalized genetic and drug interventions will be used depending on the basic genetic makeup. One such trend can be seen in current mRNA cancer research, which is driven forward by companies such as BioNTech and Moderna. BioNTech, which originally specialized in cancer research—and for many has become synonymous with ruthless profiteering at the expense of patients due to the "invention" of COVID-19 injections—is testing various mRNA therapies in clinical trials, including against lung, skin, and colorectal cancer. Moderna, on

> the other hand—whose COVID "vaccine" was even suspended for younger age groups in several countries, including Sweden and Denmark, in October 2021 because an increased risk of heart muscle and pericarditis could no longer be denied—is seeking approval for corresponding products in the USA and the EU before the end of 2025. These approaches rely heavily on personalized medicine—but again without addressing the crucial question of why cancer and other chronic diseases occur so frequently in the first place.

This three-part, self-contained narrative shapes today's cultural (mis)understanding that only technological progress can redeem and save us from suffering (through pharmaceutical medicine) and even from death (through transhumanist technology). There are probably many reasons for the success of this narrative: on the one hand, the media power of corporations, which can conjure up an "expert" for any nonsense, and credibly sell any narrative to the masses. Furthermore, it's simply too convenient, as one needn't contemplate the consequences of their actions: Why laboriously change one's accustomed lifestyle if it supposedly offers no health benefits anyway; illness is inevitable and fate. But whatever the reasons, this narrative—that chemistry and genetic engineering can rescue us from supposedly unavoidable diseases—deliberately undermines our natural health potentials. It operates both intellectually, by portraying pathology as natural and thus acceptable, and practically: instead of preventive education, symptom-masking medications dominate, often causing more harm than the easily preventable or causally treatable diseases themselves. Belief in the pharmaceutical promise of salvation has now advanced to such an extent that many people relinquish all personal responsibility to a prescription pad.

Chronic Diseases as a Self-Fulfilling Prophecy

In today's conventional medicine, insofar as it's influenced by the pharmaceutical industry's interests, the focus isn't on preventing chronic diseases. The priority is the medicinal treatment of symptoms—an approach not aimed at definitive cure, but rather promising chronic or perpetually increasing revenue, true to the motto: "A patient cured is a customer lost."[45] COVID-19 brought this into stark relief. However, the promise of prevention usually serves merely as a sales strategy for highly lucrative, but often only ostensibly disease-preventing, vaccinations. The insidious truth: in most cases, one can protect oneself against dangerous infection trajectories naturally, with no side effects. Yet, this fact—in line with the purely revenue-driven business model—is deliberately concealed or even actively combated. For these reasons, the Fundamental Laws of Life, which we discussed in Chapter 1, are not yet taught in medical schools (as of spring 2025). They would render the pharmaceutical narrative and its promises of salvation absurd, as can be easily seen from the example of essential lithium: No allopathic drug can compensate for an underlying deficiency in a disease and thereby cure it. This pursuit of profit likely largely explains why lithium is not recognized as an essential trace element—in this way, the pharmaceutical industry creates an immense market for itself and cements the associated narrative of humanity as inherently deficient.

Many People Fall Ill Due to a Chronically Dysregulated Lithiome

Earlier, we discussed how GSK-3 is active even in a resting state, without external stimuli (referred to as "constitutive activity")—a circumstance detrimental to health and thus unusual for a supposedly natural bodily process. In a comprehensive 2022 review article on hyperactive GSK-3 as a tumor activator in various forms of blood cancer, the authors summarized the problematic role of constitutively active GSK-3: "Over the last three decades, GSK-3 has also emerged as a kinase clearly implicated in the pathogenesis [the onset and development of a disease] and progression of a wide spectrum of human disorders, including cardiovascular disease, type 2 diabetes, chronic inflammation, bipolar disorder, neurodegenerative diseases (e.g., Alzheimer's disease, Parkinson's disease, Huntington's disease), and cancer."[46] However, under a more species-appropriate lifestyle that includes essential lithium, GSK-3 activity would be reduced; as a result, not only these diseases but also many other lucrative ones would lose economic significance.

Lithium Targets Dysregulated by Lithium Deficiency, Such as the (Constitutively) Highly Active GSK-3, Offer the Ideal Basis for the Development of "Magic Bullets"

A review article published in *Neuropsychopharmacology* in 2010 states: "Mood disorders may result in part from impairments in mechanisms controlling the activity of GSK3 or GSK3-regulated functions [. . .]. This substantial evidence supports the conclusion that bolstering the inhibitory control of GSK3 is an important component of the therapeutic actions of drugs used to treat mood disorders and that GSK3 is a valid target for developing new therapeutic interventions."[47]

In other words: instead of recognizing the natural lithium deficiency as the cause of the dysregulation and compensating for it, the pharmaceutical industry is concentrating on specifically inhibiting GSK-3 with medication. This strategy follows the typical pattern of not addressing a problem caused by a nutrient deficiency at its root, but rather exploiting it as a market opportunity. By presenting GSK-3 as a "valid target" for new drugs, the need for lithium as an essential trace element is deliberately obscured—in favor of patentable active ingredients that attempt to artificially mimic the same effect, often with side effects. *However, investments in the development of "magic bullets" against lithium targets only make sense as long as they don't have to compete with essential lithium, the "natural bullet."*

However, these pharmaceutical "bullets" are by no means as magically precise as their name promises. At best, they are rich in side effects and symptomatic—and even then, with very little chance of success, as a 2022 review article explains:

> It is not surprising that tremendous efforts have been invested in developing GSK-3 inhibitors as potential drugs for treating neurodegenerative and psychiatric disorders (as well as other indications not described here). Over the past two decades, a large number of highly diverse classes of GSK-3 inhibitors have been developed by pharmaceutical companies, and academic institutions. [. . .] An updated search clearly indicates that the field is 'exploding' with a growing number of new scaffolds and molecules discovered to be potent GSK-3 inhibitors. [. . .] Taken together, the remarkable increase in the number of GSK-3 inhibitors developed, undoubtedly reflects the great interest in, as well as the hope of, identifying GSK-3 inhibitors suitable for clinical use. To date, none of the GSK-3 inhibitors tested has reached the market.[48]

In short: so far, not only has the natural solution been suppressed, but a pharmaceutical alternative has been sought in vain. However, this also means that a lithium target such as GSK-3 (and this logically applies to all targets of the Lithiome) only serves the pharmaceutical industry as a promising pharmaceutical target for the development of a "magic bullet" as long as it is dysregulated due to lithium deficiency. This is the case as long as lithium is not recognized as an essential trace element and adequately supplemented. Therefore, it must be in the interest of a profit-oriented pharmaceutical industry to counteract such recognition with all the means at its disposal. From a profit-oriented point of view, the ban on lithium supplements, which is completely nonsensical from a health perspective, offered an almost irresistible financial opportunity that opened up huge markets for the pharmaceutical industry and was therefore enforced without hesitation using perfidious means. All large (and many small) pharmaceutical companies will lose the race to bring "lithium mimetics" (i.e., chemical compounds that bind to the same receptor as lithium) to market as new drugs at the very moment that lithium is recognized as essential—or at least the ban on bringing lithium to market as a dietary supplement is lifted. However, the question is: How exactly did it even happen that an element which, when consumed in trace amounts, is proven not to be harmful but even beneficial to health, was banned as a dietary supplement almost worldwide?

"Salting"—Or, How an Essential Trace Element Was Banned

Lithium was part of a healthy diet until the mid-twentieth century. For the reasons discussed above, this trace element had to be removed from the market and discredited as a foodstuff before it could be classified as essential. The problem to be solved was "the health benefits and curative powers of naturally occurring lithium in water," which had been "known for centuries," as stated by the authors of the review article "Lithium in drinking water linked with lower suicide rates."[49] Lithia Springs in Georgia, once a sacred Native American healing spring, was renowned for its naturally lithium-rich water and its health-promoting properties. The water from a spring in Londonderry, New Hampshire, was also marketed as Londonderry Lithia Water, citing numerous health benefits (see book cover from 1891, Fig. 31). As a 1903 label indicates, doctors recommended the lithium-containing beverage "for rheumatism, gout, neuralgia [chronic nerve pain], diabetes, Bright's disease [chronic kidney inflammation], eczema [skin rashes], kidney stones, catarrh [cold symptoms], and various stomach complaints."[50] These are almost all conditions for which essential lithium actually helps, as we've seen in previous chapters. Well over a hundred years ago, medical treatment was, in this regard, more advanced than it is today. In contrast to the high-dose treatment of bipolar disorder, which was only developed half a century later and can lead to kidney failure, lithium from this natural source was even considered helpful for kidney disease. A spectroscopic analysis in 1891 found that one imperial gallon (about 4.55 liters) of Londonderry Lithia Water contained about 8.620 "grains" of lithium bicarbonate.[51] One grain equals approximately 0.0648 grams, so the spring water contained 23 mg of lithium per liter. One or two glasses (about 0.2 liters each) therefore corresponded to a low-dose treatment of about 5 to 10 mg/day (see Chapter 3). The English physician Sir Alfred Baring Garrod (1819–1907), who discovered the cause of gout in 1848 and was the first physician to propose lithium as a remedy for this disease, had already established dosage guidelines back then.[52]

At the beginning of the twentieth century, lithium-containing beverages were so popular due to the trace element's then-known, diverse effects that small amounts were added to various drinks still known today. For example, the soft drink 7UP was originally called *Bib-Label Lithiated Lemon-Lime Soda*. It was launched in 1929 and, until 1950, contained about 5 mg of lithium in the form of lithium citrate per liter (see Fig. 33).[53] A 0.2-liter glass would provide the RDA of 1 mg of lithium. Incidentally, it's speculated (the man who named it took his thoughts to his grave) that the "7" in 7UP refers to lithium's atomic mass, and "Up" to the elevated mood it supposedly brought. The soft drink was marketed as a remedy for alcohol intoxication, or "hangover"—a logical association, as alcohol leads to neuroinflammation, which lithium alleviates.[54] Even an early version of Coca-Cola was available at soda fountains in pharmacies. *Lithia Coke* was a mixture of Coca-Cola syrup and lithium-containing spring water. However, the emerging pharmaceutical industry, which focused on patentable allopathic medicines, must have found ubiquitous lithium (see Fig. 34) a tremendous thorn in its side. At least until 1949, when the U.S. Food and Drug Administration (FDA) issued a ban on the use of lithium in soft drinks and beer—just a few months before John Cade published his suggestion that lithium could be essential—and the 7UP formula had to be changed. But on what basis? How was it that the public simply accepted this ban?

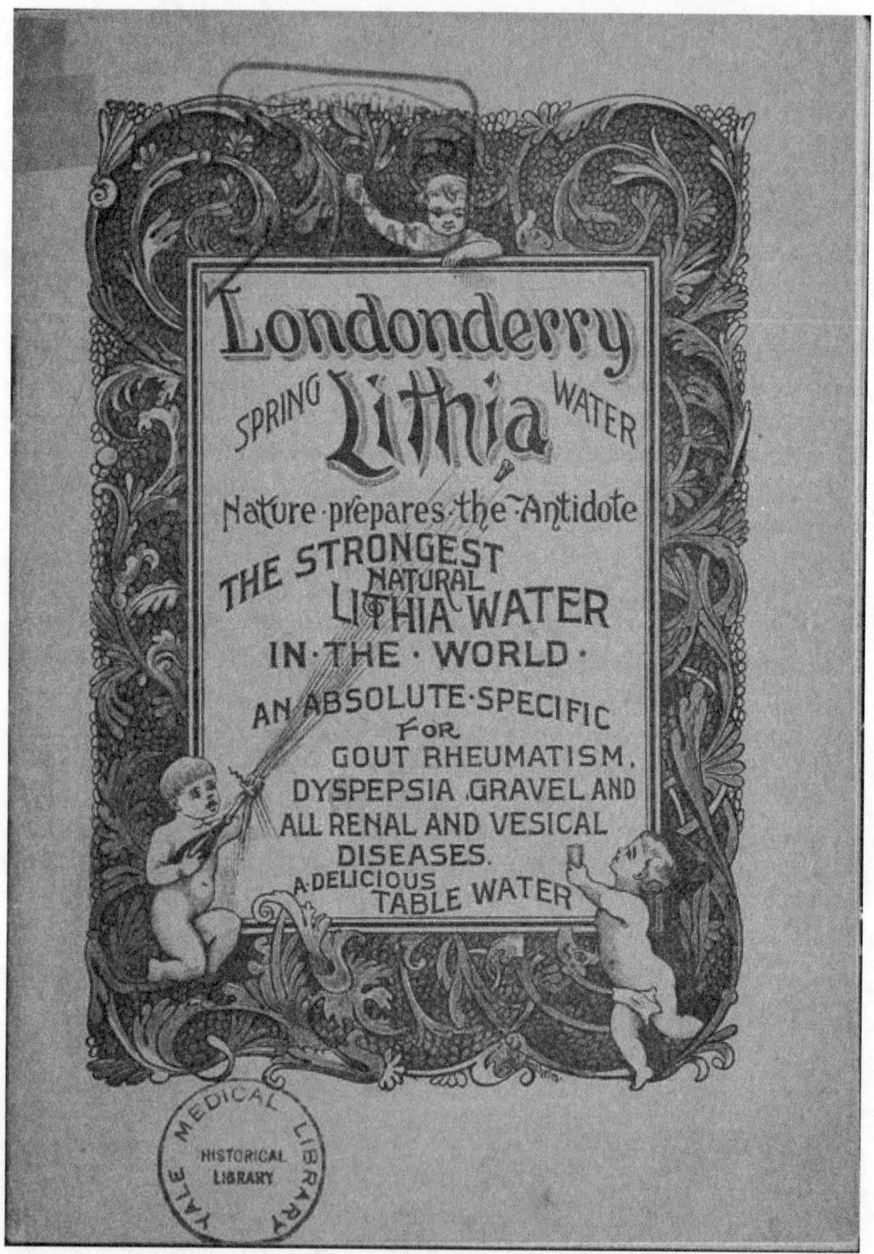
Londonderry
SPRING Lithia WATER
Nature prepares the Antidote
THE STRONGEST NATURAL LITHIA WATER
IN·THE·WORLD·
AN ABSOLUTE·SPECIFIC
FOR
GOUT RHEUMATISM,
DYSPEPSIA, GRAVEL AND
ALL RENAL AND VESICAL
DISEASES.
A·DELICIOUS
TABLE WATER
MEDICAL
HISTORICAL
LIBRARY
YALE
LIBRARY

Figure 32

Figure 33

Figure 34

The ball started rolling and the public became aware of the unfolding crisis when the US newsmagazine *Time* reported on February 28, 1949, about mysterious deaths linked to a new lithium-based salt substitute. A study published shortly thereafter explained the product's introduction: "As restricting sodium intake for heart failure and other conditions gains importance, a safe salt substitute for seasoning food was sought. Recently, a 25 percent lithium chloride solution, similar in taste to table salt, was launched. Although older literature contained isolated warnings that inorganic lithium salts could cause weakness, tremors, and visual disturbances, the solution was marketed as 'absolutely safe,' though its use 'under medical supervision' was recommended."[55] The conflicting claims of "absolutely safe" and "only to be used under medical supervision" should have immediately raised doubts about its unconditional safety. What was the rationale behind requiring medical oversight? Clearly, there was at least an awareness that lithium salt intake was far from "absolutely safe," contrary to the advertising. Perhaps they even wanted doctors to report the anticipated problems? This, at least, happened very quickly. The historical *Time* article continued: "Last spring the Foster-Milburn Co. of Buffalo thought it had found something harmless that would give food a salty flavor. The new product, Westsal®, contained lithium chloride (table salt is sodium chloride)."[56] However, by early February 1949, "doctors at a Manhattan hospital suspected that the substitute salt might have played a part in the death of a patient with heart disease." According to the *Time* report, the FDA "began experimenting, and found that heavy doses of lithium chloride killed laboratory animals." Subsequently, the FDA reportedly investigated human patients who had used lithium chloride as a salt substitute and discovered "that they were suffering variously from drowsiness, weakness, loss of appetite, nausea, tremors, blurred vision, unconsciousness."

The *Time* article suggests the lithium poisoning was accidental—perhaps some consumers simply used too much of the new table salt—and that the FDA only began investigating the issue in patients and animals after the deaths became known. But this is demonstrably not entirely correct; at least in the case of the credibility of the report by the medical team at Bellevue Hospital (Manhattan, New York), published in the *Journal of the American Medical Association* (*JAMA*) on March 12, 1949—i.e. twelve days after the *Time* article.[57] According to this report, the doctors *deliberately* administered extremely high doses of lithium chloride to a seriously ill heart patient, although they said they were aware of lithium's toxicity. In their *JAMA* article, they cite several reports of toxic effects dating back to the nineteenth century. These include a physician's self-experiment with an overdose, also published in *JAMA* in 1913, which we discussed in Chapter 3. As you'll recall, that physician had taken an enormous 1.3 g of lithium over twenty-eight hours and was on the verge of a coma due to severe toxic effects. However, the doctors at Bellevue Hospital were far more liberal with their patient, who later died, allowing him to ingest many times that amount. This was done despite their prior knowledge and their own observations of three other patients that clearly demonstrated the life-threatening nature of such high lithium doses. They soberly wrote: "On the fifty-first day of his hospitalization, the patient [suffering from severe heart failure] was given lithium chloride and instructed to flavor his food with it. He found this substance to be a pleasant substitute for salt and used it liberally. This equates to roughly 6 g, or 6,000 mg, of pure lithium (!), an intake amount that was evidently closely monitored. "On the morning of the third day after starting to take lithium, he complained of general weakness but had no other symptoms. The next morning, the weakness was so severe that he could no longer hold his eating utensils." Lithium was then discontinued, and for the next three days, the patient's decline into a coma and eventual death were meticulously documented. His blood was also regularly tested for lithium: "The highest lithium level in the blood was 4.8 [mmol/l] on the day the lithium chloride treatment was discontinued. Three days later, one day before death, the lithium level in the blood was 3.0 [mmol]." These values are clearly in the lethal range, as shown in Fig. 22 in Chapter 3. Why this patient (and the three others) was effectively intoxicated with lithium is not clarified in the article. However, it cannot be considered an oversight or medical "malpractice." Given the existing knowledge, it was evidently a deliberate act that, astonishingly, could be published without penalty.

Even more astonishingly, this wasn't the only article in that particular issue of *JAMA* detailing intentional lithium intoxication. It was one of three (supposedly) independently published submissions—though it would take compelling reasons to believe in such independence and rule out obvious collusion or a guiding hand orchestrating these events. Researchers at the Cleveland Clinic Foundation (Case Western Reserve University) report seven cases of medically supervised lithium poisoning with the "absolutely safe" (or in high doses, certainly fatal) saline substitute: "The purpose of this report is to direct attention to lithium intoxication as a complication of the use of lithium chloride as a salt substitute for flavoring in the course of low-sodium diets. We describe the syndrome as observed in 7 cases, of which in 2 cases intoxication appears to have been a contributory cause of death. The salt substitute used by these patients was westsal,* which is a solution of [25 percent] lithium chloride with citric acid and a small amount of potassium iodide. That the cause of the intoxication was lithium ion, and not sodium depletion as such, seems evident from a study of these cases, in 1 of which the intoxication was induced as a clinical trial."[58] Again, the authors brazenly admitted this was a clinical trial, meaning they deliberately poisoned their patients—despite, it cannot be stressed enough, knowing the life-threatening risks of such high doses. However, in the third *JAMA* article, from the Cedars of Lebanon Hospital, Los Angeles, physician and author Robert L. Stern attempted to evade responsibility by claiming: "At the time of the event, the possible toxic effects of lithium chloride were not known."[59] Similarly, *Time* argued: "Whether the symptoms were due to the patients' diseases or to the lithium chloride, no one could positively say—at the time." This claim—that no one knew about the dangers of an overdose—reads like a desperate defense, especially since the other two reports clearly demonstrate that this knowledge was, in fact, available!

Some obvious questions immediately arise: How likely is it that three clinics would, practically simultaneously and independently, decide to overdose severely ill people with lithium, under the completely implausible pretense that it was a healthier alternative to table salt in their diet? How likely is it that this would then be published concurrently in three independent articles in the same journal? Were these clinical "studies" orchestrated? While these crucial questions were conspicuously absent from the *Time* article, it did contain, in my opinion, a highly revealing hint about a plausible motivation. According to *Time*'s investigation, the FDA was "planning to reclassify lithium chloride as a drug instead of as a special dietary food, FDA heard of the deaths." Miraculously, these three studies conveniently supported the FDA's objective, enabling them to publicly warn against lithium via *Time*: "Stop using this dangerous poison at once."

However, any trace element, even an essential one, can be lethal if its intake exceeds a thousand times the RDA. The same tactic could be used to discredit iodine, selenium, zinc, and other supplements and have them banned. Yet, at that time, only lithium was targeted. Strangely, doctor-prescribed high-dose lithium treatments were not banned—a few months later, these treatments would be popularized by John Cade's publication, which we've cited repeatedly. Absurdly, the ban applied exclusively to the seemingly harmless, low-dose use of lithium as a dietary supplement. This had significant repercussions. A key selling point for 7UP and other beverage manufacturers' lithium-containing products vanished. They were forced to remove the very ingredient that had contributed to the popular, beneficial effects on consumers. In a perverse way, deliberately tolerated, media-staged victims were used to intentionally fuel fear of lithium. Professor Hart, in his extensively discussed "Hormesis" article, also touches on these events without delving into the underlying reasons: "It soon became apparent that such an approach [replacing sodium chloride or table salt in the diet of seriously ill people with high-dose lithium chloride] was toxic, and it led to multiple deaths. This led the FDA in the USA to declare LiCl[lithium chloride] a poison, and the practice was discontinued."[60] Three victims, who died from overdoses deliberately induced by their doctors, were enough to condemn lithium. But we must ask how many people have unnecessarily suffered and died since then because lithium, as a trace element, fell into disrepute due to this maneuver. The deliberate discrediting and unjustified demonization of lithium—instead of its recognition as an essential trace element, as John Cade hypothesized in that same year—has needlessly cost countless individuals their mental health and, consequently, their quality of life. The fact that 7UP and other popular beverage brands had to remove microdosed lithium from their formulas further solidified the public's fear of lithium. For those unaware of the deeper context, the idea that this happened without cause must seem utterly preposterous. As Hart writes, this also impacted clinical practice, at least in the US, for a period: "The second result from this [supposedly] failed [but, from the pharmaceutical industry's perspective, very successful] 'experiment' was that there was a reluctance to use Li salts in the treatment of bipolar disorder in the USA, although it was widely studied and used in Europe." Perhaps this too wasn't entirely unintended, because even then—as is still the case today—there was hope that an allopathic pharmaceutical would replace lithium.

After all, lithium was only approved and used to treat bipolar disorder because, apart from the cruel lobotomy, no other alternative existed. "More recently, prefrontal leukotomy has been performed on agitated and psychopathic mental patients to restrain their restless impulses and uncontrollable mood swings," Cade writes, "Probably lithium medication would be preferable to leukotomy."[61]

No Lithium Ban in the USA? The fact that lithium-containing dietary supplements like lithium orotate are not (or no longer) banned in the USA today might be an administrative oversight. As a *Lancet* review states: "In the early 1990s, the US Food and Drug Administration (FDA) considered introducing stricter standards for vitamins, minerals and other food additives that could have harmful effects," which led to the passage of the Dietary Supplement Health and Education Act (DSHEA) in 1994.[62] Among other things, the DSHEA stipulates that dietary supplements are to be classified as foods, not drugs. This means they no longer require FDA approval, provided certain guidelines are met. For example, before the FDA can intervene or prohibit its sale, a dietary supplement must first be proven to be harmful; this is not possible with daily doses of between 1 and 5 mg of pure lithium in the various salt compounds. Responsibility and liability rest with the manufacturer; as long as no harm comes to anyone, the supplement can remain on the market. In this respect, these "stricter" standards have paradoxically ensured lithium's availability as a dietary supplement. This path was blocked for the trace element in the EU due to opposing legal regulations (see in the Additional Information above: *Statement on the legal situation of lithium as a food supplement in Europe*) However, this doesn't mean there haven't been renewed attempts to disparage lithium or lithium orotate as a dietary supplement in the USA. In 2007, for instance, a case report was published about a woman who voluntarily took eighteen tablets, each containing 120 mg of lithium orotate, totaling approximately 68.9 mg of pure lithium "The patient complained of nausea and reported one episode of emesis. Her examination revealed normal vital signs.[. . .] After 3 hours of observation, nausea and tremor were resolved."[63] Why this case was even published is highly curious, as it's barely distinguishable from the countless reported side effects of lithium treatment for bipolar disorder. Yet, the authors' general conclusion allows for a certain inference: "Over-the-Internet dietary supplements may contain ingredients capable of causing toxicity in overdose. Chronic lithium toxicity from ingestion of this product is also of theoretical concern."

The ban on lithium as a dietary supplement is unlikely to be the result of incredible coincidence or a puzzling misinterpretation of circumstances. Rather, it appears to have been a deliberate action aimed at hindering its utility in the prevention and treatment of numerous diseases. This played directly into the hands of conventional medicine, largely established by figures like Rockefeller, which exclusively favored allopathic pharmaceuticals. Once you make this connection, it becomes crystal clear why a brain-toxic trace element like aluminum is permitted as a dietary supplement at 10 mg (EU) to 20 mg (WHO) per day, while the brain-healthy trace element lithium remains banned in most countries. The systemic abuses revealed here should serve as a wake-up call for everyone concerned about their own health and that of their children. Just imagine if iodine-containing food supplements, like iodized salt, were banned. Countless children would become cretins, as was indeed the case for a long time (and, tragically, still is in many parts of the world; see Additional Information in Chapter 2, Argument 6: *Is Mental Health Undesirable?*) Although a lithium deficiency of the kind typically seen in humans may not harm overall brain development as much as an iodine deficiency, its effects on the mental immune system are equally serious! This problem persists and underscores the urgent need for a fundamental systemic change. For many concerned individuals, this means first grappling with the comprehensive historical analysis I'm presenting to you here.

Lithium: The Stumbling Block Before Species-Appropriate Medicine

The inseparable link between mental and physical health was a cornerstone of Hippocratic teaching. Consequently, physical activity was considered vital for both bodily and mental well-being, just as a healthy diet formed a crucial foundation. Even music and theater were employed to treat illnesses and improve human behavior. Humanity was viewed as part of nature, and health care operated within a holistic framework. Prevention was paramount in medical practice—not solely because therapeutic options were limited, but because, then as now, it's easier to stay healthy than to become healthy. Thus, it's wiser to preserve your health than to rely on medicine once it's lost. Today, acknowledging all essential nutrients forms the basis for consciously maintaining physical and mental health. In this regard, recognizing lithium's essentiality is a decisive and indispensable step toward a species-appropriate medicine. After all, "essentiality" isn't an official award; it's the definition of a biological necessity. Nature has the first and last word here. Health authorities and global NGOs like the WHO, which champion improving human health, are obligated to defer to this natural truth. However, by presuming to ignore the evidence for the indispensability of essential trace elements like lithium, and thereby denying nature itself, we must ensure our voice is heard. Now that this essentiality has been proven in countless studies, we stand on firm ground to take a significant step against these self-proclaimed "authorities" and expose their harmful influence. The ban on making essential amounts of lithium readily and affordably available to everyone through dietary supplements should be a wake-up call and a summons to nothing less than a revolution in medicine. (The fact that, as of spring 2025, the forbidden sale of lithium salts as dietary supplements in some European countries is tolerated by authorities doesn't change the fundamental problem). The Latin word *revolutio* literally means "turning back." In the context of a revolution in medicine, it means restoring a natural, species-appropriate understanding of human nature—an understanding that existed before unscrupulous profiteers, obsessed with ruthless greed, wreaked havoc on the world. This isn't about backward-looking romanticism or a simplistic "everything used to be better," but about a "forward to the roots." It's about a future where we preserve what works while leaving behind everything that harms us and doesn't serve us.

In his 2002 article "Healthcare System: Trapped in Progress," published in the *Deutsches Ärzteblatt,* German psychiatrist, medical, and psychiatric historian Klaus Dörner (1933–2022) elucidated the problem of a medical system driven by profit and market power: "Competition forces the development of new markets. The aim must be to transform all healthy people into sick people—that is, into individuals who, for as long as possible, consider themselves in need of therapeutic, rehabilitative, and preventive manipulation by experts, both chemically, physically, and psychologically, in order to [supposedly] 'live healthy.'"[64] Dörner acknowledged the excellence of advanced acute medicine, both pharmaceutical and surgical. However, he argued that its fundamental concept must fail in the case of chronic illnesses, where the goal is to address the chronic underlying causes. This demands a completely different approach to medical thought and action—indeed, a fundamentally different cultural evolution toward a deeper understanding of human nature and its inherent needs. Dörner lamented the capitalist penetration of the entire medical system, which is in fact not a health-care system, but a lucrative disease system—a system that systematically makes us sick and keeps us sick: "All the described trends, which subjectively aim to promote health but in reality drain vitality from society, become even more destructive the more they are left to marketing and competition." He wasn't against competition per se:

> These principles are beneficial in the rest of the economy, but deadly in the social sector and thus in healthcare (perhaps with the exception of some areas). If health becomes a service and thus a commodity, if every medical institution is condemned to maximize profits by expanding its services, [. . .] then it should come as no surprise that artificial needs are eventually invented that promise to satisfy the customer's wishes, [. . .], that immature products and procedures are rushed to market, and that there is a tendency to retain and "milk" good customers for life.

Therefore, a crucial interim goal must be to eliminate predatory capitalism from the health-care system and completely sever the toxic influence of corporations on politics and media. People do not need advertising for their products, rather education, if possible when in school, about the Fundamental Laws of Life (see Chapter 1).

Arguments for this perspective are readily available: While preventable diseases are rapidly increasing, an estimated 882,000 people die each year in the US alone from the consequences of taking prescription drugs—making it the leading cause of death. This includes psychotropic drugs (the third leading cause of death) and painkillers.[65] These figures alone demonstrate that conventional medicine is failing and itself becoming a health hazard. However, the necessary change won't happen on its own. We'll be disappointed if we merely hope that globally operating corporations—who, with their industrial power, aim to control every detail of our lives—will abandon their profitable business with our health and suddenly act for our benefit. This fundamentally contradicts their perverted business model. If you want change, you have to take action! With a functioning mental immune system, we are ideally equipped to recognize such dangers in the macro world and protect ourselves and our children from them. We must not allow vaccines to cause more harm than good to infants, or even, as with the latest mRNA "vaccines," to genetically modify them. It is indefensible that not a single vaccine has ever been thoroughly tested for side effects; that we therefore don't know how truly dangerous they are, yet we tolerate injecting infants with such inadequately tested chemicals—as if no conceivable alternative exists (see *"Turtles all the way down: The science and myth of vaccination*).[66] The authors of a vaccination study published in 2025 succinctly summarize the issue: "Vaccine science has focused on protection against specific pathogens and specific vaccines, while the overall impact of vaccination programs on children's health has remained unexplored. The results of this study add to a growing body of evidence that raises concerns about the safety of the current vaccination program and its potential contribution to increasing rates of neurodevelopmental disorders."[67] We must not ignore or accept the numerous long-term harms that continue to be reported (see Chapter 4 on autism). Instead, we must stand united as a society to ensure that modern medical practice does not pose such a threat to us and our children. Individual concerned parents, on their own, stand little chance against the insidious power system that has secured control over our health and that of our children through vaccination passports. This system profits financially at our expense and—far more seriously—damages our long-term mental health and social cohesion.

Let's summarize and take this a step further: Modern medicine has made enormous strides, particularly in acute pharmacological treatment and surgical traumatology. Yet, it remains unable to solve the vast and ever-growing problem of so-called diseases of civilization. The reason lies in a fundamental, global systemic flaw: multicausal problems cannot be solved with a single-cause approach, and certainly not through symptomatic interventions. These interventions actually worsen the problem by merely masking it. Additionally, this creates an incentive to ignore, deny, or even promote the root causes out of profit-driven self-interest. The solution can only be found in a multicausal approach: guidance toward a return to a species-appropriate way of life under modern conditions and with modern methods. I've attempted to outline what such a system might look like and what to consider in Chapter 4 (see *Systemic Prevention and Therapy of Multicausal Chronic Diseases*). Furthermore, special priority must be given to fostering a health-conscious society where personal responsibility is encouraged and supported. Healthy child development and a healthy life require stable homeostasis based on fundamental natural laws. If essential vital substances like lithium are withheld, it leads to long-term illness—with severe consequences for society. What's more, since these developmental disorders and illnesses stem from dysregulated targets of this essential vital substance, these very vital substance targets can be misappropriated to become targets for drugs with numerous side effects. This paradoxically benefits those who are partly responsible for the deficiency. The consequences: psychological deficits, unnecessary and excessive illnesses, countless medications with side effects (i.e., substances that trigger new treatable diseases), and an enormous financial burden on all affected. Since these problems result from an inappropriate lifestyle, a fundamental restructuring of the current health system and a focus on the main causes are necessary.

Returning to species-appropriate living conditions within a modern context is crucial for preventing (and treating) multicausal diseases of civilization. Healthy behavior and recovery cannot be delegated; they are the responsibility of each individual. The aim of any new health policy program should therefore be to guide the population toward this self-understanding and empower them to take personal responsibility, without being patronizing (you can find a more detailed discussion under "Info" on my website).[68] Even though people are capable of learning, change takes a long time. The indoctrination that leads many to see themselves as drug-dependent "misconstructions" must be overcome. Instead, it's about correcting our self-perception so that the natural interaction between humans (as part of nature, not separate from it) and nature becomes conscious and tangible again. In this light, it becomes self-evident that humans have natural and essential needs, without the satisfaction of which health is impossible. There is hope that more and more people will develop a lost awareness of a more natural way of life, if only we promote the skills for self-healing and maintaining health—under species-appropriate conditions—in both schools and adult education. According to the Law of the Maximum, this also includes teaching genuine media literacy to distinguish facts from opinions and recognize manipulative representations. Change will not come from "above" but, like almost all revolutions, from the grassroots (or "from below")—from the majority of people who recognize that there is no good alternative to a natural, species-appropriate life that respects fundamental human needs. And this hope is growing: At least since the COVID-19 events, people are becoming more aware of national, global, systemic, and specific health grievances.

Thus, one of the biggest crises is also becoming the biggest opportunity for change. Ultimately, this could naturally lead to a cultural shift and even contribute to a more peaceful global community.

Lithium: A Symbol for a More Peaceful Future?

They say without health, you have nothing, but health isn't everything. Humans need community, space for creativity, and, crucially, peace. Yet, just as health is more than the absence of illness, "Peace is not an absence of war, it is a virtue, a state of mind, a disposition for benevolence, confidence, justice," as the Dutch philosopher Baruch Spinoza (1632–1677) once recognized. The American journalist Dorothy Celene Thompson (1893–1961) offered a similar insight: "Peace is not the absence of conflict but the presence of creative alternatives for responding to conflict—alternatives to passive or aggressive responses, alternatives to violence. Peace must be created to be maintained." In brief: peace is a lifelong, proactive process that engages every function of our mental immune system.

When I was invited by Swiss Foreign Minister Ignazio Cassis to deliver a keynote speech at the International Cooperation Forum (IC-Forum) 2024 on April 12, 2024, to approximately 1,500 representatives from governments and non-governmental organizations (NGOs) across over a hundred nations, I was asked what the brain needs to think peace. My—to me, obvious—answer was: a healthy mental immune system, which, in turn, is inconceivable without essential lithium.[69]

The profiteers of this agenda have been aware of this longer than we have. Their intentions are brazenly obvious; they have confidently unveiled their plans themselves, and many even proudly display their ideological allegiance to this future blueprint. COVID-19 was the prelude to a final upheaval, designed to implant a new antisocial operating system into the minds of humanity, so that people would willingly accept their increasing loss of freedom (without which peace is impossible). It was, therefore, far more than just a cleverly orchestrated medical crisis. It was, in essence, a central part of a global political agenda that its protagonists have not abandoned—quite the contrary: we still face this threat. A brief look at events since 2020 shows how we are being steered toward total dependence and control by a global power complex, in accordance with the goals of Agenda 2030. I've analyzed this in detail in my book *The Indoctrinated Brain*.[70] The ruthless instrumentalization of health issues reveals the system's utter perfidy: the pharmaceutical industry did everything in its power to conceal the option of causally remedying winter vitamin D3 deficiency as a natural prophylaxis against life-threatening infections. Humanity was deliberately left unprotected against respiratory viruses, including artificially created ones, in order to present an artificial solution to self-made problems. The entire population was meant to submit to life-threatening genetic manipulation in the form of an mRNA injection. With a sufficient vitamin D level, an infection with SARS-CoV-2 would have been completely harmless (See Additional Information, Chapter 1, *Vitamin D Deficiency—A Deadly Example of Ignoring the Law of the Minimum.*), but the solution presented to us—genetically modifying humanity—created an even greater problem.[71] The manipulation of the masses was skillfully orchestrated, well-conceived from the outset, and executed according to a master plan. Independent French scientist Fabien Deruelle reconstructed this plan in a 2022 peer-reviewed article as follows:

> Since the beginning of COVID-19, we can list the following methods of information manipulation which have been used: falsified clinical trials and inaccessible data; fake or conflict-of-interest studies; concealment of vaccines' short-term side effects and total lack of knowledge of the long-term effects of COVID-19 vaccination; doubtful composition of vaccines; inadequate testing methods; governments and international organizations under conflicts of interest; bribed physicians; the denigration of renowned scientists; the banning of all alternative effective treatments; unscientific and liberticidal social methods; government use of behavior modification and social engineering techniques to impose confinements, masks, and vaccine acceptance; scientific censorship by the media.[72]

His conclusion is stark:

> By supporting and selecting only the one side of science information while suppressing alternative viewpoints, and with obvious conflicts of interest revealed by this study, governments and the media constantly disinform the public. Consequently, the unscientifically validated vaccination laws, originating from industry-controlled medical science, led to the adoption of social measures for the supposed protection of the public but which became serious threats to the health and freedoms of the population.

Indeed, the "war" against an artificial virus was an operation against humanity executed with military precision, as Dutch Minister of Health Fleur Agema publicly revealed on November 24, 2024, before cameras: "COVID-19 was a NATO military operation."[73]

The neurological assault on the world's mental immune system, via genetically modified spike protein, originated from the medical sector—but it was by no means the only vector of attack. Deruelle points to further developments that will continue to psychologically and mentally impair us and our children. In a 2024 article titled "Microwave radiofrequencies, 5G, 6G, graphene nanomaterials Technologies used in neurological warfare," we read: "Scientific literature, with no conflicts of interest, shows that even below the limits defined by the International Commission on Non-Ionizing

Radiation Protection, microwaves from telecommunication technologies cause numerous health effects: neurological, oxidative stress, carcinogenicity, deoxyribonucleic acid and immune system damage, electro-hypersensitivity. The majority of these biological effects of non-thermal microwave radiation have been known since the 1970s."[74] Yet, there is no interest in protecting our health—quite the opposite: "Despite reports and statements from the authorities presenting the constant deployment of new wireless communication technologies, as well as medical research into nanomaterials, as society's ideal future, in-depth research into these scientific fields shows, above all, an objective linked to the current cognitive war. It could be hypothesized that, in the future, this aim will correspond to the control of humanity by machines."

What Deruelle describes here is the ultimate objective of Rockefeller's plan, which the technocratic power complex he helped create continues to pursue. Agenda 2030 is designed to usher humanity into a *Brave New World*, where a self-appointed elite—"world controllers" in the spirit of Aldous Huxley's dystopian novel—monitors an indoctrinated and effectively will-less humanity.

The Technocratic Power Complex Per John F. Kennedy (1917–1963) Speech to the American Newspaper Publishers Association on April 27, 1961 — a moment in which the U.S. President likely knew exactly what he was talking about:

"The very word 'secrecy' is repugnant in a free and open society; and we are as a people inherently and historically opposed to secret societies, to secret oaths and to secret proceedings. We decided long ago that the dangers of excessive and unwarranted concealment of pertinent facts far outweighed the dangers which are cited to justify it. [. . .] For we are opposed around the world by a monolithic and ruthless conspiracy that relies primarily on covert means for expanding its sphere of influence—on infiltration instead of invasion, on subversion instead of elections, on intimidation instead of free choice, on guerrillas by night instead of armies by day. It is a system which has conscripted vast human and material resources into the building of a tightly knit, highly efficient machine that combines military, diplomatic, intelligence, economic, scientific and political operations."[75]

Now, sixty-four years later, his nephew, Robert F. Kennedy, who became the U.S. Secretary of Health and Human Services on February 13, 2025, has both the opportunity and the mandate to rein in this power complex. As reported by the *Epoch Times* on February 13, 2025, "Kennedy has announced sweeping changes to the department. As part of his *Make America Healthy Again* initiative, he specifically aims to combat what he calls the chronic disease epidemic. To achieve this, he plans to curb what he views as excessive corporate influence on federal health agencies and ban toxic chemicals from the U.S. food supply. Kennedy is convinced that the health of Americans will hardly improve as long as the influence of large corporations on the FDA, the CDC, and the Department of Agriculture is not critically questioned."[76] Ever since the announcement that Robert F. Kennedy Jr. would take office, I have been watching closely to see whether the CDC would revise its most troubling vaccination guidelines, particularly the one that, as of March 27, 2025, still advised infants to receive multiple COVID "vaccinations" with genetically modified spike mRNA during their first year of life. That this recommendation no longer appears on the CDC website (as of July 2025) in that form is, in my view, a hopeful first sign. I sincerely hope these recommendations will soon be removed entirely, as anything less would fail to acknowledge the true extent of harm caused by this undeclared gene therapy. May Mr. Kennedy succeed in implementing his ambitious agenda—and in fulfilling the deep and diverse expectations placed in him by those who believe in integrity, transparency, and real health protection.[77]

Every war robs people of their freedom, depletes valuable societal resources, and further empowers the already powerful. Those who profit from conflict will inevitably view a peaceful society as a threat to their self-serving interests. Beyond the obvious profits the pharmaceutical industry reaps from lithium-deficiency-related diseases, others benefit from a suboptimal mental immune system, without which peace is difficult to imagine. Since the social immune system also suffers as a result, a more peaceful society is hard to envision in the presence of widespread lithium deficiency. Perhaps the recorded history of humanity is dominated by wars and mass destruction precisely because there has always been a lack of vital lithium since the "expulsion from the Garden of Eden"? The thinking and behavior of individuals with a weakened mental immune system, due to their reduced psychological resilience, have always been easily controlled by fear or fear-inducing narratives, even to the point of instigating the next war.

In the 2016 scientific article, "Is violence in part a lithium deficiency state?" the authors point to "interestingly, lithium, in trace amounts, as occurs in some drinking water, has been inversely related to aggression, and suicidal and homicidal violence."[78] By strengthening our mental immune system, lithium acts as an antidote to hatred, violence, and power-driven interests. Therefore, a political motive to deprive people of an essential micronutrient, to make them more pliable with war rhetoric, cannot be entirely ruled out. Without a functioning mental immune system, and thus without vital lithium, genuine social and inner peace are almost unthinkable. This trace element is far more than just the long-lost key to healthy mental development and a long life. I addressed the profound problem this presents for humanity in my speech at the

IC-Forum 2024: "The global community does not acknowledge the importance [of lithium.] One of my major projects for the next few years is to show the world that lithium [. . .] is essential for an efficient mental immune system." True, lasting peace, both on a personal level and on the global stage, requires creativity, courage, mental resilience, the ability to reflect and compromise, and, not least, the capacity for empathy—all functions of a healthy and efficient mental immune system. In this sense, lithium could be something akin to the sought-after "philosopher's stone," which, instead of turning lead into gold, might be capable of "refining" people in precisely the way so urgently needed today. Therefore—and I have no doubt—a global community where the majority possesses a healthy mental immune system would form the foundation for comprehensive positive change. If more people had a stable mental immune system, it could also create the basis for societal development where empathy, cooperation, and rational compassion play a greater role. Whether this will lead to global change, of course, depends on many other factors—but we at least know which biological prerequisite could play a key role and establish the fundamental conditions for such a transformation. So perhaps not much is missing, and that gives us hope. So maybe there isn't much left to do, and that gives us hope. Because perhaps what the French sculptor Jean-Élie Chaponnière (1801–1835) once put on paper is true.

Optimism is the true philosopher's stone that turns into gold whatever it touches.

ACKNOWLEDGMENTS

The many pieces of the puzzle that I have put together in this book with the aim of proving the essentiality of lithium were developed by numerous doctors and scientists, for which I am infinitely grateful, especially as in some cases it took considerable courage. In the case of the physician Hans Alfred Niepers (1928–1998), for example, his commitment to orthomolecular medicine (which Wikipedia calls "controversial")—he also addressed lithium as an essential trace element—led to him being publicly attacked. Influential circles knew how to prevent the Law of the Minimum, on which species-appropriate medicine is based, from being applied to humans.

I am also grateful to the pediatrician Dr. Christian Schellenberg, who repeatedly encouraged me to push forward with this book project with many moving and positive case reports from his own practice. My heartfelt thanks also go to Dr. rer. nat. Katrin Huesker and her medical colleague, Andrea Thiem, from IMD Berlin. They selflessly compiled and provided much of the data on the pharmacokinetics of lithium orotate needed for this book—data that were previously unavailable in the medical literature.

The fact that this book could be realized at all without having to refer to ultimately illegal sources of lithium orotate is thanks to the pharmacist Sabine Bäumer, who was the first to agree to serve prescriptions for lithium orotate in essential quantities; this also made the public aware of the absurdity of the current legal situation. My statements and assessments of the legal situation were thankfully reviewed by Dr. Brigitte Röhrig, a lawyer; with her support, I became clear about the path we must take to recognize lithium as an essential trace element. In this respect, I would also like to thank health expert Patric Heizmann, whose foundation provided financial support for the initial legal review of the legal situation. Like many other freethinkers at home and abroad, he has helped me a lot to make this essential knowledge available to countless people via social platforms. We would also like to take this opportunity to thank all the other unmentioned supporters.

My special thanks go to my editor, Corvin P. Rabenstein, who previously edited *The Indoctrinated Brain* with sensitivity and understanding. He and my wife, Sabine, have worked hard to ensure that the relationships presented in this book are as comprehensible as possible for the medical laypersonwhich was no easy task given my love of scientific detail. It is unavoidable that the descriptions remain complex in places; after all, the functioning of the lithium atom is a fascinatingly complex symphony of physical, physiological, biochemical and psychological processes which, in their orchestrated interplay, make our lives possible in an almost miraculous way.

Special thanks also go to David Ter-Avanesyan for his creative cover design. I would like to thank my longtime family friend Bettina Simonis (Dipl.-Biol.), Patrick Detta, my daughter Sarah, and Grit Graefe, who did the professional final proofreading.

Last but not least, I would like to thank my family for their support and understanding as I immersed myself for months in this project, which, set to prove the essentiality of lithium, is much more than just another nonfiction book. Without them, I wouldn't have been able to take this path.

IMAGE CREDITS

Figure 1: https://pixabay.com/ de/illustrations/lithium-atom-freigestellt-atomar-2784853/, "Saat des Lebens" Nehls, mod.

Figure 2: Nehls: https://commons.wikimedia.org/ wiki/User:DooFi, mod.

Figure 3: https://de.freepik.com/vektoren-kostenlos/schiesssport_23807144.htm, mod.

Figure 7: https://www. shotshop.com/stockphoto/dp59529179, mod.;

Figure 10: https://link.springer.com/article/10.1007/s00232-017-9998-2, mod.;

Figure 15: doi:10.1038/srep40726, mod.

Figure 16: https://link.springer.com/ article/10.1007/s00232-017-9998-2, mod.

Figure 19: @brgfx, https://de.freepik.com/vektoren-kostenlos/medizinische-handveranschaulichung_221749881.htm, https://pixabay.com/de/illustrations/lithium-atom-freigestellt-atomar-2784853/, mod.

Figure 23: https://commons.wikimedia.org/wiki/File:BurjKhalifaHeight.svg, mod.; Woman with Child, Designed by kjpargeter / Freepik; mouse, https://de.freepik.com/autor/freepik.

Figure 24: Nehls, https://de.freepik.com/freie-psd/erstaunliches-blaues-dna-helix-3dmodell-genetischer-code-wissenschaft-biologie_410549674.htm, mod., https://www.shotshop.com/stockphoto/dp59529179, mod.

Figures 26–27: Nehls, mod. nach https://publichealthpolicyjournal.com/vaccination-and-neurodevelopmental-disorders-a-study of-nine-year-old-children-enrolled-in-medicaid/.

Figure 28: doi: 10.2174/ 1567205011310010014. mod.

Figure 29: https://de.freepik.com/vektorenkostenlos/tornado-himmel-illustration_3886432.htm mod.

Figure 32: https://wellcomecollection.org/works/tqd6fp3r.

Figure 33: https://science.howstuffworks.com/environmental/earth/geology / lithium.htm.

Figure 34: doi: 10.3390/ijms19072143.

*Figures 4–6, 8–9, 11–14, 17–18, 20–22, and 25 were created by the author.

NOTES

ESSENTIAL OPENING REMARKS

1 Nyncke G: Weggefährten, *Gedanken und Aphorismen.* 1990; ISBN-13: 9783874090438.

2 Haage BD, *Alchemie im Mittelalter: Ideen und Bilder, von Zosimos bis Paracelsus.* Artemis & Winkler 1996, pg. 84.

3 M. Nehls, "Unified theory of Alzheimer's disease (UTAD): implications for prevention and curative therapy," *J Mol Psychiatry*, 2016, doi: 10.1186/s40303-016-0018-8.

4 C. Spuch et al., "Efficacy and Safety of Lithium Treatment in SARS-CoV-2 Infected Patients," *Front Pharmacol*, 2022, https://www.frontiersin.org/articles/10.3389/fphar.2022.850583/full.

CHAPTER 1

1 Allert-Wybranietz K, *Best Wishes.* New Gift Texts, Lucy Körner Verlag. 1982.

2 C. F. Kleisiaris et al., "Health care practices in ancient Greece: The Hippocratic ideal," *J Med Ethics Hist Med.* 2014, https://pmc.ncbi.nlm.nih.gov/articles/PMC4263393/.

3 Von Haller A, The Power and Mystery of Food: The Dramatic Discoveries of the Foundations of Life and Health. Unikat-Verlag Ingo F. Rittmeyer; 4th Edition, October 1, 1995, pg. 52.

4 Ibid. Pg. 43, 44.

5 Ibid. Pg. 53.

6 Schmiedel V, The microbe is nothing, the environment is everything! *Empirical Medicine* 2020, www.thieme-connect.com/products/ejournals/abstract/10.1055/a-1158-4256.

7 Nehls M, Herd health: The way out of the Corona crisis and the natural alternative to the global vaccination program. Mental Enterprises 2022.

8 Wendt G, *Carl Sprengel and the mineral theory he created as the foundation of the new plant nutrition theory*, Commission Publishing House Ernst Fischer 1950, 1st Edition, Pg. 129.

9 M. F. Luxwolda et al., "Vitamin D Status Indicators in Indigenous Populations in East Africa," *Eur J Nutr.*, 2013, 52:1115-1125. doi: 10.1007/s00394-012-0421-6.;

M. F. Luxwol-da et al., "Traditionally Living Populations in East Africa Have a Mean Serum 25-hydroxyvitamin D Concentration of 115 nmol/l," *Br J Nutr.,* 2012, 108:1557-1561. doi: 10.1017/S0007114511007161.

10 Nehls M, "Herd health: The way out of the Corona crisis and the natural alternative to the global vaccination program," *Mental Enterprises*, 2022; L. Borsche et al., "COVID-19 Mortality Risk Correlates Inversely with Vitamin D3 Status, and a Mortality Rate Close to Zero Could Theoretically Be Achieved at 50 ng/mL 25(OH) D3: Results of a Systematic Review and Meta-Analysis," *Nutrients*, 2021, doi: 10.3390/nu13103596.; M. Sartini et al., "Preventive Vitamin D Supplementation and Risk for COVID-19 Infection: A Systematic Review and Meta-Analysis," *Nutrients*, 2024, doi: 10.3390/nu16050679.

11 Sartini M, et al., "Preventive Vitamin D Supplementation and Risk for COVID-19 Infection: A Systematic Review and Meta-Analysis." *Nutrients*. 2024 Feb 28;16(5):679. doi: 10.3390/nu16050679.

12 A. Rubio-Casillas et al., "Review: N1-methyl-pseudouridine (m1Ψ): Friend or foe of cancer?" *Int J Biol Macromol.* 2024, doi: 10.1016/j.ijbiomac.2024.131427. Erratum in: doi: 10.1016/j.ijbiomac.2024.132447.; P. Nordström et al., "Risk of infection, hospitalisation, and death up to 9 months after a second dose of COVID-19 vaccine: a retrospective, total population cohort study in Sweden," *Lancet*, 2022, 399:814823. doi: 10.1016/S0140-6736(22)00089-7.

13 Nehls, M. *The Indoctrinated Brain: How to Successfully Fend Off the Global Attack on Your Mental Freedom*. Skyhorse Publishing, 2023.

14 https://michael-nehls.de/infos/biowaffe-gegen-das-ungeborene-kind/.

15 Ken Sakura, "Big Data Analysis Suggests COVID Vaccination Increases Excess Mortality Of Highly vaccinated North Temperate Zone and North Frigid Zone Countries" (2024)10.31219/osf.io/zv6j8.

16 J. Reboul et al., "The gene number dilemma: Direct evidence for at least 19,000 protein-encoding genes in C. elegans and implications for the human genome," *Nat Genet* 27 (Suppl 4), 82 (2001). https://doi.org/10.1038/87264; P. Amaral, S. Carbonell-Sala, F. M. De La Vega et al., "The status of the human gene catalogue," *Nature*, 2023, 622:41-47, doi: org/10.1038/s41586-023-06490-x.

17 S. R. Cummings & C. Rosen, "VITAL Findings: A Decisive Verdict on Vitamin D Supplementation," *N Engl J Med*, 2022, 387:368-370. doi: 10.1056/NEJMe2205993.

18 Nehls, M., *The Formula Against Alzheimer's*. Heyne 2018.

19 H. A. Bischoff-Ferrari et al., "Combined Vitamin D, Omega-3 Fatty Acids, and a Simple Home Exercise Program May Reduce Cancer Risk Among Active Adults Aged 70 and Older: A Randomized Clinical Trial," *Front Aging*, 2022, doi: 10.3389/fragi.2022.852643.; siehe auch hier: https://www.dr-schmiedel.de/helfen-omega-3-und-vitamin-d-doch-gegen-krebs/ (15.05.2022, zuletzt abgerufen am 8.12.2024).

20 P. C. Gøtzsche, "Prescription Drugs Are the Leading Cause of Death. And psychiatric drugs are the third leading cause of death," *Mad in America* 2024,

https://www. madinamerica.com/2024/04/prescription-drugs-are-the-leading-cause-of-death/.

21 J. Y. Ho, "Life Course Patterns of Prescription Drug Use in the United States. Demography," 2023, 60:1549-1579. doi: 10.1215/00703370-10965990.

22 M. Gandhi, *A Guide to Health*, CreateSpace Independent Publishing Platform, 2012, https://www.gutenberg.org/files/40373/40373-h/40373-h.htm, S. 12.

23 E. Yirmiya et al., "Phages overcome bacterial immunity via diverse anti-defence proteins," *Nature,* 2024, 625:352-359. doi: 10.1038/s41586-023-06869-w.

24 Nehls, M., et al., "Pillars article: new member of the winged-helix protein family disrupted in mouse and rat nude mutations," *Nature*, 1994. 372: 103-107. *J Immunol.* 2015, 194:849-853.

25 Nehls, M., *The Methuselah strategy: Avoiding what prevents us from growing older healthily and wiser.*

26 Nehls, M. *The Indoctrinated Brain: How to Successfully Fend Off the Global Attack on Your Mental Freedom.* Skyhorse Publishing, 2023.

27 https://www.nobelprize.org/prizes/economic-sciences/2002/summary/.

28 F. Crick & C. Koch, "A framework for consciousness," *Nat Neurosci,* 2003, 6: 119-126. doi: 10.1038/nn0203-119.

29 Nehls, M. *The Indoctrinated Brain: How to Successfully Fend Off the Global Attack on Your Mental Freedom.* Skyhorse Publishing, 2023.

30 M. Lahdenperä et al., "Fitness benefits of prolonged post-reproductive lifespan in women," *Nature*, 2004, 428:178-181. doi: 10.1038/nature02367.

31 K. Hawkes, "Colloquium paper: how grandmother effects plus individual variation in frailty shape fertility and mortality: guidance from human-chimpanzee comparisons," *Proc Natl Acad Sci* USA. 2010, 107:8977-8984, doi: 10.1073/pnas.0914627107.

32 Nehls, M., *The Alzheimer's lie: The truth about a preventable disease.* Heyne 2017, pg. 389.

33 L. J. N. Brent et al., "Ecological Knowledge, Leadership, and the Evolution of Menopause in Killer Whales," *Curr Biol.*, 2015, 25:746-750. doi: 10.1016/j.cub.2015.01.037.; S. Nattrass et al., "Postreproductive Killer Whale Grandmothers Improve the Survival of Their Grand offspring," *Proc Natl Acad Sci* USA. 116:26669-26673. doi: 10.1073/pnas.1903844116.

34 M. Gurven & H. Kaplan, "Longevity among hunter-gatherers: a cross-cultural examination," *Population and Development Review*, 2007, 33:321-365. https://www.researchgate.net/publication/4780476.

35 G. Kempermann et al., "Human Adult Neurogenesis: Evidence and Remaining Questions," *Cell Stem Cell* 23:25-30, 2018, https://doi.org/10.1016/j.stem.2018.04.004.

36 S. Simard. et al., "Spatial transcriptomic analysis of adult hippocampal neurogenesis in the human brain," *J Psychiatry Neurosci.*, 2024, 49:319-333. doi: 10.1503/jpn.240026.

37 M. Alonso et al., "The impact of adult neurogenesis on affective functions: of mice and men," *Mol Psychiatry*, 2024, 29:2527-2542. doi: 10.1038/s41380-024-02504-w.

38 E. P. Moreno-Jiménez et al., "Adult hippocampal neurogenesis is abundant in neurologically healthy subjects and drops sharply in patients with Alzheimer's disease," *Nat Med.*, 2019, 25:554-560. doi: 10.1038/s41591-019-0375-9.

39 M. Flor-García et al., "Unraveling human adult hippocampal neurogenesis," *Nat Protoc.*, 2020, 15:668-693. doi: 10.1038/s41596-019-0267-y.

40 E. Steiner et al., "A fresh look at adult neurogenesis," *Nat Med.*, 2019, 25:542-543. doi: 10.1038/s41591-019-0408-4.

41 M. K. Tobin et al., "Human Hippocampal Neurogenesis Persists in Aged Adults and Alzheimer's Disease Patients," *Cell Stem Cell.*, 2019, 24:974-982.e3. doi: 10.1016/j.stem.2019.05.003.

42 Moreno-Jiménez, "Adult hippocampal neurogenesis."

43 M. Nehls, "Unified theory of Alzheimer's disease (UTAD): implications for prevention and curative therapy," *J Mol Psychiatry*, 2016, doi: 10.1186/s40303-016-0018-8.

44 S. H. Choi & R. E. Tanzi, "Is Alzheimer's Disease a Neurogenesis Disorder?" *Cell Stem Cell*, 2019, 25:7-8. doi: 10.1016/j.stem.2019.06.001.

45 A. Montagrin et al., "The hippocampus dissociates present from past and future goals," *Nat Commun*, 2024, doi: 10.1038/s41467-024-48648-9.

46 J. Freund et al., "Emergence of individuality in genetically identical mice," *Science*, 2013, 340:756-759. doi: 10.1126/science.1235294.; S. Zocher et al., "Early-life environmental enrichment generates persistent individualized behavior in mice," *Sci Adv.* 2020, doi: 10.1126/sciadv.abb1478.

47 N. A. DeCarolis & A. J. Eisch, "Hippocampal neurogenesis as a target for the treatment of mental illness: a critical evaluation," *Neuropharmacology*, 2010, 58:884-893. doi: 10.1016/j.neuropharm.2009.12.013.

48 C. A. Denny et al., "4-to 6-week-old adult-born hippocampal neurons influence novelty-evoked exploration and contextual fear conditioning," *Hippocampus*, 2012, 22:1188-201. doi: 10.1002/hipo.20964.; V. Lemaire et al., "Behavioural trait of reactivity to novelty is related to hippocampal neurogenesis," *Eur J Neurosci.*, 1999, 11:4006-4014. doi: 10.1046/j.1460-9568.1999.00833.x.

49 C. S. S. Weeden et al., "A role for hippocampal adult neurogenesis in shifting attention toward novel stimuli," *Behav Brain Res.*, 2019, doi: 10.1016/j.bbr.2019.112152.

50 J. R. Ryu et al., "Control of adult neurogenesis by programmed cell death in the mammalian brain," *Mol Brain*, 2016, doi: 10.1186/s13041-016-0224-4.

51 A. Besnard & A. Sahay, "Adult Hippocampal Neurogenesis, Fear Generalization, and Stress," *Neuropsychopharmacology*, 2016, 41:24-44. doi: 10.1038/Npp.2015.167.

CHAPTER 2

1 World Health Organization, "Trace elements in human nutrition and health," 1996, https://www.who.int/publications/i/item/9241561734.

2 A. W. Voors, "Minerals in the municipal water and atherosclerotic heart death," *Am J Epidemiol,* 1971, 93:259-266. https://pubmed.ncbi.nlm.nih.gov/5550342/.

3 E. B. Dawson et al., "The mathematical relationship of drinking water lithium and rainfall to mental hospital admission," *Dis Nerv Syst.*, 1970, 31:811-820.

4 World Health Organization, "Trace elements in human nutrition and health," 1996, https://www.who.int/publications/i/item/9241561734, S. 225.

5 A. Mehri "Trace Elements in Human Nutrition (II) An Update," *Int J Prev Med.*, 2020, doi: 10.4103/ijpvm.IJPVM_48_19.

6 J. F. Cade, "Lithium salts in the treatment of psychotic excitement," *Med J Aust.,* 1949, 2:349-352. doi: 10.1080/j.1440-1614.1999.06241.x.

7 Fels, "Should We All Take a Bit of Lithium?" *The New York Times,* 13.07.2014, https://www.nytimes.com/2014/09/14/opinion/sunday/should-we-all-take-a-bit-of-lithium.html?referrer&_r=1.

8 M. Anke, B. Groppel and H. Kronemann, "Significance of the Newer Essential Trace Elements (like Si, Ni, As, Al, Li, V, . . .) for the Nutrition of Man and Animal" In *Vol 3 Proceedings of the Third International Workshop Neuherberg, Federal Republic of Germany, April 1984* edited by Pter Bratter and Peter Schramel, 421–464. Berlin, Boston: De Gruyter, 1984. https://doi.org/10.1515/9783112417188-049M.

9 Nehls, M., *Das erschöpfte Gehirn: Der Ursprung unserer mentalen Energie – und warum sie schwindet Willenskraft, Kreativität und Fokus zurückgewinnen.* Heyne 2022.

10 E. B. Dawson et al., "The mathematical relationship of drinking water lithium and rainfall to mental hospital admission," *Dis Nerv Syst.,* 1970, 31:811-820. https://pubmed.ncbi.nlm.nih.gov/5497853/.

11 E. B. Dawson et al., "Relationship of lithium metabolism to mental hospital admission and homicide," *Dis Nerv Syst.*, 1972, 33:546-556. https://pubmed.ncbi.nlm.nih. gov/4648454/.

12 A. W. Voors, "Drinking-water lithium and mental hospital admission in North Carolina," *N C Med J.,* 1972, 33:597-602. https://pubmed.ncbi.nlm.nih.gov/4504412/.

13 S. Shimodera et al., "Lithium levels in tap water and psychotic experiences in a general population of adolescents," *Schizophr Res.*, 2018, 201:294-298. doi: 10.1016/j. schres.2018.05.019.

14 Eyre-Watt B, et all, "The association between lithium in drinking water and neuropsychiatric outcomes: A systematic review and meta-analysis from across 2678 regions containing 113 million people," *Aust N Z J Psychiatry.* 2021 Feb; 55(2):139-152. doi: 10.1177/0004867420963740.

15 E. Isometsä, "Suicidal behaviour in mood disorders—who, when, and why?" *Can J Psychiatry*, 2014, 59:120-130. doi: 10.1177/070674371405900303; N. D. Kapusta

et al., "Lithium in drinking water and suicide mortality," *British Journal of Psychiatry*, 2011, 198:346-350. doi:10.1192/bjp.bp.110.091041.

16 Kapusta ND et al., "Lithium in drinking water and suicide mortality," *British Journal of Psychiatry*. 2011, 198:346-350. doi:10.1192/bjp.bp.110.091041.

17 F. Barjasteh-Askari et al., "Relationship between suicide mortality and lithium in drinking water: A systematic review and meta-analysis," *J Affect Disord.*, 2020, 264:234-241. doi: 10.1016/j.jad.2019.12.027.

18 Bschor, T., *Mehr Lithium im Trinkwasser, weniger Suizide. InFo Neurologie + Psychiatrie*. 2021, doi:10.1007/s15005-021-1833-8.

19 S. Ando et al., "Comparison of lithium levels between suicide and non-suicide fatalities: Cross-sectional study," *Transl Psychiatry*, 2022, doi: 10.1038/s41398-02202238-9.

20 Pichler EM et al., "Too early to add lithium to drinking water? No association between lithium and suicides in a pre-registered Swiss study," *J Affect Disord.* 2024, 367:598-605. doi: 10.1016/j.jad.2024.08.239.

21 https://www.geog.uni-heidelberg.de/forschung/gis_lithiuminwater_en.html (2.10.2013, zuletzt abgerufen am 2.10.2024).

22 M. Helbich et al., "Lithium in drinking water and suicide mortality: interplay with lithium prescriptions," *Br J Psychiatry*, 2015, 207:64-71. doi: 10.1192/bjp.bp.114.152991.

23 H. Matsuzaki et al., "Re-analysis of the association of temperature or sunshine with hyperthymic temperament using lithium levels of drinking water," *J Affect Disord*, 2017, 223:126-129. doi: 10.1016/j.jad.2017.07.039.; N. Ishii et al., "Lithium in drinking water may be negatively associated with depressive temperament in the nonclinical population," *CNPT*, 2017, 8:7-11. https://www.jstage.jst.go.jp/article/cnpt/8/0/8_7/_pdf.

24 W. F. Parker et al., "Association Between Groundwater Lithium and the Diagnosis of Bipolar Disorder and Dementia in the United States," *JAMA Psychiatry*, 2018, 75:751-754. doi: 10.1001/jamapsychiatry.2018.1020.

25 S. Ando et al., "Lithium Levels in Tap Water and the Mental Health Problems of Adolescents: An Individual-Level Cross-Sectional Survey," *J Clin Psychiatry*, 2017, 78:252-256. doi: 10.4088/JCP.15m10220.

26 G. N. Schrauzer & K. P. Shrestha, "Lithium in drinking water and the incidences of crimes, suicides, and arrests related to drug addictions," *Biol Trace Elem Res*, 1990, 25:105-113. doi: 10.1007/BF02990271.

27 G. N. Schrauzer et al., "Lithium in scalp hair of adults, students, and violent criminals. Effects of supplementation and evidence for interactions of lithium with vitamin B12 and with other trace elements," *Biol Trace Elem Res.*, 1992, 34:161-176. doi: 10.1007/BF02785244.

28 Schrauzer GN, Shrestha KP. "Lithium in drinking water and the incidences of crimes, suicides, and arrests related to drug addictions," *Biol Trace Elem Res*. 1990:105-13. doi: 10.1007/BF02990271.

29 O. Giotakos et al., "A negative association between lithium in drinking water and the incidences of homicides, in Greece," *Biol Trace Elem Res.*, 2015, 164:165-168. doi: 10.1007/s12011-014-0210-6.

30 O. Giotakos et al., "Lithium in the public water supply and suicide mortality in Greece," *Biol Trace Elem Res*, 2013, 156:376-379. doi: 10.1007/s12011-013-9815-4.

31 K. Kohno et al., "Lithium in drinking water and crime rates in Japan: cross-sectional study," *BJPsych Open.*, 2020, doi: 10.1192/bjo.2020.63.

32 T. M. Marshall, "Lithium as a Nutrient," J*ournal of American Physicians and Surgeons*, 2015, https://www.jpands.org/vol20no4/marshall.pdf.

33 M. Nehls, "Unified theory of Alzheimer's disease (UTAD): implications for prevention and curative therapy," *J Mol Psychiatry,* 2016, doi: 10.1186/s40303-016-0018-8.

34 Nehls, M., *Alzheimer ist heilbar: Rechtzeitig zurück in ein gesundes Leben.* Heyne 2017.

35 L. V. Kessing et al., "Association of Lithium in Drinking Water With the Incidence of Dementia," *JAMA Psychiatry*, 2017, 74:1005-1010. doi: 10.1001/jamapsychiatry.2017.2362.

36 L. B. Zahodne et al., "Depressive symptoms precede memory decline, but not vice versa, in non-demented older adults," *J Am Geriatr Soc.*, 2014, 62:130-134. doi: 10.1111/jgs.12600.

37 M. A. Rapp et al., "Increased neurofibrillary tangles in patients with Alzheimer disease with comorbid depression," *Am J Geriatr Psychiatry*, 2008, 16:168-174. doi: 10.1097/JGP.0b013e31816029ec.; G. Spalletta et al., "The role of persistent and incident major depression on rate of cognitive deterioration in newly diagnosed Alzheimer's disease patients," *Psychiatry Res.*, 2012, 198:263-268. doi: 10.1016/j.psychres.2011.11.018.

38 H. Zhou et al., "Prevention of Keshan Disease by Selenium Supplementation: a Systematic Review and Meta-analysis," *Biol Trace Elem Res*, 2018, 186:98-105. doi: 10.1007/s12011-018-1302-5.

39 K. Rampon, "Anemia: Microcytic Anemia," *FP Essent.*, 2023, 530:12-16. https://pubmed.ncbi.nlm.nih.gov/37390396/.

40 M. B. Zimmermann, "The effects of iodine deficiency in pregnancy and infancy," *Paediatr Perinat Epidemiol.*, 2012, 1:108-117. doi: 10.1111/j.1365-3016.2012.01275.x.

41 M. Adida et al., "Lithium might be associated with better decision-making performance in euthymic bipolar patients," *Eur Neuropsychopharmacol.*, 2015, 25:788797. doi: 10.1016/j.euroneuro.2015.03.003.

42 G. Martone, "Nutritional Lithium," *J Clin Psychiatry Neurosci*, 2017, 1:3-4, https://www.pulsus.com/scholarly-articles/nutritional-lithium.pdf.

43 J. M. Greenblatt & K. Grossmann, *Nutritional Lithium: A Cinderella Story: The Untold Tale of a Mineral That Transforms Lives and Heals the Brain* (CreateSpace Independent Publishing Platform, 2016), S. 193/194.

44 E. Shorter, "The history of lithium therapy," *Bipolar Disord*, 2009, 2:4-9. doi: 10.1111/j.1399-5618.2009.00706.x.

45 J. F. Cade, "Lithium salts in the treatment of psychotic excitement," *Med J Aust.*, 1949, 2:349-352. doi: 10.1080/j.1440-1614.1999.06241.x.

46 Ibid.

47 J. K. Rybakowski, "Response to lithium in bipolar disorder: clinical and genetic findings," *ACS Chem Neurosci*, 2014, 5:413-421. doi: 10.1021/cn5000277.

48 H. Sugawara et al., "Predictors of efficacy in lithium augmentation for treatment-resistant depression," *J Affect Disord.*, 2010, 125:165-168. doi: 10.1016/j.jad.2009.12.025.

49 N. Embi et al., "Glycogen synthase kinase-3 from rabbit skeletal muscle. Separation from cyclic-AMP-dependent protein kinase and phosphorylase kinase," *Eur J Biochem.*, 1980, 107:519-527. https://pubmed.ncbi.nlm.nih.gov/6249596/.

50 P. S. Klein & D. A. Melton, "A molecular mechanism for the effect of lithium on development," *Proc Natl Acad Sci USA* 1996, 93:8455-8459. doi: 10.1073/pnas.93.16.8455.

51 L. Wang et al., "Glycogen synthesis and beyond, a comprehensive review of GSK3 as a key regulator of metabolic pathways and a therapeutic target for treating metabolic diseases," *Med Res Rev.*, 2022, 42:946-982. doi: 10.1002/med.21867.

52 J. Lee & M. S. Kim, "The role of GSK3 in glucose homeostasis and the development of insulin resistance," *Diabetes Res Clin Pract.*, 2007, 77:49-57. doi: 10.1016/j.diabres.2007.01.033.; S. E. Nikoulina et al., "Potential role of glycogen synthase kinase-3 in skeletal muscle insulin resistance of type 2 diabetes," *Diabetes*, 2000, 49:263-271. doi: 10.2337/diabetes.49.2.263.

53 E. Beurel et al., "Glycogen synthase kinase-3 (GSK3): regulation, actions, and diseases," *Pharmacol Ther,* 2015, 148:114-131. doi: 10.1016/j.pharmthera.2014.11.016.

54 E. Jakobsson et al., "Towards a Unified Understanding of Lithium Action in Basic Biology and its Significance for Applied Biology," *J Membr Biol.*, 2017, 250:587-604. doi: 10.1007/s00232-017-9998-2.

55 R. Khairova et al., "Effects of lithium on oxidative stress parameters in healthy subjects," *Mol Med Rep.*, 2012, 5:680-682. doi: 10.3892/mmr.2011.732.

56 M. Roux & A. Dosseto, "From direct to indirect lithium targets: a comprehensive review of omics data," *Metallomics*, 2017, 9:1326-1351. doi: 10.1039/c7mt00203c.

57 J. M. Beaulieu et al., "A beta-arresting 2 signaling complex mediates lithium action on behavior," *Cell*, 2008, 132:125-136. doi: 10.1016/j.cell.2007.11.041.; D. Chatterjee & J. M. Beaulieu, "Inhibition of glycogen synthase kinase 3 by lithium, a mechanism in search of specificity," *Front Mol Neurosci.*, 2022, doi: 10.3389/fnmol.2022.1028963.

58 Roux & Dosseto, "From direct to indirect lithium targets: a comprehensive review of omics data."

59 B. W. Doble & J. R. Woodgett, "GSK-3: tricks of the trade for a multi-tasking kinase,"*J Cell Sci.* (2003), 116:1175-1186. doi: 10.1242/jcs.00384.; J. L. Stamos et al., "Structural basis of GSK-3 inhibition by N-terminal phosphorylation and by

the Wnt receptor LRP6," *Elif.* (2014) doi: 10.7554/eLife.01998.; I. Rippin et al., "Discovery and Design of Novel Small Molecule GSK-3 Inhibitors Targeting the Substrate Binding Site," *Int J Mol Sci.* (2020), doi: 10.3390/ijms21228709.

60 M. E. Snitow et al., "Lithium and Therapeutic Targeting of GSK-3," *Cells* (2021), doi: 10.3390/cells10020255.

61 P. S. Klein & D. A. Melton, "A molecular mechanism for the effect of lithium on development," *Proc Natl Acad Sci USA* (1996), 93:8455-8459. doi: 10.1073 /pnas.93.16.8455.; https://de.wikipedia.org/wiki/Liste_der_Ionenradien.

62 P. S. Klein & D. A. Melton, "A molecular mechanism for the effect of lithium on development," *Proc Natl Acad Sci USA.* (1996), 93:8455-8459. doi: 10.1073 /pnas.93.16.8455.; V. Stambolic et al., "Lithium inhibits glycogen synthase kinase-3 activity and mimics wingless signaling in intact cells," *Curr Biol.* (1996), 6:16641668. doi: 10.1016/s0960-9822(02)70790-2.

63 C. W. Hsu et al., "Lithium concentration and recurrence risk during maintenance treatment of bipolar disorder: Multicenter cohort and meta-analysis," *Acta Psychiatr Scand.* (2021), 144:368-378. doi: 10.1111/acps.13346.

64 W. J. Ryves & A. J. Harwood, "Lithium inhibits glycogen synthase kinase-3 by competition for magnesium," *Biochem Biophys Res Commun.* (2001), 280:720-725. doi: 10.1006/bbrc.2000.4169.; T. D. Gould, H. K. Manji, "Glycogen synthase kinase-3: a putative molecular target for lithium mimetic drugs," *Neuropsychopharmacology* (2005), Jul;30(7):1223-37. doi: 10.1038/sj.npp.1300731 .PMID: 15827567.

65 A. Bortolozzi et al., "New Advances in the Pharmacology and Toxicology of Lithium: A Neurobiologically Oriented Overview," *Pharmacol Rev* (2024), 76:323-357. doi: 10.1124/pharmrev.120.000007.

66 E. Chalecka-Franaszek & D. M. Chuang, "Lithium activates the serine/threonine kinase Akt-1 and suppresses glutamate-induced inhibition of Akt-1 activity in neurons," *Proc Natl Acad Sci USA* (1999), 96:8745-8750. doi: 10.1073/pnas .96.15.8745.; N. Kirshenboim et al., "Lithium-mediated phosphorylation of glycogen synthase kinase-3beta involves PI3 kinase-dependent activation of protein kinase C-alpha," *J Mol Neurosci.* (2004), 24:237-245. doi: 10.1385/JMN:24:2:237.

67 F. Zhang et al., "Inhibitory phosphorylation of glycogen synthase kinase-3 (GSK-3) in response to lithium. Evidence for autoregulation of GSK-3," *J Biol Chem.* (2003), 278:33067-33077. doi: 10.1074/jbc.M212635200.

68 E. Beurel et al., "Glycogen synthase kinase-3 (GSK3): regulation, actions, and diseases," *Pharmacol Ther.* (2015), 148:114-131. doi: 10.1016/j.pharmthera. 2014.11.016.

69 L. Freland & J. M. Beaulieu, "Inhibition of GSK3 by lithium, from single molecules to signaling networks," *Front Mol Neurosci.* (2012), doi: 10.3389 /fnmol.2012.00014.

70 K. Yang et al., "The Key Roles of GSK-3β in Regulating Mitochondrial Activity," *Cell Physiol Biochem* (2017), 44:1445-1459, https://karger.com/cpb/articlepdf

/44/4/1445/2446568/000485580.pdf; I. T. Struewing et al., "Lithium increases PGC-1alpha expression and mitochondrial biogenesis in primary bovine aortic endothelial cells," *FEBS J.* (2007), 274:2749-2765. doi: 10.1111/j.1742-4658 .2007.05809.x.

71 H. Y. Pan & M. Valapala, "Regulation of Autophagy by the Glycogen Synthase Kinase-3 (GSK-3) Signaling Pathway," *Int J Mol Sci.* (2022), doi: 10.3390/ijms 23031709.

72 S. Sarkar et al., "Lithium induces autophagy by inhibiting inositol monophosphatase," *J Cell Biol.* (2005), 170:1101-1111. doi: 10.1083/jcb .200504035.

73 T. T. Nguyen et al., "Mitochondria-associated programmed cell death as a therapeutic target for age-related disease," *Exp Mol Med.* (2023), 55:1595-1619. doi: 10.1038 /s12276-023-01046-5.

74 C. Wang & R. J. Youle, "The role of mitochondria in apoptosis*," *Annu Rev Genet.* (2009), 43:95-118. doi: 10.1146/annurev-genet-102108-134850.

75 S. I. Hamstra et al., "Beyond Its Psychiatric Use: The Benefits of Low-dose Lithium Supplementation," *Curr Neuropharmacol.* (2023), 21:891-910. doi : 10.2174/1570159X 20666220302151224.

76 Nehls M: Die Alzheimer-Lüge: Die Wahrheit über eine vermeidbare Krankheit. Heyne 2014, pg. 87/88.

77 F. Léveillé et al., "Suppression of the intrinsic apoptosis pathway by synaptic activity," *J Neurosci.* (2010), 30:2623-2635. doi: 10.1523/JNEUROSCI.5115-09.2010.

78 K. J. Christie et al., "Adult hippocampal neurogenesis, Rho kinase inhibition and enhancement of neuronal survival," *Neuroscience.* (2013), 247:75-83. doi: 10.1016 /j.neuroscience.2013.05.019.

79 G. Chen et al., "Enhancement of hippocampal neurogenesis by lithium," *J Neurochem.* (2000), 75:1729-1734. doi: 10.1046/j.1471-4159.2000.0751729.x.; K. J. Christie et al., "Adult hippocampal neurogenesis, Rho kinase inhibition and enhancement of neuronal survival," *Neuroscience.* (2013), 247:75-83. doi: 10.1016/j. neuroscience.2013.05.019.

80 P. A. Zunszain et al., "Interleukin-1β: a new regulator of the kynurenine pathway affecting human hippocampal neurogenesis," *Neuropsychopharmacology* (2012), 37:939-949. doi: 10.1038/npp.2011.277.; A. Borsini et al., "Rescue of IL-1β-induced reduction of human neurogenesis by omega-3 fatty acids and antidepressants," B*rain Behav Immun.* (2017), 65:230-238. doi: 10.1016/j. bbi.2017.05.006.

81 X. Kong et al., "JAK2/STAT3 signaling mediates IL-6-inhibited neurogenesis of neural stem cells through DNA demethylation/methylation," *Brain, Behavior, and Immunity* (2019), 79:159-173. doi: 10.1016/j.bbi.2019.01.027.

82 H. Neumann et al., "Tumor necrosis factor inhibits neurite outgrowth and branching of hippocampal neurons by a rho-dependent mechanism," *J Neurosci.* (2002), 22:854-862. doi: 10.1523/JNEUROSCI.22-03-00854.2002.

83 M. Nehl, "Unified theory of Alzheimer's disease (UTAD): implications for prevention and curative therapy," *J Mol Psychiatry* (2016), doi: 10.1186/s40303-016-0018-8.
84 A. B. Engin & A. Engin, "The Connection Between Cell Fate and Telomere," *Adv Exp Med Biol.* (2021), 1275:71-100. doi: 10.1007/978-3-030-49844-3_3.
85 J. W. Shay & W. E. Wright, "Hayflick, his limit, and cellular ageing," *Nat Rev Mol Cell Biol.* (2000), 1:72-76. doi: 10.1038/35036093.
86 L. A. Tucker & C. J. Bates, "Telomere Length and Biological Aging: The Role of Strength Training in 4814 US Men and Women," *Biology (Basel.)* (2024), doi: 10.3390/biology13110883.
87 M. Chen et al., "Association between modifiable lifestyle factors and telomere length: a univariable and multivariable Mendelian randomization study," *J Transl Med.* (2024), doi: 10.1186/s12967-024-04956-8.
88 F. Rossiello et al., "Telomere dysfunction in ageing and age-related diseases," *Nat Cell Biol.* (2022), 24:135-147. doi: 10.1038/s41556-022-00842-x.
89 F. Coutts et al., "The polygenic nature of telomere length and the anti-ageing properties of lithium," *Neuropsychopharmacology* (2019), 44:757-765. doi: 10.1038/s41386-018-0289-0.; L. Martinsson et al., "Long-term lithium treatment in bipolar disorder is associated with longer leukocyte telomeres," *Transl Psychiatry.* (2013), doi: 10.1038/tp.2013.37.
90 T. Viel et al., "Microdose lithium reduces cellular senescence in human astrocytes – a potential pharmacotherapy for COVID-19?" *Aging* (2020), 12:1003510040. doi:10.18632/aging.103449.
91 Y. B. Wei et al., "Telomerase dysregulation in the hippocampus of a rat model of depression: normalization by lithium," *Int J Neuropsychopharmacol.* (2015), doi: 10.1093/ijnp/pyv002.; G. Saretzki & T. Wan, "Telomerase in Brain: The New Kid on the Block and Its Role in Neurodegenerative Diseases," *Biomedicines* (2021) Apr 29;9(5):490. doi: 10.3390/biomedicines9050490.
92 A. B. Palmos et al., "Telomere length and human hippocampal neurogenesis," *Neuropsychopharmacology* (2020), 45:2239-2247. doi: 10.1038/s41386-020-00863-w.
93 Y. B. Wei et al. "Telomerase dysregulation in the hippocampus of a rat model of depression: normalization by lithium," *Int J Neuropsychopharmacol.* (2015), doi: 10.1093/Ijnp/pyv002.
94 S. Hägg et al., "Short telomere length is associated with impaired cognitive performance in European ancestry cohorts," *Transl Psychiatry* (2017), doi: 10.1038/tp.2017.73.
95 A. M. Staffaroni et al., "Telomere attrition is associated with declines in medial temporal lobe volume and white matter microstructure in functionally independent older adults," *Neurobiol Aging* (2018), 69:68-75. doi: 10.1016/j.neurobiolaging.2018.04.021.

96 T. R. Powell et al., "Telomere Length and Bipolar Disorder," *Neuropsychopharmacology* (2018), 43:445-453. doi: 10.1038/npp.2017.125. Erratum in: doi: 10.1038/npp.2017.239.

97 T. R. Powell et al., "Telomere length as a predictor of emotional processing in the brain," *Hum Brain Mapp.* (2019), 40:1750-1759. doi: 10.1002/hbm.24487.

98 G. Zanni et al., "Lithium Accumulates in Neurogenic Brain Regions as Revealed by High Resolution Ion Imaging," *Sci Rep.* (2017), doi: 10.1038/srep40726.

99 S. A. Kaladchibachi et al., "Glycogen synthase kinase 3, circadian rhythms, and bipolar disorder: a molecular link in the therapeutic action of lithium," *J Circadian Rhythms* (2007), doi: 10.1186/1740-3391-5-3.

100 C. Fernandes et al., "Detrimental role of prolonged sleep deprivation on adult neurogenesis," *Front Cell Neurosci.* (2015), doi: 10.3389/fncel.2015.00140.; I. Koyanagi et al., "Memory consolidation during sleep and adult hippocampal neurogenesis," *Neural Regen Res.* (2019), 14:20-23. doi: 10.4103/1673-5374.243695.

101 Y. Takaesu, "Circadian rhythm in bipolar disorder: A review of the literature," *Psychiatry Clin Neurosci.* (2018), 72:673-682. doi: 10.1111/pcn.12688.

102 A. Palmos et al., "Lithium treatment and human hippocampal neurogenesis," *Translational Psychiatry* (2021), doi: 10.1038/s41398-021-01695-y.; W. Y. Kim et al., "GSK-3 is a master regulator of neural progenitor homeostasis," *Nat Neurosci.* (2009), 12:1390-1397. doi: 10.1038/nn.2408.; R. Hellweg et al., "Subchronic treatment with lithium increases nerve growth factor content in distinct brain regions of adult rats," *Mol Psychiatry* (2002), 7:604-608. doi: 10.1038/sj.mp.4001042.

103 Hellweg R et al., "Subchronic treatment with lithium increases nerve growth factor content in distinct brain regions of adult rats," *Mol Psychiatry*. 2002, 7:604-608. doi: 10.1038/sj.mp.4001042.

104 V. J. De-Paula et al., "Long-term lithium treatment increases intracellular and extracellular brain-derived neurotrophic factor (BDNF) in cortical and hippocampal neurons at subtherapeutic concentrations," *Bipolar Disord.* (2016), 18:692-695. doi: 10.1111/bdi.12449.

105 T. Numakawa et al., "Actions of Brain-Derived Neurotrophin Factor in the Neurogenesis and Neuronal Function, and Its Involvement in the Pathophysiology of Brain Diseases," *Int J Mol Sci.* (2018), doi: 10.3390/ijms19113650.; H. Frielingsdorf et al., "Nerve growth factor promotes survival of new neurons in the adult hippocampus," *Neurobiol Dis.* (2007), 26:47-55. doi: 10.1016/j.nbd.2006.11.015.; T. Siddiqui et al., "Nerve growth factor receptor (Ngfr) induces neurogenic plasticity by suppressing reactive astroglial Lcn2/Slc22a17 signaling in Alzheimer's disease," *NPJ Regen Med.* (2023) Jul 10;8(1):33. doi: 10.1038/s41536-023-00311-5.

106 A. J. Valvezan & P. S. Klein, "GSK-3 and Wnt Signaling in Neurogenesis and Bipolar Disorder," *Front Mol Neurosci.* (2012), doi: 10.3389/fnmol.2012.00001.;

W. Young, "Review of lithium effects on brain and blood," *Cell Transplant* (2009), 18:951-975, doi: 10.3727/096368909X471251; Palmos AB et al., "Lithium treatment and human hippocampal neurogenesis," *Transl Psychiatry*. 2021, doi: 10.1038/s41398-021-01695-y.

107 S. Brady & G. Morfini, "A perspective on neuronal cell death signaling and neurodegeneration," *Mol Neurobiol.* (2010), 42:25-31. doi: 10.1007/s12035-010-8128-2.

108 G. N. Schrauzer, "Lithium: occurrence, dietary intakes, nutritional essentiality," *J Am Coll Nutr.* (2002), 21:14-21. doi: 10.1080/07315724.2002.10719188.

109 W. E. Wright et al., "Telomerase activity in human germline and embryonic tissues and cells," *Dev Genet. (*1996), 18:173-179. doi: 10.1002/(SICI)1520-6408(1996)18:2<173::AID-DVG10>3.0.CO;2-3. PMID: 8934879.

110 Lundberg M. et al., "Lithium and the Interplay Between Telomeres and Mitochondria in Bipolar Disorder," *Front Psychiatry*. 2020, doi: 10.3389/fpsyt.2020.586083.; Fernandez RJ et al., "GSK3 inhibition rescues growth and telomere dysfunction in dyskeratosis congenita iPSC-derived type II alveolar epithelial cells," *Elife*. 2022, doi: 10.7554/eLife.64430.

111 Gallicchio VS & Chen MG: Modulation of murine pluripotential stem cell proliferationin vivo by lithium car nate. *Blood.* 1980, 56:1150-1152. https://pubmed.ncbi.nlm.nih.gov/7437517/.

112 J. Huang et al., "Pivotal role for glycogen synthase kinase-3 in hematopoietic stem cell homeostasis in mice," *J Clin Invest.* (2009), 119:3519-3529. doi: 10.1172 /JCI40572.

113 G. N. Schrauzer, "Lithium: occurrence, dietary intakes, nutritional essentiality," *J Am Coll Nutr.* (2002), 21:14-21. doi: 10.1080/07315724.2002.10719188.; M. Anke et al., "Recent progress in exploring the essentiality of the ultratrace element lithium to the nutrition of animals and man," *Biomed Res Trace Elements* (2005), 16:169-176, https://www.jstage.jst.go.jp/article/brte/16/3/16_3_169/_pdf.

114 E. Jakobsson et al., "Towards a Unified Understanding of Lithium Action in Basic Biology and its Significance for Applied Biology," *J Membr Biol.* (2017), 250: 587-604. doi: 10.1007/s00232-017-9998-2.

115 Y. Tamari & K. Tsuchiya, "Lithium Content of Fish in the Ocean: Investigation of raw, drying and canned fish available," *Biomedical Research on Trace Elements* (2004), 15:248-258, https://doi.org/10.11299/brte.15.248.

116 G. Zanni et al., "Lithium Accumulates in Neurogenic Brain Regions as Revealed by High Resolution Ion Imaging," *Sci Rep.* (2017), doi: 10.1038/srep40726.

117 V. Donega et al., "Single-cell profiling of human subventricular zone progenitors identifies SFRP1 as a target to re-activate progenitors," *Nat Commun.* (2022), doi: 10.1038/s41467-022-28626-9.

118 J. Stout et al., "Accumulation of Lithium in the Hippocampus of Patients With Bipolar Disorder: A Lithium-7 Magnetic Resonance Imaging Study at 7 Tesla," *Biol Psychiatry* 2020, 88:426-433. doi: 10.1016/j.biopsych.2020.02.1181.

119 E. Jakobsson et al., "Towards a Unified Understanding of Lithium Action in Basic Biology and its Significance for Applied Biology," *J Membr Biol.* (2017), 250:587-604. doi: 10.1007/s00232-017-9998-2.; H. Luo et al., "Sodium Transporters Are Involved in Lithium Influx in Brain Endothelial Cells," *Mol Pharm.* (2018), 15:2528-2538. doi: 10.1021/acs.molpharmaceut.8b00018.

120 M. Blum & T. Ott, "Animal left-right asymmetry," *Curr Biol.* (2018), 28:301-304. doi: 10.1016/j.cub.2018.02.073.

121 L. J. Rogers, "Brain Lateralization and Cognitive Capacity," *Animals* (Basel.) (2021), doi: 10.3390/ani11071996.

122 G. Lawton, "Mind tricks: ways to explore your brain," *New Scientist* (2007), 195:34-41, doi: 10.1016/S0262-4079(07)62411-7.

123 O. A. Shipton et al., "Left-right dissociation of hippocampal memory processes in mice," *Proc Natl Acad Sci USA* (2014), 111:15238-15243. doi: 10.1073/pnas.1405648111.

124 N. Burgess et al., "The human hippocampus and spatial and episodic memory," *Neuron.* (2002), 35:625-41. doi: 10.1016/s0896-6273(02)00830-9.

125 A. Ezzati et al., "Differential association of left and right hippocampal volumes with verbal episodic and spatial memory in older adults," *Neuropsychologia* (2016), 93:380-385. doi: 10.1016/j.neuropsychologia.2016.08.016.

126 S. S. Nemati et al., "Lateralization of the hippocampus: A review of molecular, functional, and physiological properties in health and disease," *Behav Brain Res.* (2023), doi: 10.1016/j.bbr.2023.114657.

127 M. L. Phillips & H. A. Swartz, "A critical appraisal of neuroimaging studies of bipolar disorder: toward a new conceptualization of underlying neural circuitry and a road map for future research," *Am J Psychiatry* (2014), 171:829-843. doi: 10.1176/appi.ajp.2014.13081008.

128 Zanni et al., "Lithium Accumulates in Neurogenic Brain Regions."

129 M. Lundberg et al., "Expression of telomerase reverse transcriptase positively correlates with duration of lithium treatment in bipolar disorder," *Psychiatry Res.* (2020), doi: 10.1016/j.psychres.2020.112865.

130 F. Coutts et al., "The polygenic nature of telomere length and the anti-ageing properties of lithium," *Neuropsychopharmacology* (2019), 44:757-765. doi: 10.1038/s41386018-0289-0.

131 A. B. Palmos et al., "Telomere length and human hippocampal neurogenesis," *Neuropsychopharmacology* (2020), 45:2239-2247. doi: 10.1038/s41386-020-00863-w.

132 T. W. Moorhead et al., "Progressive gray matter loss in patients with bipolar disorder," *Biol Psychiatry* (2007), 62:894-900. doi: 10.1016/j.biopsych.2007.03.005.; S. J. Quigley et al., "Volume and shape analysis of subcortical brain structures and ventricles in euthymic bipolar I disorder," *Psychiatry Res.* (2015), 233:324-330. doi: 10.1016/j.pscychresns.2015.05.012.

133 T. Hajek et al., "Hippocampal volumes in bipolar disorders: opposing effects of illness burden and lithium treatment," *Bipolar Disord.* (2012), 14:261-270. doi: 10.1111/j.1399-5618.2012.01013.x.; S. Zung et al., "The influence of lithium on hippocampal volume in elderly bipolar patients: a study using voxel-based morphometry," *Transl Psychiatry.* (2016), doi: 10.1038/tp.2016.97.

134 K. Brosch et al., "Reduced hippocampal gray matter volume is a common feature of patients with major depression, bipolar disorder, and schizophrenia spectrum disorders," *Mol Psychiatry* (2022), 27:4234-4243. doi: 10.1038/s41380-022-01687-4.

135 D. Sussman et al., "The autism puzzle: Diffuse but not pervasive neuroanatomical abnormalities in children with ASD," *Neuroimage Clin.* (2015), 8:170-179. doi: 10.1016/j.nicl.2015.04.008.

136 T. Huebner et al., "Morphometric brain abnormalities in boys with conduct disorder," *J Am Acad Child Adolesc Psychiatry* (2008), 47:540-547. doi: 10.1097/CHI.0b013e3181676545.; A. Abdolalizadeh et al., "Larger left hippocampal presubiculum is associated with lower risk of antisocial behavior in healthy adults with childhood conduct history," *Sci Rep.* (2023), doi: 10.1038/s41598-023-33198-9.

137 Y. Luo et al., "Decreased left hippocampal volumes in parents with or without posttraumatic stress disorder who lost their only child in China," *J Affect Disord.* (2016), 197:223-30. doi: 10.1016/j.jad.2016.03.003.; H. Xie et al., "Relationship of Hippocampal Volumes and Posttraumatic Stress Disorder Symptoms Over Early Post-trauma Periods," *Biol Psychiatry Cogn Neurosci Neuroimaging.* (2018), 3:968-975. doi: 10.1016/j.bpsc.2017.11.010.

138 E. Jakobsson et al., "Towards a Unified Understanding of Lithium Action in Basic Biology and its Significance for Applied Biology," *J Membr Biol.* (2017), 250:587-604. doi: 10.1007/s00232-017-9998-2.

139 P. G. Higgs & N. Lehman, "The RNA World: molecular cooperation at the origins of life," *Nat Rev Genet.* (2015), 16:7-17. doi: 10.1038/nrg3841.

140 P. C. Joshi & M. F. Aldersley, "Significance of mineral salts in prebiotic RNA synthesis catalyzed by montmorillonite," *J Mol Evol.* (2013), 76:371-319. doi: 10.1007/s00239013-9568-x.

141 T. Tsuruta, "Removal and recovery of lithium using various microorganisms," *J Bioscience and Bioengineering.* (2005), 100:562-566. doi: 10.1263/jbb.100.562.

142 R. Koishi et al., "A superfamily of voltage-gated sodium channels in bacteria," *J Biol Chem.* (2004), 279:9532-9538. doi: 10.1074/jbc.M313100200.; E. Richelson, "Lithium ion entry through the sodium channel of cultured mouse neuroblastoma cells: a biochemical study," *Science* (1977), 196:1001-1002. doi: 10.1126/science.860126.; H. Krishnamurthy et al., "Unlocking the molecular secrets of sodium-coupled transporters," *Nature* (2009), 459:347-355. doi: 10.1038/nature08143.

143 H. Luo et al., "The role of brain barriers in the neurokinetics and pharmacodynamics of lithium," *Pharmacol Res.* (2021), doi: 10.1016/j.phrs.2021.105480.

144 H. Luo et al., "Sodium Transporters Are Involved in Lithium Influx in Brain Endothelial Cells," *Mol Pharm. (*2018), 15:2528-2538. doi: 10.1021/acs.molpharmaceut.8b00018.

145 E. Jakobsson et al., "Towards a Unified Understanding of Lithium Action in Basic Biology and its Significance for Applied Biology," *J Membr Biol.* (2017), 250:587-604. doi: 10.1007/s00232-017-9998-2.

146 C. De Duve, *Vital Dust: The Origin And Evolution Of Life On Earth* (Basic Books, 1995).

147 Y. Oba et al., "Identifying the wide diversity of extraterrestrial purine and pyrimidine nucleobases in carbonaceous meteorites," *Nat Commun.* (2022), doi: 10.1038/s41467-022-29612-x.; S. A. Krasnokutski et al., "Formation of extraterrestrial peptides and their derivatives," *Sci Adv.* (2024), doi: 10.1126/sciadv.adj7179.; A. A. Sharov, "Genome increase as a clock for the origin and evolution of life," *Biol Direct.* (2006), doi: 10.1186/1745-6150-1-17.

148 P. Starokadomskyy & K. V. Dmytruk, "A bird's-eye view of autophagy," *Autophagy* (2013), 9:1121-1126. doi: 10.4161/auto.24544.

149 S. Sarkar et al., "Lithium induces autophagy by inhibiting inositol monophosphatase," *J Cell Biol.* (2005), 170:1101-1111. doi: 10.1083/jcb.200504035.

150 D. E. Dollins et al., "A structural basis for lithium and substrate binding of an inositide phosphatase," *J Biol Chem.* (2021), doi: 10.1074/jbc.RA120.014057.

151 Boonekamp FJ et al., "Full humanization of the glycolytic pathway in Saccharomyces cerevisiae," *Cell Rep*. 2022, doi: 10.1016/j.celrep.2022.111010.

152 F. J. J. Chain & R. Assis, "BLAST from the Past: Impacts of Evolving Approaches on Studies of Evolution by Gene Duplication," *Genome Biol Evol.* (2021), doi: 10.1093/gbe/evab149.

153 S. D. Copley, "Evolution of new enzymes by gene duplication and divergence," *FEBS J.* (2020), 287:1262-1283. doi: 10.1111/febs.15299.

154 Y. Saidi et al., "Function and evolution of 'green' GSK3/Shaggy-like kinases," *Trends Plant Sci.* (2012), 17:39-46. doi: 10.1016/j.tplants.2011.10.002.

155 M. J. Novacek, "Mammalian evolution: an early record bristling with evidence," *Curr Biol.* (1997), 7:489-491. doi: 10.1016/s0960-9822(06)00245-4.

156 Y. X. Ma et al., "Differential Roles of Glycogen Synthase Kinase 3 Subtypes Alpha and Beta in Cortical Development," *Front Mol Neurosci.* (2017), doi: 10.3389/fnmol.2017.00391.; M. Pardo et al., "GSK3β isoform-selective regulation of depression, memory and hippocampal cell proliferation," *Genes Brain Behav.* (2016), 15:348-355. doi: 10.1111/gbb.12283.

157 A. K. Brown et al., "Dimerization of inositol monophosphatase Mycobacterium tuberculosis SuhB is not constitutive, but induced by binding of the activator Mg2+," *BMC Struct Biol.* (2007) doi: 10.1186/1472-6807-7-55.

158 Ibid.
159 A. Dutta et al., "Structural elucidation of the binding site and mode of inhibition of Li(+) and Mg(2+) in inositol monophosphatase," *FEBS J.* (2014), 281:5309-5324. doi: 10.1111/febs.13070.
160 A. Haimovich et al., "Determination of the lithium binding site in inositol monophosphatase, the putative target for lithium therapy, by magic-angle-spinning solid-state NMR," *J Am Chem Soc.* (2012), 134:5647-51. doi: 10.1021/ja211794x.
161 E. R. R. Moody et al., "The nature of the last universal common ancestor and its impact on the early Earth system," *Nat Ecol Evol.* (2024), 8:1654-1666. doi: 10.1038/s41559024-02461-1.
162 D. Piovesan et al., "The human 'magnesome': detecting magnesium binding sites on human proteins," *BMC Bioinformatics.* (2012), doi: 10.1186/1471-2105-13-S14S10.; P. S. Klein & D. A. Melton, "A molecular mechanism for the effect of lithium on development," *Proc Natl Acad Sci USA.* (1996), 93:8455-8459. doi: 10.1073/pnas.93.16.8455.
163 E. L. Patt et al., "Effect of dietary lithium levels on tissue lithium concentrations, growth rate, and reproduction in the rat," *Bioinorganic Chemistry* (1978), 9:299-310, https://doi.org/10.1016/S0006-3061(00)80024-1
164 E. E. Pickett & B. L. O'Dell, "Evidence for dietary essentiality of lithium in the rat," *Biol Trace Elem Res.* (1992), 34:299-319. doi: 10.1007/BF02783685.
165 M. Anke et al., "Recent progress in exploring the essentiality of the ultratrace element lithium to the nutrition of animals and man," *Biomed Res Trace Elements* (2005), 16:169-176, https://www.jstage.jst.go.jp/article/brte/16/3/16_3_169/_pdf.
166 T. Ono & O. Wada, "Effects of lithium deficient diet on avoidance behavior," Nihon Eiseigaku Zasshi. (1989), 44:748-455. doi: 10.1265/jjh.44.748.
167 Anke et al., "Recent progress in exploring the essentiality."
168 G. N. Schrauzer, "Lithium: occurrence, dietary intakes, nutritional essentiality," *J Am Coll Nutr.* (2002), 21:14-21. doi: 10.1080/07315724.2002.10719188. PMID: 11838882.
169 G. N. Schrauzer & E. de Vroey, "Effects of nutritional lithium supplementation on mood. A placebo-controlled study with former drug users," *Biol Trace Elem Res.* (1994), 40:89-101. doi: 10.1007/BF02916824.
170 https://www.youtube.com/watch?v=tP3dXSEKaNE&t (11/22/2024, last accessed on 1/14/2025).
171 A. Cipriani et al., "Lithium in the prevention of suicide in mood disorders: updated systematic review and meta-analysis," *BMJ* (2013), doi: 10.1136/bmj.f3646.
172 E. Araldi et al., "Lithium treatment extends human lifespan: findings from the UK Biobank," *Aging* (Albany NY.) (2023), 15:421-440. doi: 10.18632/aging.204476.
173 E. Beurel & R. S. Jope, "Inflammation and lithium: clues to mechanisms contributing to suicide-linked traits," *Translational Psychiatry* (2014), doi: 10.1038/tp.2014.129.

174 J. K. N. Chan et al., "Life expectancy and years of potential life lost in people with mental disorders: a systematic review and meta-analysis," *EClinicalMedicine* (2023), doi: 10.1016/j.eclinm.2023.102294.

175 https://qbi.uq.edu.au/article/2019/10/life-expectancy-mapped-people-mental-disorders (10/25/2019, last accessed on 09/12/2024); O. Plana-Ripoll et al., "A comprehensive analysis of mortality-related health metrics associated with mental disorders: a nationwide, register-based cohort study," *The Lancet.* (2019), 394:1827-1835, https://www.thelancet.com/journals/lancet/article/PIIS0140-6736(19)32316-5/.

176 V. A. Fajardo et al., "Examining the Relationship between Trace Lithium in Drinking Water and the Rising Rates of Age-Adjusted Alzheimer's Disease Mortality in Texas." *J Alzheimer's Dis.* (2018), 61:425-434. doi: 10.3233/JAD-170744.

177 U. Schönfelder, Lithium – Jungbrunnen aus der Wasserleitung. Stabsstelle Kommunikation/Pressestelle Friedrich-Schiller-Universität Jena, https://idw-online.de/de/news408654 (11.02.2011, zuletzt abgerufen am 12.09.2024).

178 K. Zarse et al., "Low-dose lithium uptake promotes longevity in humans and metazoans," *Eur J Nutr.* (2011), 50:387-389. doi: 10.1007/s00394-011-0171-x.

179 V. A. Fajardo et al., "Trace lithium in Texas tap water is negatively associated with allcause mortality and premature death," *Appl Physiol Nutr Metab.* (2018), 43:412-414. doi: 10.1139/apnm-2017-0653.

180 Zarse et al., "Low-dose lithium uptake."

181 G. McColl et al., "Pharmacogenetic analysis of lithium-induced delayed aging in *Caenorhabditis elegans*," *J Biol Chem.* (2008), 283:350-357. doi: 10.1074/jbc.M705028200.

182 Zarse et al., "Low-dose lithium uptake."

183 Y. B. Wei et al., "Telomerase dysregulation in the hippocampus of a rat model of depression: normalization by lithium," I*nt J Neuropsychopharmacol.* (2015), doi: 10.1093/Ijnp/pyv002.

184 Z. Y. Tam et al., "Effects of Lithium on Age-related Decline in Mitochondrial Turnover and Function in *Caenorhabditis elegans*," T*he Journals of Gerontology* (2014), 69: 810–820, https://doi.org/10.1093/gerona/glt210, https://www.researchgate.net/publication/259626839.

185 A. R. Konopka et al., "Markers of human skeletal muscle mitochondrial biogenesis and quality control: effects of age and aerobic exercise training," *J Gerontol A Biol Sci Med Sci.* (2014), 69:371-378. doi: 10.1093/gerona/glt107.

186 I. T. Struewing et al., "Lithium increases PGC-1alpha expression and mitochondrial biogenesis in primary bovine aortic endothelial cells," *FEBS J.* (2007), 274:2749-2765. doi: 10.1111/j.1742-4658.2007.05809.x.

187 O. Sofola-Adesakin et al., "Lithium suppresses Aβ pathology by inhibiting translation in an adult Drosophila model of Alzheimer's disease," *Front Aging Neurosci.* (2014), doi: 10.3389/fnagi.2014.00190.

188 J. I. Castillo-Quan et al., "Lithium Promotes Longevity through GSK3 /NRF2-Dependent Hormesis," *Cell Rep.* (2016), 15:638-650. doi: 10.1016/j .celrep.2016.03.041.

189 T. Nespital et al., "Lithium can mildly increase health during ageing but not lifespan in mice," *Aging Cell.* (2021), doi: 10.1111/acel.13479.

190 K. Zarse et al., "Low-dose lithium uptake promotes longevity in humans and metazoans," *Eur J Nutr.* (2011), 50:387-389. doi: 10.1007/s00394-011-0171-x.

191 Sofola-Adesakin et al., "Lithium suppresses Aβ pathology."

192 M. Gandhi et al., "Rediscovering a Forgotten Disease," *Diseases* (2023), doi: 10.3390/diseases11020078.

193 Arnold D: British India and the "beriberi problem", 1798-1942. *Med Hist.* 2010, 54:295-314. doi: 10.1017/s0025727300004622.

194 Marean CW et al., "Als die Menschen fast ausstarben," *Spektrum der Wissenschaft.* 2010, 12:59-65, https://www.spektrum.de/magazin/als-die-menschen -fast-ausstarben/1050007.

195 Behar DM et al., "The dawn of human matrilineal diversity," *Am J Hum Genet.* 2008, 82:1130-40. doi: 10.1016/j.ajhg.2008.04.002.

196 Marean CW et al., "Early human use of marine resources and pigment in South Africa during the Middle Pleistocene," *Nature* 2007, 449:905-908, doi: 10.1038 /nature06204.

197 A. Gibbons, "Humans' head start: new views of brain evolution," *Science* (2002), 296:835-837. doi: 10.1126/science.296.5569.835.

198 S. B. Eaton & M. Konner, "Paleolithic nutrition. A consideration of its nature and current implications," *N Engl J Med.* (1985), 312: 283-289, doi: 10.1056 /NEJM198501313120505.

199 B. Stetka, "By Land or by Sea: How Did Early Humans Access Key Brain-Building Nutrients?" *Scientific American* (2016); www.scientificamerican.com/article /by-land-or-by-sea-how-did-early-humans-access-key-brain-building-nutrients.

200 Ibid.

201 A. Gibbons, "Humans' head start: new views of brain evolution," *Science* (2002), 296:835-837. doi: 10.1126/science.296.5569.835.

202 Stetka, "By Land or by Sea."

203 M. A. Crawford & C. L. Broadhurst, "The role of docosahexaenoic and the marine food web as determinants of evolution and hominid brain development: the challenge for human sustainability," *Nutr Health* (2012), 21:17-39, doi: 10.1177/0260106012437550.

204 C. B. Ruff et al., "Body mass and encephalization in *Pleistocene Homo*," *Nature* (1997), 387:173-176, https://pubmed.ncbi.nlm.nih.gov/9144286/.

205 Marean et al., "Early human use of marine resources and pigment."

206 T. F. Strasser et al., "Stone Age Seafaring in the Mediterranean: Evidence from the Plakias Region for Lower Palaeolithic and Mesolithic Habitation of Crete," *Hesperia* (2010), 79:145-190, doi: 10.2972/hesp.79.2.145.

207 C. W. Marean, "The origins and significance of coastal resource use in Africa and Western Eurasia," *J Hum Evol.* (2014), 77:17-40. doi: 10.1016 /j.jhevol.2014.02.025.

208 Ibid.

209 Bregman R: Im Grunde gut: Eine neue Geschichte der Menschheit. Rowohlt 2021, Pg. 69.

210 L. A. Dugatkin & L. Trut, "How to Tame A Fox and Build a Dog," *Scientific American* (2017), https://www.americanscientist.org/article/how-to-tame -a-fox-and-build-a-dog.

211 Bregman R: Im Grunde gut: Eine neue Geschichte der Menschheit. Rowohlt 2021, Pg. 85.

212 N. Alenina & F. Klempin, "The role of serotonin in adult hippocampal neurogenesis," *Behav Brain Res.* (2015), 277:49-57. doi: 10.1016/j. bbr.2014.07.038.; Y. T. Lin et al., "Oxytocin stimulates hippocampal neurogenesis via oxytocin receptor expressed in CA3 pyramidal neurons," *Nat Commun.* (2017), doi: 10.1038/s41467-017-00675-5.

213 Bregman R: Im Grunde gut: Eine neue Geschichte der Menschheit. Rowohlt 2021, Pg. 89.

214 Ibid. Pg. 85.

215 P. Rakic, "Limits of neurogenesis in primates," *Science.* (1985), 227:1054-1056. doi: 10.1126/science.3975601.; A. Duque et al., "An assessment of the existence of adult neurogenesis in humans and value of its rodent models for neuropsychiatric diseases," *Mol Psychiatry.* (2022), 27:377-382. doi: 10.1038/s41380-021-01314-8.

216 G. Kempermann et al., "Human Adult Neurogenesis: Evidence and Remaining Questions," *Cell Stem Cell.* 23:25-30, (2018), https://doi.org/10.1016/j. stem.2018.04.004.

217 D. N. Abrous et al., "A Baldwin interpretation of adult hippocampal neurogenesis: from functional relevance to physiopathology," *Mol Psychiatry.* (2022), 27:383-402. doi: 10.1038/s41380-021-01172-4.

218 M. K. Skinner, "Environmental Epigenetics and a Unified Theory of the Molecular Aspects of Evolution: A Neo-Lamarckian Concept that Facilitates Neo-Darwinian Evolution," *Genome Biol Evol.* (2015), 7:1296-302. doi: 10.1093/gbe/evv073.

219 L. Bonfanti & I. Amrein, "Editorial: Adult Neurogenesis: Beyond Rats and Mice," *Front Neurosci.* (2018), doi: 10.3389/fnins.2018.00904.

220 Abrous et al., "A Baldwin interpretation."

221 Marean, "The origins and significance of coastal resource use."

222 Wilson EO: Die soziale Eroberung der Erde: Eine biologische Geschichte des Menschen. C.H.Beck 2013.

223 M. R. Goldstein & L. Mascitelli, "Is violence in part a lithium deficiency state? *Med Hypotheses.* (2016), doi: 10.1016/j.mehy.2016.02.002.

224 J. Greenblatt, "Irritability, Anger, and Rage: Lithium Deficiency Syndrome," *Townsend Letter.* (2022), 20-25. https://www.jamesgreenblattmd.com/wp-content

/uploads/2022/10/Irritability-Anger-and-Rage_GREENBLATT-Townsend-Letter Oct.-2022.pdf.

225 Nehls M: Herdengesundheit: Der Weg aus der Corona-Krise und die natürliche Alternative zum globalen Impfprogramm. Mental Enterprises 2022.

226 S. Ye et al., "Higher oxygen content and transport characterize high-altitude ethnic Tibetan women with the highest lifetime reproductive success," *Proc Natl Acad Sci USA*. (2024), doi: 10.1073/pnas.2403309121.

227 O. Bar-Yosef, "The role of western Asia in modern human origins," *Philos Trans R Soc Lond B Biol Sci.* (1992), 337:193-200, doi: 10.1098/rstb.1992.0097.

228 M. P. Richards et al., "Isotope evidence for the intensive use of marine foods by Late Upper Palaeolithic humans," *J Hum Evol.* (2005), 49:390-394, doi: 10.1016 /j.jhevol.2005.05.002.

229 H. Bocherens et al., "Isotopic evidence for diet and subsistence pattern of the Saint Césaire I Neanderthal: review and use of a multi-source mixing model," *J Hum Evol.* (2005), 49:71-87, doi: 10.1016/j.jhevol.2005.03.003.

230 H. Choudhry & M. Nasrullah, "Iodine consumption and cognitive performance: Confirmation of adequate consumption," *Food Sci Nutr* (2018), 6:1341-1351 doi: 10.1002/fsn3.694. Erratum in: doi: 10.1002/fsn3.2123.

231 E. P. Wiedmer & Woermann UJ: Erinnerung an ein vergessen gegangenes Krankheitsbild. https://www.iml.unibe.ch/themen/uebersichten/artikel/kretinis-mus (2017, last accessed on 9 October 2024).

232 https://www.jod.de/zwangsjodierung/die-geschichte-der-jodmangelprophylaxe-in-deutschland (last accessed on 9 October 2024).

233 https://www.degam.de/files/Inhalte/Leitlinien-Inhalte/Dokumente/DEGAM-S2-Leitlinien/053-046_Erhoehter%20TSH-Wert%20in%20der%20Hausarztpraxis /oeffentlich/S2k%20LL%20TSH%20ver%C3%B6ffentlicht%202023/DEGAM _LL_TSH_2023_Kurz_RZ_010723.pdf (2023, zuletzt abgerufen am 27.12.2024).

234 F. Delange, "The role of iodine in brain development," *Proc Nutr Soc.* (2000), 59:75-79. doi: 10.1017/s0029665100000094.

235 B. G. Biban & C. Lichiardopol, "Iodine Deficiency, Still a Global Problem?" *Curr Health Sci J.* (2017), 43:103-111. doi: 10.12865/CHSJ.43.02.01.

236 W. W. Takele et al., "Two-thirds of pregnant women attending antenatal care clinic at the University of Gondar Hospital are found with subclinical iodine deficiency, 2017," *BMC Res Notes* (2018), https://www.ncbi.nlm.nih.gov/pmc/articles/ PMC6192361/.

237 C. B. Businge et al., "Iodine nutrition status in Africa: potentially high prevalence of iodine deficiency in pregnancy even in countries classified as iodine sufficient," *Public Health Nutr* (2021), 24:3581-3586. https://www.ncbi.nlm.nih.gov/pmc /articles/PMC8369456/.

238 M. J. Orlich & G. E. Fraser, "Vegetarian diets in the Adventist Health Study 2: a review of initial published findings," *Am J Clin Nutr.* (2014), 1:353-138.

doi: 10.3945/ajcn.113.071233.; https://adventistreview.org/lifestyles/well-being/adventisthealth-update-2/ (08/21/2024, last accessed on 11/13/2024).

239 Nehls, M., *Die Algenöl-Revolution: Lebenswichtiges Omega-3 – Das pflanzliche Lebenselixier aus dem Meer*. Heyne 2022.

240 B. Stetka, "By Land or by Sea: How Did Early Humans Access Key Brain-Building Nutrients?" *Scientific American*. (2016); www.scientificamerican.com/article/by-land-or-by-sea-how-did-early-humans-access-key-brain-building-nutrients.

241 https://ods.od.nih.gov/factsheets/Iodine-Consumer/ (5/1/2024, last accessed on 10/22/2024).

242 M. Sprague et al., "Iodine Content of Wild and Farmed Seafood and Its Estimated Contribution to UK Dietary Iodine Intake," *Nutrients*. (2021), doi: 10.3390/nu14010195.; . K. M. Eckhoff & A. Maage, "Iodine Content in Fish and Other Food Products from East Africa Analyzed by ICP-MS," *Journal of Food Composition and Analysis*. (1997), 10:270-282, https://doi.org/10.1006/jfca.1997.0541.

243 H. Barbosa et al., "Lithium: A review on concentrations and impacts in marine and coastal systems," *Sci Total Environ*. (2023), doi: 10.1016/j.scitotenv.2022.159374.

244 F. Thibon et al., "Bioaccumulation of Lithium Isotopes in Mussel Soft Tissues and Implications for Coastal Environments," *ACS Earth and Space Chemistry* (2021), doi: 10.1021/acsearthspacechem.1c00045.

245 Marean et al., "Early human use of marine resources and pigment."

246 Y. Tamari & K. Tsuchiya, "Lithium Content of Fish in the Ocean: Investigation of raw, drying and canned fish available," *Biomedical Research on Trace Elements*. (2004), 15:248-258, https://doi.org/10.11299/brte.15.248.

247 G. N. Schrauzer & E. de Vroey, "Effects of nutritional lithium supplementation on mood. A placebo-controlled study with former drug users," *Biol Trace Elem Res*. (1994), 40:89-101. doi: 10.1007/BF02916824.

CHAPTER 3

1 G. N. Schrauzer, "Lithium: occurrence, dietary intakes, nutritional essentiality," *J Am Coll Nutr*. (2002), 21:14-21. doi: 10.1080/07315724.2002.10719188.

2 Verbraucherzentrale: Lithiumorotat: Welche Wirkung hat dieses Nahrungsergänzungsmittel und weshalb ist es in Deutschland so schwer und teuer zu bekommen? https://www.verbraucherzentrale.de/faq/projekt-klartext-nem/lithiumorotat-40582.

3 E. B. Dawson et al., "Relationship of lithium metabolism to mental hospital admission and homicide," *Dis Nerv Syst. (*1972), 33:546-556. https://pubmed.ncbi.nlm.nih. gov/4648454/.

4 G. N. Schrauzer & K. P. Shrestha, "Lithium in drinking water and the incidences of crimes, suicides, and arrests related to drug addictions," *Biol Trace Elem Res*. (1990), 25:105-113. doi: 10.1007/BF02990271.

5 Schrauzer & de Vroey, "Effects of nutritional lithium supplementation on mood."

6 M. A. Nunes et al., "Microdose lithium treatment stabilized cognitive impairment in patients with Alzheimer's disease," *Curr Alzheimer Res* (2013), 10:104-107, www.ncbi.nlm.nih.gov/pubmed/22746245.
7 V. A. Fajardo et al., "Trace lithium in Texas tap water is negatively associated with allcause mortality and premature death," *Appl Physiol Nutr Metab.* (2018), 43:412-414. doi: 10.1139/apnm-2017-0653.
8 Schrauzer, "Lithium: occurrence, dietary intakes, nutritional essentiality."
9 T. M. Marshall, "Lithium as a nutrient." *J Am Phys Sur.* (2015), 20:104-109. https://www. jpands.org/vol20no4/marshall.pdf.
10 J. Enderle et al., "Plasma Lithium Levels in a General Population: A Cross-Sectional Analysis of Metabolic and Dietary Correlates," *Nutrients.* (2020), doi: 10.3390/nu12082489.
11 https://www.imd-berlin.de/fachinformationen/diagnostikinformationen/lithium-Ein-essentielles-spurenelement (2025, last accessed on 12/31/2024).
12 https://www.mlhb.de/labor/profil (zuletzt abgerufen am 31.12.2024).
13 U. Seidel et al., "Lithium-Rich Mineral Water is a Highly Bioavailable Lithium Source for Human Consumption," *Mol Nutr Food Res.* (2019), doi: 10.1002/mnfr.201900039.
14 N. M. de Roos et al., "Serum lithium as a compliance marker for food and supplement intake," *Am J Clin Nutr.* (2001), 73:75-79. doi: 10.1093/ajcn/73.1.75.
15 A. Post et al., "Dietary lithium intake, graft failure and mortality in kidney transplant recipients," *Nephrol Dial Transplant.* (2023), 38:1867-1879. doi: 10.1093/ndt/gfac340.
16 M. Długaszek et al., "Lithium supply in the daily food rations of students," *Probl Hig Epidemiol.* (2012), 93:867-870, http://www.phie.pl/phe.php?opc=AR&lng=en&art=858.
17 R. Van Cauwenbergh et al., "Daily dietary lithium intake in Belgium using duplicate portion sampling," *Z Lebensm Unters Forsch.* 1999, 208:153-155, doi: 10.1007/s002170050393.
18 Schrauzer, "Lithium: occurrence, dietary intakes, nutritional essentiality."
19 G. N. Schrauzer et al., "Lithium in scalp hair of adults, students, and violent criminals. Effects of supplementation and evidence for interactions of lithium with vitamin B12 and with other trace elements," *Biol Trace Elem Res.* (1992), 34:161-176. doi: 10.1007/BF02785244.
20 World Health Organization, "Trace elements in human nutrition and health," (1996), https://www.who.int/publications/i/item/9241561734, S. 224.
21 O. I. Sobolev et al., "Lithium in the natural environment and its migration in the trophic chain," *Ukr. J. Ecol.* (2019), 9:195–203. https://cyberleninka.ru/article/n/lithium-in-the-natural-environment-and-its-migration-in-the-trophic-chain.
22 https://www.imd-berlin.de/fachinformationen/diagnostikinformationen/lithium-Ein-essentielles-spurenelement (2025, last accessed on 12/31/2024).
23 https://echa.europa.eu/registration-dossier/-/registered-dossier/15034/7/6/1.

24 G. J. Fosmire, "Zinc toxicity," *Am J Clin Nutr.* (1990), 51:225-227. doi: 10.1093/ajcn/51.2.225.
25 https://ods.od.nih.gov/factsheets/Zinc-HealthProfessional/ (09/28/2022, last ac-cessed on 01/08/2025).
26 https://www.gesundheit.gv.at/leben/ernaehrung/vitamine-mineralstoffe/spurenele-mente/selen.html.
27 U. Tinggi, "Essentiality and toxicity of selenium and its status in Australia: a review," *Toxicol Lett.* (2003), 137:103-110. doi: 10.1016/s0378-4274(02)00384-3.
28 https://www.msdmanuals.com/professional/nutritional-disorders/mineral-deficien-cy-and-toxicity/selenium-toxicity.
29 https://www.scientificamerican.com/article/strange-but-true-drinking-too-much-water-can-kill/ (06/21/2007, last accessed 12/29/2024); https://www.nbcnews.com/id/wbna16614865 (1/14/2007, last accessed on 12/29/2024).
30 Young W: Review of lithium effects on brain and blood. Cell Transplant. 2009, 18:951-75. doi: 10.3727/096368909X471251.
31 S. E. Cleaveland, "A case of poisoning by lithium. Presenting some new features," *JAMA.* (1913), 60:722. doi: 10.1001/jama.1913.04340100014005.
32 Grafik inspiriert durch Dr. med. Christian Schellenberg.
33 V. S. Gallicchio & M. G. Chen, "Modulation of murine pluripotential stem cell proliferation in vivo by lithium carbonate," *Blood.* (1980), 56:1150-1152. https://pubmed.ncbi.nlm.nih.gov/7437517/; V. S. Gallicchio & M. G. Chen, Influence of lithium on proliferation of hematopoietic stem cells," *Exp Hematol.* (1981), 9:804-810, https://pubmed.ncbi.nlm.nih.gov/7318982/; Q. Wang et al., "Lithium, an anti-psychotic drug, greatly enhances the generation of induced pluripotent stem cells," *Cell Res.* (2011), 21:14241435. doi: 10.1038/cr.2011.108.
34 Schrauzer, "Lithium: occurrence, dietary intakes, nutritional essentiality."
35 Gesundheitskrise bei Kindern: Kleine Änderung verbessert alles! (Praxis-Erfolg!):
36 https://www.mayoclinic.org/drugs-supplements/lithium-oral-route/description/drg-20064603#drug-proper-use, (1/8/2024, last accessed on 2/2/2025).
37 D. Janiri et al., "Use of Lithium in Pediatric Bipolar Disorders and Externalizing Childhood related Disorders: A Systematic Review of Randomized Controlled Trials," *Curr Neuropharmacol.* (2023), 21:1329-1342. doi: 10.2174/1570159X21666230126153105.
38 Gesundheitskrise bei Kindern: Kleine Änderung verbessert alles! (Praxis-Erfolg!): https://www.youtube.com/watch?v=sMEWisa-zZc (07/27/2024).
39 Fallberichte, Echte, "Wie Lithium Großartiges bei Kindern und Jugendlichen bewirkt (Erstaunlich!)," https://www.youtube.com/watch?v=tP3dXSEKaNE&t=187s (09/22/2024).
40 Nehls M: Lithium ist essentiell, auch für Kinder. https://michael-nehls.de/lithiumfuer-kinder/ (June 2024).

41 F. Harari et al., "Environmental exposure to lithium during pregnancy and fetal size: a longitudinal study in the Argentinean Andes," *Environ Int.* (2015), 77:48-54. doi: 10.1016/j.envint.2015.01.011.

42 E. M. Poels et al., "Lithium exposure during pregnancy increases fetal growth," *J Psychopharmacol.* (2021), 35:178-183. doi: 10.1177/0269881120940914.

43 S. J. Jacobson et al., "Prospective multicentre study of pregnancy outcome after lithium exposure during first trimester," *Lancet.* (1992), 339:530-533. doi: 10.1016/01406736(92)90346-5.

44 G. Concha et al., "High-level exposure to lithium, boron, cesium, and arsenic via drinking water in the Andes of northern Argentina," *Environ Sci Technol.* (2010), 44:68756980. doi: 10.1021/es1010384.

45 J. Hu et al., "Prenatal metal mixture exposure and birth weight: A two-stage analysis in two prospective cohort studies," *Eco Environ Health.* (2022), 1:165-171. doi: 10.1016/j. eehl.2022.09.001.

46 R. Quansah et al., "Association of arsenic with adverse pregnancy outcomes/infant mortality: a systematic review and meta-analysis." *Environ Health Perspect.* (2015), 123:412-421. doi: 10.1289/ehp.1307894.

47 Uran in Mineralwasser (Stand Mai 2009), https://www.foodwatch.org/fileadmin /foodwatch.de/news/Uran-in-Mineralwasser_20090518_01.pdf; Birke M. et al., "Distribution of uranium in German bottled and tap water," *J Geochemical Exploration.* 2010, 107: 272-282, doi: 10.1016/j.gexplo.2010.04.003.

48 W. Zhang et al., "Association of adverse birth outcomes with prenatal uranium exposure: A population-based cohort study," *Environ Int.* (2020), doi: 10.1016 /j.envint.2019.105391.

49 N. M. de Roos et al., "Serum lithium as a compliance marker for food and supplement intake," *Am J Clin Nutr.* (2001), 73:75-79. doi: 10.1093/ajcn/73.1.75.

50 S. Hassan et al., "Lithium toxicity in the setting of nonsteroidal anti-inflammatory medications," *Case Rep Nephrol.* (2013), doi: 10.1155/2013/839796.; G. S. Malhi et al., "Lithium therapy and its interactions," *Aust Prescr.* (2020), 43:91-93. doi: 10.18773/austprescr.2020.024. Erratum in: doi: 10.18773/austprescr.2020.041.

51 B. Shine et al., "Long-term effects of lithium on renal, thyroid, and parathyroid function: a retrospective analysis of laboratory data," *Lancet.* (2015), 386:461-468. doi: 10.1016/S0140-6736(14)61842-0.

52 K. Broberg et al., "Lithium in drinking water and thyroid function," *Environ Health Perspect.* (2011), 119:827-830. doi: 10.1289/ehp.1002678.

53 E. V. Popova et al., "Boron A potential goiterogen?" *Med Hypotheses.* (2017), 104:63-67. doi: 10.1016/j.mehy.2017.05.024.

54 A. Esform et al., "Environmental arsenic exposure and its toxicological effect on thyroid function: a systematic review," *Rev Environ Health.* (2021), 37:281-289. doi: 10.1515/reveh-2021-0025.

55 M. Alda, "Lithium in the treatment of bipolar disorder: pharmacology and pharmacogenetics," *Mol Psychiatry.* (2015), 20:661-670. doi: 10.1038/mp.2015.4.

56 W. J. Scotton et al., "Serotonin Syndrome: Pathophysiology, Clinical Features, Management, and Potential Future Directions," *Int J Tryptophan Res.* (2019), doi: 10.1177/1178646919873925.

57 B. D. Fields & K. A. Olive, "Big bang nucleosynthesis." *Nuclear Physics A.* (2006), 777:208-225. doi: 10.1016/j.nuclphysa.2004.10.033.

58 H. Aral & A. Vecchio-Sadus, "Toxicity of lithium to humans and the environment—a literature review," *Ecotoxicol Environ Saf.* (2008), 70:349-356. doi: 10.1016/j.ecoenv.2008.02.026.

59 T. Ushikubo et al., "Lithium in Jack Hills zircons: evidence for extensive weathering of Earth's earliest crust," *Earth Planet Sci Lett.* (2008), 272:666–676. doi: 10.1016 /j.epsl.2008.05.032.

60 K. K. Turekian & K. H. Wedepohl, "Distribution of the Elements in Some Major Units of the Earth's Crust," *GSA Bulletin.* (1961), 72:175-192. doi: 10.1130/0016-7606(1961)72[175:DOTEIS]2.0.CO;2.

61 A. Gutiérrez-Ravelo et al., "Toxic Metals (Al, Cd, Pb) and Trace Element (B, Ba, Co, Cu, Cr, Fe, Li, Mn, Mo, Ni, Sr, V, Zn) Levels in Sarpa Salpa from the North-Eastern Atlantic Ocean Region," *Int J Environ Res Public Health.* (2020), doi: 10.3390 /ijerph17197212.; L. Alberghini et al., "Microplastics in Fish and Fishery Products and Risks for Human Health: A Review," *Int J Environ Res Public Health.* (2022), doi: 10.3390/ijerph20010789.

62 Nehls, M., *Die Algenöl-Revolution: Lebenswichtiges Omega-3 – Das pflanzliche Lebenselixier aus dem Meer.* Heyne 2022.

63 Nehls, M., *Herdengesundheit: Der Weg aus der Corona-Krise und die natürliche Alternative zum globalen Impfprogramm.* Mental Enterprises 2022.

64 A. M. Iordache et al., "Lithium Content and Its Nutritional Beneficence, Dietary Intake, and Impact on Human Health in Edibles from the Romanian Market," *Foods.* (2024), doi: 10.3390/foods13040592.

65 Gonzalez-Weller D. et al., "Dietary intake of barium, bismuth, chromium, lithium, and strontium in a Spanish population (Canary Islands, Spain)," *Food Chem Toxicol.* 2013, 62:856-858. https://pubmed.ncbi.nlm.nih.gov/24416776/.

66 T. Filippini et al., "Dietary Estimated Intake of Trace Elements: Risk Assessment in an Italian Population," *Exposure and Health.* (2020), doi: 10.1007/s12403 -019-00324-w.

67 A. Naeem et al., "Lithium: Perspectives of nutritional beneficence, dietary intake, biogeochemistry, and biofortification of vegetables and mushrooms," *Sci Total Environ.* (2021), doi: 10.1016/j.scitotenv.2021.149249.

68 M. Kalinowska et al., "The influence of two lithium forms on the growth, L-ascorbic acid content and lithium accumulation in lettuce plants," *Biol Trace Elem Res.* (2013), 152:251-257. doi: 10.1007/s12011-013-9606-y.; B. Hawrylak-Nowak et al., "A study on selected physiological parameters of plants grown under lithium supplementation," *Biol Trace Elem Res.* (2012), 149:425-430. doi: 10.1007/s12011-012-9435-4.

69 N. Gayathri et al., "Effect of lithium on seed germination and plant growth of Amaranthus viridis," *J. Appl. Nat. Sci.* (2022), 14:133-139. doi: 10.31018/JANS.V14I1.3165.

70 B. Shahzad et al., "Toxicity in plants: Reasons, mechanisms and remediation possibilities. A review," *Plant Physiol Biochem.* (2016), 107:104-115. doi: 10.1016/j.plaphy.2016.05.034.

71 L. Kavanagh et al., "Induced Plant Accumulation of Lithium," *Geosciences.* (2018), doi: 10.3390/geosciences8020056.

72 T. G. Ammari et al., "The occurrence of lithium in the environment of the Jordan Valley and its transfer into the food chain," *Environ Geochem Health.* (2011), 33:427-437. doi: 10.1007/s10653-010-9343-5.

73 U. Seidel et al., "Lithium (Li) concentrations [μg/L] in 381 German mineral and medicinal waters," *ResearchGate* (2020), https://www.researchgate.net/publication/341287028_Lithium_Li_concentrations_gL_in_381_German_mineral_and_medicinal_waters.

74 Spitzer M & Graf H: Lithium im Trinkwasser – Lithium ins Trinkwasser? Nervenheilkunde 2010, 29:157-158, https://oparu.uni-ulm.de/items/b0382f07-6498-47f99b60-19800bb9aed4; P. Araya et al., "Lithium in Drinking Water as a Public Policy for Suicide Prevention: Relevance and Considerations," *Front Public Health.* (2022), doi: 10.3389/fpubh.2022.805774.; G. N. Schrauzer & K. P. Shrestha, "Lithium in drinking water and the incidences of crimes, suicides, and arrests related to drug addictions," *Biol Trace Elem Res.* (1990), 25:105-113. doi: 10.1007/BF02990271.

75 https://michael-nehls.de/infos/apothekenverzeichnis-fuer-lithium/.

76 https://michael-nehls.de/infos/therapeutenliste-orthomolekulare-medizin/.

77 Kostengünstiges Lithium eine Anleitung. (06/28/2024) https://www.youtube.com/watch?v=4dz2oilGh-s.

78 T. M. Marshall, "Lithium as a Nutrient," *J American Physicians and Surgeons.* (2015), 20:104-109.

79 C. Moisa et al., "Aspects Regarding the Relationship Between the Stability of Six Magnesium Compounds and their Cellular Uptake in Mice," *Revista de Chimie.* (2020), 71:193-204. doi: 10.37358/RC.20.10.8364.

80 T. M. Marshall, "Lithium as a Nutrient," *J American Physicians and Surgeons* (2015), 20:104-109.

81 H. A. Nieper, "The clinical applications of lithium orotate. A two years study," *Agressologie.* (1973), 14:407-411. https://pubmed.ncbi.nlm.nih.gov/4607169/.

82 Y. Shinoda et al., "Functional characterization of human organic anion transporter 10 (OAT10/SLC22A13) as an orotate transporter," *Drug Metab Pharmacokinet.* (2022), doi: 10.1016/j.dmpk.2021.100443.

83 F. Bardanzellu et al., "The Human Breast Milk Metabolome in Overweight and Obese Mothers," *Front Immunol.* (2020), doi: 10.3389/fimmu.2020.01533.

84 Wehrmuller K. et al., "Orotsauregehalt in Kuh-, Schaf- und Ziegenmilch," https://www.agrarforschungschweiz.ch/2008/07/orotsaeuregehalt-in-kuh-schaf-und-ziegenmilch/.

85 F. Karatas, "An investigation of orotic acid levels in the breastmilk of smoking and non-smoking mothers," *Eur J Clin Nutr.* (2002), 56:958-960. doi: 10.1038/sj.ejcn.1601420.

86 C. Schiopu et al., "Magnesium Orotate and the Microbiome-Gut-Brain Axis Modulation: New Approaches in Psychological Comorbidities of Gastrointestinal Functional Disorders," *Nutrients.* (2022), doi: 10.3390/nu14081567.; L. D. Wright et al., "Orotic acid, a growth factor for lactobacillus bulgaricus," *J. Am. Chem. Soc.* (1950), 72:2312-2313. doi: 10.1021/ja01161a544.

87 M. A. Kling et al., "Rat brain and serum lithium concentrations after acute injections of lithium carbonate and orotate," *J Pharm Pharmacol.* (1978), 30:368-370. doi: 10.1111/j.2042-7158.1978.tb13258.x.

88 A. G. Pacholko & L. K. Bekar, "Different pharmacokinetics of lithium orotate inform why it is more potent, effective, and less toxic than lithium carbonate in a mouse model of mania," *J Psychiatr Res.* (2023), 164:192-201. doi: 10.1016/j.jpsychires.2023.06.012.

89 C. Henry, "Lithium side-effects and predictors of hypothyroidism in patients with bipolar disorder: sex differences," *J Psychiatry Neurosci.* (2002), 27:104-107. https://pmc. ncbi.nlm.nih.gov/articles/PMC161639/.

90 D. F. Smith & M. Schou, "Kidney function and lithium concentrations of rats given an injection of lithium orotate or lithium carbonate," *J Pharm Pharmacol.* (1979), 31:161-163. doi: 10.1111/j.2042-7158.1979.tb13461.x.

91 Murbach T. S. et al., "A toxicological evaluation of lithium orotate," *Regul Toxicol Pharmacol.* 2021, doi: 10.1016/j.yrtph.2021.104973.

92 H. E. Sartori, "Lithium orotate in the treatment of alcoholism and related conditions," *Alcohol.* (1986), 3:97-100. doi: 10.1016/0741-8329(86)90018-2. J. Fawcett et al., "A double-blind, placebo-controlled trial of lithium.

93 J. Fawcett et al., "A double-blind, placebo-controlled trial of lithium carbonate therapy for alcoholism," *Arch Gen Psychiatry.* (1987), 44:248-256. doi: 10.1001/archpsyc.1987.01800150060008.; W. Dorus et al., "Lithium treatment of depressed and nondepressed alcoholics," *JAMA.* (1989), 262:1646-1652. https://pubmed.ncbi.nlm.nih. gov/2504944/.

94 J. R. de la Fuente et al., "A controlled study of lithium carbonate in the treatment of alcoholism," *Mayo Clin Proc.* (1989), 64:177-180. doi: 10.1016/s0025-6196(12)65672-9.

95 A. G. Pacholko & L. K. Bekar, "Lithium orotate: A superior option for lithium therapy?" *Brain Behav.* (2021), doi: 10.1002/brb3.2262.

96 S. P. Tyrer et al., "Bioavailability of lithium carbonate and lithium citrate: a comparison of two controlled-release preparations," *Pharmatherapeutica.* (1982), 3:243-246. https://pubmed.ncbi.nlm.nih.gov/6815664/.

97 R. Olbrich et al., "Lithium in the treatment of chronic alcoholic patients with brain damage—a controlled study," *Nervenarzt.* 1991, 62:182-186. https://pubmed.ncbi.nlm.nih.gov/2052117/.

98 M. Daunderer, "Lithium aspartate in drug dependence," *Fortschr Med.* (1982), 100:15001502. https://pubmed.ncbi.nlm.nih.gov/7129311/.

99 J. M. Morrison Jr et al., "Plasma and brain lithium levels after lithium carbonate and lithium chloride administration by different routes in rats," *Proc Soc Exp Biol Med.* (1971), 137:889-892. doi: 10.3181/00379727-137-35687.

100 A. Edström & G. Persson, "Comparison of side effects with coated lithium carbonate tablets and lithium sulphate preparations giving medium-slow and slow-release," *Acta Psychiatr Scand.* (1977), 55:153-158. doi: 10.1111/j.1600-0447.1977.tb00152.x.

101 W. Sim et al., "Antimicrobial Silver in Medicinal and Consumer Applications: A Patent Review of the Past Decade (2007-2017)," *Antibiotics* (Basel.) (2018), doi: 10.3390/antibiotics7040093.

102 J. Mauermann & W. Bauer: Kolloide Mineralien & Spurenelemente in kolloidaler Form. https://alternativgesund.de/media/4c/b2/8e/1699756401/eBook-Jutta-Mauermann-Kolloide-6-Auflage.pdf, S.63.

103 T. M. Marshall, "Lithium as a nutrient," *J Am Phys Sur.* (2015), 20:104-109. https://www. jpands.org/vol20no4/marshall.pdf.

CHAPTER 4

1 C. Zipfel, "Plant pattern-recognition receptors," *Trends Immunol.* (2014), 35:345-351. doi: 10.1016/j.it.2014.05.004.

2 D. Li & M. Wu, "Pattern recognition receptors in health and diseases," *Signal Transduct Target Ther.* (2021), doi: 10.1038/s41392-021-00687-0.

3 Y. Zhao et al., "SARS-CoV-2 spike protein interacts with and activates TLR4," *Cell Res.* (2021), 31:818-820. doi: 10.1038/s41422-021-00495-9. Erratum in: doi: 10.1038/s41422-021-00501-0.

4 R. Ko & S. Y. Lee, "Glycogen synthase kinase 3β in Toll-like receptor signaling," *BMB Rep.* (2016), 49:305-310. doi: 10.5483/bmbrep.2016.49.6.059.

5 K. Shirato & T. Kizaki, "SARS-CoV-2 spike protein S1 subunit induces pro-inflammatory responses via toll-like receptor 4 signaling in murine and human macrophages," *Heliyon.* (2021), doi: 10.1016/j.heliyon.2021.e06187.

6 R. Hanamsagar et al., "Inflammasome activation and IL-1β/IL-18 processing are influenced by distinct pathways in microglia," *J Neurochem.* (2011), 119:736-748. doi: 10.1111/j.1471-4159.2011.07481.x.

7 H. Dong et al., "Lithium ameliorates lipopolysaccharide-induced microglial activation via inhibition of toll-like receptor 4 expression by activating the PI3K/Akt/FoxO1 pathway," *J Neuroinflammation.* (2014), doi: 10.1186/s12974-014-0140-4.

8 H. Yu et al., "Targeting NF-κB pathway for the therapy of diseases: mechanism and clinical study," *Signal Transduct Target Ther.* (2020), doi: 10.1038/s41392-020-00312-6.

9 T. Liu et al., NF-κB signaling in inflammation, *Signal Transduct Target Ther.* (2017), doi: 10.1038/sigtrans.2017.23.

10 R. C. Bransfield, "Suicide and Lyme and associated diseases," *Neuropsychiatr Dis Treat.* (2017), 13:1575-1587. doi: 10.2147/NDT.S136137.

11 L. Butler & K. A. Walker, "The Role of Chronic Infection in Alzheimer's Disease: Instigators, Co-conspirators, or Bystanders?" *Curr Clin Microbiol Rep.* (2021), 8:199-212. doi: 10.1007/s40588-021-00168-6.

12 M. Nehls, *Das indoktrinierte Gehirn: Wie wir den globalen Angriff auf unsere mentale Freiheit erfolgreich abwehren*. Mental Enterprises 2023, S. 168-201.

13 Zhao et al., "SARS-CoV-2 spike protein interacts with and activates TLR4," *Cell Res.* (2021), 31:818-820. doi: 10.1038/s41422-021-00495-9. Erratum in: doi: 10.1038/s41422-021-00501-0.; K. Shirato & T. Kizaki, "SARS-CoV-2 spike protein S1 subunit induces pro-inflammatory responses via toll-like receptor 4 signaling in murine and human macrophages."

14 J. S. Roh & D. H. Sohn, "Damage-Associated Molecular Patterns in Inflammatory Diseases," *Immune Netw.* (2018), doi: 10.4110/in.2018.18.e27.

15 D. Nikolopoulos et al., "Microglia activation in the presence of intact blood-brain barrier and disruption of hippocampal neurogenesis via IL-6 and IL-18 mediate early diffuse neuropsychiatric lupus," *Ann Rheum Dis.* (2023), 82:646-657. doi: 10.1136/ard2022-223506.

16 S. S. Shim & G. E. Stutzmann, "Inhibition of Glycogen Synthase Kinase-3: An Emerging Target in the Treatment of Traumatic Brain Injury," *J Neurotrauma.* (2016), 33:20652076. doi: 10.1089/neu.2015.4177.

17 P. R. Leeds et al., "A new avenue for lithium: intervention in traumatic brain injury," *ACS Chem Neurosci.* (2014), 5:422-433. doi: 10.1021/cn500040g.

18 Y. P. Liu et al., "Tumor necrosis factor-alpha and interleukin-18 modulate neuronal cell fate in embryonic neural progenitor culture," *Brain Res.* (2005), 1054:152-158. doi: 10.1016/j.brainres.2005.06.085.; M. D. Wu et al., "Adult murine hippocampal neurogenesis is inhibited by sustained IL-1β and not rescued by voluntary running," *Brain Behav Immun.* (2012), 26:292-300. doi: 10.1016/j.bbi.2011.09.012.; X. Kong et al., "JAK2/ STAT3 signaling mediates IL-6-inhibited neurogenesis of neural stem cells through DNA demethylation/methylation," *Brain Behav Immun.* (2019), 79:159-173. doi: 10.1016/j.bbi.2019.01.027.

19 Z. R. Patterson & M. R. Holahan, "Understanding the neuroinflammatory response following concussion to develop treatment strategies," *Front Cell Neurosci.* (2012), doi: 10.3389/fncel.2012.00058.

20 Hill AS, Sahay A, Hen R. Increasing Adult Hippocampal Neurogenesis is Sufficient to Reduce Anxiety and Depression-Like Behaviors. *Neuropsychopharmacology.* 2015 Sep;40(10):2368-78. doi: 10.1038/npp.2015.85. Epub 2015 Apr 2. PMID: 25833129; PMCID: PMC4538351.

21 S. J. Lupien et al., "Cortisol levels during human aging predict hippocampal atrophy and memory deficits," *Nat Neurosci.* (1998), 1:69-73. doi: 10.1038/271.

22 A. L. Roberts et al., "Exposure to American Football and Neuropsychiatric Health in Former National Football League Players: Findings From the Football Players Health Study," *Am J Sports Med.* (2019) Oct;47(12):2871-2880. doi: 10.1177/0363546519868989.

23 E. J. Lehman et al., "Neurodegenerative causes of death among retired National Football League players," *Neurology.* (2012), 79:1970-1974. doi: 10.1212 /WNL.0b013e31826daf50.

24 H. J. Bruce et al., "American Football Play and Parkinson Disease Among Men," *JAMA Netw Open.* (2023), doi: 10.1001/jamanetworkopen.2023.28644.

25 G. D. Batty et al., "Dementia in former amateur and professional contact sports participants: population-based cohort study, systematic review, and meta-analysis," *EClinicalMedicine.* (2023), doi: 10.1016/j.eclinm.2023.102056.

26 R. S. Jope et al., "Stressed and Inflamed, Can GSK3 Be Blamed?" *Trends Biochem Sci.* (2017), 42:180-192. doi: 10.1016/j.tibs.2016.10.009.

27 M. Fleshner et al., "Danger Signals and Inflammasomes: Stress-Evoked Sterile Inflammation in Mood Disorders," *Neuropsychopharmacology.* (2017), 42:36-45. doi: 10.1038/npp.2016.125.

28 Y. Cheng et al., "Stress-induced neuroinflammation is mediated by GSK3-dependent TLR4 signaling that promotes susceptibility to depression-like behavior," *Brain Behav Immun.* (2016), 53:207-222. doi: 10.1016/j.bbi.2015.12.012.; M. Fleshner et al., "Cat exposure induces both intra and extracellular Hsp72: the role of adrenal hormones," *Psychoneuroendocrinology.* (2004), 29:1142-1152. doi: 10.1016/j. psyneuen.2004.01.007.

29 G. M. Slavich & M. R. Irwin, "From stress to inflammation and major depressive disorder: a social signal transduction theory of depression," *Psychol Bull.* (2014), 140:774815. doi: 10.1037/a0035302.; A. H. Miller & C. L. Raison, "The role of inflammation in depression: from evolutionary imperative to modern treatment target," *Nat Rev Immunol.* (2016), 16:22-34. doi: 10.1038/nri.2015.5.

30 A. Danese et al., "Elevated inflammation levels in depressed adults with a history of childhood maltreatment," *Arch Gen Psychiatry.* (2008), 65:409-415. doi: 10.1001 /archpsyc.65.4.409.; M. Fleshner et al., "Danger Signals and Inflammasomes: Stress-Evoked Sterile Inflammation in Mood Disorders," *Neuropsychopharmacology.* (2017), 42:36-45. doi: 10.1038/npp.2016.125.

31 J. C. Felger et al., "Inflammation and immune function in post-traumatic stress disorder: mechanisms, consequences and translational implications," In: K. J. Ressler, I. Liberzon (eds) *Neurobiology of Posttraumatic Stress Disorder* (Oxford Press, 2016).

32 H. Xie et al., "Relationship of Hippocampal Volumes and Posttraumatic Stress Disorder Symptoms Over Early Posttrauma Periods," *Biol Psychiatry Cogn Neurosci Neuroimaging.* (2018), 3:968-975. doi: 10.1016/j.bpsc.2017.11.010.; M. W. Logue et al., "Smaller Hippocampal Volume in Posttraumatic Stress Disorder: A Multisite

ENIGMA-PGC Study: Subcortical Volumetry Results From Posttraumatic Stress Disorder Consortia," *Biol Psychiatry* (2018), 83:244-253. https://www.ncbi.nlm.nih.gov/pmc/articles/PMC5951719/.

33 D. Berntsen & D. C. Rubin "Pretraumatic Stress Reactions in Soldiers Deployed to Afghanistan," *Clin Psychol Sci* (2015), 3:663-674. https://www.ncbi.nlm.nih.gov/pmc/articles/PMC4564108.

34 M. W. Logue et al., "Smaller Hippocampal Volume in Posttraumatic Stress Disorder: A Multisite ENIGMA-PGC Study: Subcortical Volumetry Results From Posttraumatic Stress Disorder Consortia," *Biol Psychiatry.* (2018), 83:244-253. doi: 10.1016/j.biopsych.2017.09.006.

35 S. Prieto et al., "Posttraumatic stress symptom severity predicts cognitive decline beyond the effect of Alzheimer's disease biomarkers in Veterans," *Transl Psychiatry.* (2023), doi: 10.1038/s41398-023-02354-0.

36 C. R. Schultze-Florey et al., "When grief makes you sick: bereavement induced systemic inflammation is a question of genotype," *Brain Behav Immun.* (2012), 26:1066-1071. doi: 10.1016/j.bbi.2012.06.009.; R. L. Brown, A. S. LeRoy et al., "Grief Symptoms Promote Inflammation During Acute Stress Among Bereaved Spouses," *Psychological Science.* (2022), 33:859-873. doi 10.1177/09567976211059502.

37 A. L. Marsland et al., "Interleukin-6 covaries inversely with hippocampal grey matter volume in middle-aged adults," *Biol Psychiatry.* (2008), 64:484-490. doi: 10.1016/j.biopsych.2008.04.016.

38 Nehls M: Die Methusalem-Strategie: Vermeiden, was uns daran hindert, gesund älter und weiser zu werden. Mental Enterprises 2011.

39 Nehls M: Die Formel gegen Alzheimer: Die Gebrauchsanweisung für ein gesundes Leben Ganz einfach vorbeugen und rechtzeitig heilen. Heyne 2018.

40 Nehls M: Das indoktrinierte Gehirn: Wie wir den globalen Angriff auf unsere mentale Freiheit erfolgreich abwehren. Mental Enterprises 2023.

41 A. R. Sutin et al., "Purpose in life and stress: An individual-participant meta-analysis of 16 samples," *J Affect Disord.* (2024) Jan 15;345:378-385. doi: 10.1016/j.jad.2023.10.149.

42 S. Hassamal, "Chronic stress, neuroinflammation, and depression: an overview of pathophysiological mechanisms and emerging anti-inflammatories," *Front Psychiatry.* (2023), doi: 10.3389/fpsyt.2023.1130989.

43 A. J. Guimond et al., "Sense of purpose in life and inflammation in healthy older adults: A longitudinal study," *Psychoneuroendocrinology.* (2022), doi: 10.1016/j.psyneuen.2022.105746.

44 E. S. Kim et al., "Sense of Purpose in Life and Subsequent Physical, Behavioral, and Psychosocial Health: An Outcome-Wide Approach," *Am J Health Promot.* (2022), 36:137-147. doi: 10.1177/08901171211038545.

45 K. C. Britt et al., "The association between religious beliefs and values with inflammation among Middle-age and older adults," *Aging Ment Health.* (2024), 28:1343-1350. doi: 10.1080/13607863.2024.2335390.; K. M. Vagnini et al.,

"Multidimensional Religiousness and Spirituality Are Associated With Lower Interleukin-6 and C-Reactive Protein at Midlife: Findings From the Midlife in the United States Study," *Ann Behav Med.* (2024), 58:552-562. doi: 10.1093/abm /kaae032.

46 K. A. Muscatell et al., "Socioeconomic status and inflammation: a meta-analysis," *Mol Psychiatry.* (2020), 25:2189-2199. doi: 10.1038/s41380-018-0259-2.

47 A. J. Al Omran et al., "Social Isolation Induces Neuroinflammation And Microglia Overactivation, While Dihydromyricetin Prevents And Improves Them." *J Neuroinflammation.* (2022) Jan, doi: 10.1186/s12974-021-02368-9.

48 J. H. Roh et al., "A potential association between COVID-19 vaccination and development of Alzheimer's disease," *QJM.* (2024), 117:709-716. doi: 10.1093 /qjmed/hcae103.

49 A. Du Preez et al., "Chronic stress followed by social isolation promotes depressive-like behaviour, alters microglial and astrocyte biology and reduces hippocampal neurogenesis in male mice." *Brain Behav Immun.* (2021), 91:24-47. doi: 10.1016/j. bbi.2020.07.015.

50 L. A. Duffner et al., "Associations between social health factors, cognitive activity and neurostructural markers for brain health A systematic literature review and meta-analysis," *Ageing Res Rev.* (2023), doi: 10.1016/j.arr.2023.101986.

51 T. Charlton et al., "Brain-derived neurotrophic factor (BDNF) has direct an-ti -inflammatory effects on microglia," *Front Cell Neurosci.* (2023), doi: 10.3389 /fncel.2023.1188672.; L. Yuan et al., "Oxytocin inhibits lipopolysaccharide -induced inflammation in microglial cells and attenuates microglial activation in lipopolysaccharide-treated mice," *J Neuroinflammation.* (2016), doi: 10.1186 /s12974-016-0541-7.; B. Leuner et al., "Oxytocin stimulates adult neurogenesis even under conditions of stress and elevated glucocorticoids," *Hippocampus.* (2012), 22:861-868. doi: 10.1002/hipo.20947.

52 B. R. Goodin et al., Oxytocin a multifunctional analgesic for chronic deep tissue pain," *Curr Pharm Des.* (2015), 21:906-913. doi: 10.2174/1381612820666141027 111843.

53 M. Nehls, "Unified theory of Alzheimer's disease (UTAD): implications for prevention and curative therapy," *J Mol Psychiatry* (2016), doi: 10.1186/s40303 -016-0018-8.

54 O. Mee-Inta et al., "Physical Exercise Inhibits Inflammation and Microglial Activation," *Cells.* (2019), doi: 10.3390/cells8070691.

55 B. Mahalakshmi et al., "Possible Neuroprotective Mechanisms of Physical Exercise in Neurodegeneration," *Int J Mol Sci.* 2020, doi: 10.3390/ijms21165895.

56 M. V. Lourenco et al., "Exercise-linked FNDC5/irisin rescues synaptic plasticity and memory defects in Alzheimer's models," *Nat Med.* (2019), 25:165-175. doi: 10.1038/s41591-018-0275-4.; M. R. Islam, S. Valaris, M. F. Young, et al., "Exercise hormone irisin is a critical regulator of cognitive function," *Nat Metab.* (2021), 3:1058-1070. doi: 10.1038/s42255-021-00438-z.

57 J. J. Slate-Romano et al., "Irisin reduces inflammatory signaling pathways in inflammation-mediated metabolic syndrome," *Mol Cell Endocrinol.* (2022), doi: 10.1016/j. mce.2022.111676.; M. R. Islam et al., "Exercise hormone irisin is a critical regulator of cognitive function," *Nat Metab.* (2021), 3:1058-1070. doi: 10.1038/s42255-021-00438-z. Erratum in doi: 10.1038/s42255-021-00476-7.

58 P. K. R. Kalluru et al., "Role of erythropoietin in the treatment of Alzheimer's disease: the story so far," *Ann Med Surg* (Lond.) (2024), 86:3608-3614. doi: 10.1097/MS9.0000000000002113.

59 K. I. Erickson et al., "Exercise training increases size of hippocampus and improves memory," *Proc Natl Acad Sci USA.* (2011), 108:3017-3022. doi: 10.1073/pnas.1015950108.

60 E. Kip & L. C. Parr-Brownlie, "Healthy lifestyles and wellbeing reduce neuroinflammation and prevent neurodegenerative and psychiatric disorders," *Front Neurosci.* (2023), doi: 10.3389/fnins.2023.1092537.

61 S. Song et al., "Does Exercise Affect Telomere Length? A Systematic Review and MetaAnalysis of Randomized Controlled Trials," *Medicina* (Kaunas.) (2022), doi: 10.3390/medicina58020242.

62 S. von Holstein-Rathlou et al., "Voluntary running enhances glymphatic influx in awake behaving, young mice," *Neurosci Lett. (*2018), 662:253-258. doi: 10.1016 /j.neulet.2017.10.035.

63 J. P. Karl et al., "Effects of Psychological, Environmental and Physical Stressors on the Gut Microbiota," *Front Microbiol. (*2018), doi: 10.3389/fmicb.2018.02013.; C. Gubert et al., "Exercise, diet and stress as modulators of gut microbiota: Implications for neurodegenerative diseases," *Neurobiol Dis.* (2020), doi: 10.1016 /j.nbd.2019.104621.

64 J. Park et al., "Effects of Long-Term Endurance Exercise and Lithium Treatment on Neuroprotective Factors in Hippocampus of Obese Rats," *Int J Environ Res Public Health. (*2020), doi: 10.3390/ijerph17093317.

65 O. M. Buxton et al., "Acute and delayed effects of exercise on human melatonin secretion," *J Biol Rhythms* (1997), 12:568-574. doi: 10.1177/074873049701200611.; R. J. Reiter et al., "Melatonin and its relation to the immune system and inflammation, *Ann N Y Acad Sci.* (2000), 917:376-386. doi: 10.1111/j.1749-6632.2000.tb05402.x.; M. M. Masternak & A. Bartke, "Growth hormone, inflammation and aging," *Pathobiol Aging Age Relat Dis.* (2012), doi: 10.3402/pba.v2i0.17293.; B. Soler Palacios et al., "Growth Hormone Reprograms Macrophages toward an Anti-Inflammatory and Reparative Profile in an MAFB-Dependent Manner," *J Immunol.* (2020), 205:776-788. doi: 10.4049 /jimmunol.1901330.

66 L. Jiang et al.,: "Association of Sedentary Behavior With Anxiety, Depression, and Suicide Ideation in College Students," *Front Psychiatry.* (2020), doi: 10.3389 /fpsyt.2020.566098.

67 G. Hurtado-Alvarado et al., "Brain Barrier Disruption Induced by Chronic Sleep Loss: Low-Grade Inflammation May Be the Link," *J Immunol Res.* (2016), doi: 10.1155/2016/4576012.

68 Xie L et al., "Sleep drives metabolite clearance from the adult brain," *Science.* 2013, 342:373-377. doi: 10.1126/science.1241224.; Hablitz LM & Nedergaard M: The Glymphatic System: A Novel Component of Fundamental Neurobiology. *J Neurosci.* 2021, 41:7698-7711. doi: 10.1523/JNEUROSCI.0619-21.2021.

69 Nehls M: Das erschöpfte Gehirn: Der Ursprung unserer mentalen Energie – und warum sie schwindet Willenskraft, Kreativität und Fokus zurückgewinnen. Heyne 2022.

70 T. J. LaRocca et al., "Amyloid beta acts synergistically as a pro-inflammatory cytokine," *Neurobiol Dis.* (2021), doi: 10.1016/j.nbd.2021.105493.

71 *Cell* Press Release: "How deep sleep clears a mouse's mind, literally" 8. January 2025, https://www.eurekalert.org/news-releases/1069271.

72 N. L. Hauglund et al., "Norepinephrine-mediated slow vasomotion drives glymphatic clearance during sleep," *Cell.* (2025), doi: 10.1016/j.cell.2024.11.027.

73 S. Billioti de Gage et al., "Benzodiazepine use and risk of Alzheimer's disease: casecontrol study," *BMJ.* (2014), doi: 10.1136/bmj.g5205.

74 L. C. Walker & H. LeVine, "The cerebral proteopathies: neurodegenerative disorders of protein conformation and assembly," *Mol Neurobiol.* (2000), 21:83-95. doi: 10.1385/MN:21:1-2:083.

75 A. M. Mahalakshmi et al., "Impact of Sleep Deprivation on Major Neuroinflammatory Signal Transduction Pathways," *Sleep Vigilance.* (2022), 6:101-114. doi: 10.1007/s41782-022-00203-6.

76 R. Stickgold, "How do I remember? Let me count the ways," *Sleep Med Rev.* (2009), 13:305-308. doi: 10.1016/j.smrv.2009.05.004.

77 V. Oldridge and Dr. Robert Stickgold, "'We Are What We Sleep,' and Rethinking Going to Bed Angry," *Truffld.* (2021), https://truffld.com/what-really-happens-when-wesleep-and-how-memories-form-our-sense-of-self/.

78 M. C. Pascoe et al., "Mindfulness mediates the physiological markers of stress: Systematic review and meta-analysis" *J Psychiatr Res.* (2017), 95:156-178. doi: 10.1016/j.jpsychires.2017.08.004.; W. O. Twal et al., "Yogic breathing when compared to attention control reduces the levels of pro-inflammatory biomarkers in saliva: a pilot randomized controlled trial," *BMC Complement Altern Med.* (2016), doi: 10.1186/s12906-016-1286-7.

79 I. Buric et al., "What Is the Molecular Signature of Mind-Body Interventions? A Systematic Review of Gene Expression Changes Induced by Meditation and Related Practices," *Front Immunol.* (2017), doi: 10.3389/fimmu.2017.00670.

80 E. Kip & L. C. Parr-Brownlie, "Healthy lifestyles and wellbeing reduce neuroinflammation and prevent neurodegenerative and psychiatric disorders," *Front Neurosci.* (2023), doi: 10.3389/fnins.2023.1092537.

81 M. J. Orlich et al., "Vegetarian dietary patterns and mortality in Adventist Health Study 2," *JAMA Intern Med.* (2013), 173:1230-1238. doi: 10.1001/jamainternmed.2013.6473.

82 D. Buettner, "The Secrets of a Long Life," *National Geographic (*2005), https://web. archive.org/web/20071211162254/http://ngm.nationalgeographic.com/ngm/0511/feature1/.

83 Schrauzer, "Lithium: occurrence, dietary intakes, nutritional essentiality."

84 Nehls, M., *Die Algenöl-Revolution: Lebenswichtiges Omega-3 – Das pflanzliche Lebenselixier aus dem Meer*. Heyne 2022.

85 M. Rekatsina et al., "Pathophysiology and Therapeutic Perspectives of Oxidative Stress and Neurodegenerative Diseases: A Narrative Review," *Adv Ther*. (2020), 37:113-139. doi: 10.1007/s12325-019-01148-5.

86 C. Itsiopoulos et al., "The anti-inflammatory effects of a Mediterranean diet: a review," *Curr Opin Clin Nutr Metab Care*. (2022), 25:415-422. doi: 10.1097/MCO.0000000000000872.

87 D. Valera-Gran et al., "The Impact of Foods, Nutrients, or Dietary Patterns on Telomere Length in Childhood and Adolescence: A Systematic Review," *Nutrients*. (2022), doi: 10.3390/nu14193885.

88 K. Gabel, A. Hamm et al., "A Narrative Review of Intermittent Fasting With Exercise," *J Acad Nutr Diet*. (2025), 125:153-171. doi: 10.1016/j.jand.2024.05.015.

89 P. Regmi & L. K. Heilbronn, "Time-Restricted Eating: Benefits, Mechanisms, and Challenges in Translation," *iScience*. (2020), doi: 10.1016/j.isci.2020.101161.

90 M. F. McCarty et al., "Ketosis may promote brain macroautophagy by activating Sirt1 and hypoxia-inducible factor-1," *Med Hypotheses*. (2015), 85:631-639. doi: 10.1016/j.mehy.2015.08.002.

91 S. S. Madhavan et al., "β-hydroxybutyrate is a metabolic regulator of proteostasis in the aged and Alzheimer disease brain," *Cell Chem Biol. (*2025), 32:174-191.e8. doi: 10.1016/j.chembiol.2024.11.001.

92 I. Irfannuddin et al., "The effect of ketogenic diets on neurogenesis and apoptosis in the dentate gyrus of the male rat hippocampus," *J Physiol Sci*. (2021), doi: 10.1186/s12576-020-00786-7.

93 J. Gudden et al., "The Effects of Intermittent Fasting on Brain and Cognitive Function," *Nutrients*. (2021), doi: 10.3390/nu13093166.

94 A. R. Vasconcelos et al., "Effects of intermittent fasting on age-related changes on Na,K-ATPase activity and oxidative status induced by lipopolysaccharide in rat hippocampus," *Neurobiol Aging*. (2015), 36:1914-1923. doi: 10.1016/j.neurobiolaging.2015.02.020.

95 D. Di Majo et al., "Ketogenic and Modified Mediterranean Diet as a Tool to Counteract Neuroinflammation in Multiple Sclerosis: Nutritional Suggestions," *Nutrients*. (2022), doi: 10.3390/nu14122384.

96 U. Ojha et al., "Intermittent fasting protects the nigral dopaminergic neurons from MPTP-mediated dopaminergic neuronal injury in mice," *J Nutr Biochem*. (2023), doi: 10.1016/j.jnutbio.2022.109212.

97 A. Pinto et al., "Anti-Oxidant and Anti-Inflammatory Activity of Ketogenic Diet: New Perspectives for Neuroprotection in Alzheimer's Disease," *Antioxidants* (Basel.) (2018), doi: 10.3390/antiox7050063.

98 X. Song et al., "Advanced glycation in D-galactose induced mouse aging model," *Mech Ageing Dev.* (1999), 108:239-251. doi: 10.1016/s0047-6374(99)00022-6.

99 L. Yan L et al., "RAGE-TLR4 Crosstalk Is the Key Mechanism by Which High Glucose Enhances the Lipopolysaccharide-Induced Inflammatory Response in Primary Bovine Alveolar Macrophages," *Int J Mol Sci.* (2023), doi: 10.3390/ijms24087007.

100 N. S. Rahim et al., "Virgin Coconut Oil-Induced Neuroprotection in Lipopolysaccharide-Challenged Rats is Mediated, in Part, Through Cholinergic, Anti-Oxidative and Anti-Inflammatory Pathways," *J Diet Suppl.* (2021), 18:655-681. doi: 10.1080/19390211.2020.1830223.

101 Nehls, M., *Das erschöpfte Gehirn: Der Ursprung unserer mentalen Energie – und warum sie schwindet Willenskraft, Kreativität und Fokus zurückgewinnen.* Heyne 2022.

102 F. Cignarella et al., "Intermittent Fasting Confers Protection in CNS Autoimmunity by Altering the Gut Microbiota," *Cell Metab.* (2018), 27:1222-1235. doi: 10.1016 /j.cmet.2018.05.006.

103 Kip E, Parr-Brownlie, "Healthy lifestyles and wellbeing reduce neuroinflammation and prevent neurodegenerative and psychiatric disorders," *Front Neurosci.* 2023 Feb 15;17:1092537. doi: 10.3389/fnins.2023.1092537.

104 Y. Amagase et al., "Peripheral Regulation of Central Brain-Derived Neurotrophic Factor Expression through the Vagus Nerve," *Int J Mol Sci.* (2023), doi: 10.3390 /ijms24043543.

105 Y. T. Fang et al., "Neuroimmunomodulation of vagus nerve stimulation and the therapeutic implications," *Front Aging Neurosci.* (2023), doi: 10.3389/fnagi.2023 .1173987.

106 S. Sean Cuthbert, "The Soul Nerve and Trauma Recovery," https://www .seancuthbert. com/post/the-vagus-and-trauma-recovery (7/21/2024, last accessed on 1/28/2025); R. Menakem, *My Grandmother's Hands: Racialized Trauma and the Pathway to Mending Our Hearts and Bodies.* (Central Recovery Press 2017).

107 O. F. O'Leary et al., "The vagus nerve modulates BDNF expression and neurogenesis in the hippocampus," *Eur Neuropsychopharmacol.* (2018), 28:307-316. doi: 10.1016/j.euroneuro.2017.12.004.; M. Molska et al., "The Influence of Intestinal Microbiota on BDNF Levels," *Nutrients.* (2024), doi: 10.3390/nu16172891.

108 Kip & Parr-Brownlie, "Healthy lifestyles and wellbeing."

109 J. F. Cryan & T. G. Dinan, "Mind-altering microorganisms: the impact of the gut microbiota on brain and behaviour," *Nat Rev Neurosci.* (2012), 13:701-712. doi: 10.1038/nrn3346.

110 E. Siopi et al., "Gut microbiota changes require vagus nerve integrity to promote depressive-like behaviors in mice," *Mol Psychiatry.* (2023), 28:3002-3012. doi: 10.1038/s41380-023-02071-6.; Y. Amagase et al., "Peripheral Regulation of Central Brain-Derived Neurotrophic Factor Expression through the Vagus Nerve," *Int J Mol Sci.* (2023), doi: 10.3390/ijms24043543.

111 Nehls, M., *Das erschöpfte Gehirn: Der Ursprung unserer mentalen Energie – und warum sie schwindet Willenskraft, Kreativität und Fokus zurückgewinnen*. Heyne 2022.

112 N. Cresto et al., "Pesticides at brain borders: Impact on the blood-brain barrier, neuroinflammation, and neurological risk trajectories," *Chemosphere*. (2023), doi: 10.1016/j.chemosphere.2023.138251.; B. Vellingiri et al., "Neurotoxicity of pesticides: A link to neurodegeneration," *Ecotoxicol Environ Saf. (*2022), doi: 10.1016/j.ecoenv.2022.113972.; A. Arab & S. Mostafalou, "Neurotoxicity of pesticides in the context of CNS chronic diseases," *Int. J. Environ. Health* Res. (2022), 32:2718-2755. doi: 10.1080/09603123.2021.1987396.; J. Xiao et al., "Pesticides exposure and dopaminergic neurodegeneration," *Expos. Health*. (2021), 13:295-306. doi: 10.1007/s12403-02100384-x.

113 J. H. Moon, "Health effects of electromagnetic fields on children," *Clin Exp Pediatr*. (2020), 63:422-428. doi: 10.3345/cep.2019.01494.; S. R. Acharya et al., "Electromagnetic Field Exposure in Kindergarten Children: Responsive Health Risk Concern," *Front Pediatr.* (2021), doi: 10.3389/fped.2021.694407.; https://michael-nehls.de/infos/5gund-mobilfunk-im-allgemeinen/.

114 Ł. Dudek et al., "Silicon in prevention of atherosclerosis and other age-related diseases," *Front Cardiovasc Med. (*2024), doi: 10.3389/fcvm.2024.1370536.

115 S. Sripanyakorn et al., "The comparative absorption of silicon from different foods and food supplements," *Br J Nutr.* (2009), 102:825-834. doi: 10.1017 /S0007114509311757.; A. Prescha et al., "Dietary Silicon and Its Impact on Plasma Silicon Levels in the Polish Population," *Nutrients*. (2019), doi: 10.3390 /nu11050980.

116 Nehls, M., *Die Formel gegen Alzheimer: Die Gebrauchsanweisung für ein gesundes Leben Ganz einfach vorbeugen und rechtzeitig heilen*. Heyne 2018, S. 161/162.

117 F. Vafaee et al., "Alpha-lipoic acid, as an effective agent against toxic elements: a review," Naunyn Schmiedebergs Arch Pharmacol. (2024), doi: 10.1007/s00210-024-03576-9.; H. M. Saleh et al., "Efficacy of α-lipoic acid against cadmium toxicity on metal ion and oxidative imbalance, and expression of metallothionein and antioxidant genes in rabbit brain," Environ Sci Pollut Res Int. (2017), 24:24593-24601. doi: 10.1007/s11356017-0158-0.

118 B. Salehi et al., "Insights on the Use of α-Lipoic Acid for Therapeutic Purposes," *Biomolecules*. (2019), doi: 10.3390/biom9080356.

119 E. E. Brown et al., "Psychiatric benefits of lithium in water supplies may be due to protection from the neurotoxicity of lead exposure," *Med Hypotheses*. (2018), 115:94-102. doi: 10.1016/j.mehy.2018.04.005.

120 P. Bhalla et al., "Protective role of lithium during aluminium-induced neurotoxicity," *Neurochem Int.* (2010), 56:256-262. doi: 10.1016/j.neuint.2009.10.009.

121 C. A. Lazzara & Y. H. Kim, "Potential application of lithium in Parkinson's and other neurodegenerative diseases," *Front Neurosci.* (2015), doi: 10.3389 /fnins.2015.00403.

122 "How our autistic ancestors played an important role in human evolution," The Conversation (2017), https://theconversation.com/how-our-autistic-ancestors-playedan-important-role-in-human-evolution-73477; Nehls, M., *Die Algenöl-Revolution: Lebenswichtiges Omega-3 – Das pflanzliche Lebenselixier aus dem Meer.* Heyne 2022, Pg. 110.

123 J. Long et al., "Insights into the structure and function of the hippocampus: implications for the pathophysiology and treatment of autism spectrum disorder," *Front Psychiatry.* (2024), doi: 10.3389/fpsyt.2024.1364858.

124 S. M. Banker et al., "Hippocampal contributions to social and cognitive deficits in autism spectrum disorder," *Trends Neurosci.* (2021), 44:793-807. doi: 10.1016/j.tins.2021.08.005.

125 H. C. Hazlett et al., "Early brain development in infants at high risk for autism spectrum disorder," *Nature.* (2017), 542:348-351. doi: 10.1038/nature21369.

126 D. Sussman et al., "The autism puzzle: Diffuse but not pervasive neuroanatomical abnormalities in children with ASD," *Neuroimage Clin.* (2015), 8:170-179. doi: 10.1016/j. nicl.2015.04.008.

127 H. K. Hughes et al., "Innate immune dysfunction and neuroinflammation in autism spectrum disorder (ASD.)," *Brain Behav Immun.* (2023), 108:245-254. doi: 10.1016/j. bbi.2022.12.001.; U. T. T. Than et al., "Inflammatory mediators drive neuroinflammation in autism spectrum disorder and cerebral palsy," *Sci Rep.* (2023), doi: 10.1038/s41598023-49902-8.

128 M. J. Maenner et al., "Prevalence and Characteristics of Autism Spectrum Disorder Among Children Aged 8 Years Autism and Developmental Disabilities Monitoring Network, 11 Sites, United States, 2020," *MMWR Surveill Summ.* (2023), 72:1-14. doi: 10.15585/mmwr.ss7202a1.

129 https://www.cdc.gov/autism/data-research/index.html (16.05.2024, zuletzt aufgeru-on 2/04/2025).

130 A. B. Hill, "The environment and disease: association or causation?" *Proc R Soc Med.* (1965), 58:295-300. doi: 10.1177/003591576505800503.

131 L. Tomljenovic & C. A. Shaw, "Do aluminum vaccine adjuvants contribute to the rising prevalence of autism?" *J Inorg Biochem.* (2011), 105:1489-1499. doi: 10.1016/j.jinorgbio.2011.08.008.

132 J. Asín et al., "Cognition and behavior in sheep repetitively inoculated with aluminum adjuvant-containing vaccines or aluminum adjuvant only," *J Inorg Biochem.* (2020), doi: 10.1016/j.jinorgbio.2019.110934.

133 A. R. Mawson & J. Binu, "Vaccination and Neurodevelopmental Disorders: A Study of Nine-Year-Old Children Enrolled in Medicaid," *Science, Public Health Policy, and the Law.* (2025), 6:2019-2025. https://publichealthpolicyjournal.com/vaccination-and-neurodevelopmental-disorders-a-study-of-nine-year-old-children-enrolled-in-medicaid/.

134 M. Gandhi, *A Guide to Health*, (S. Ganesan Publisher 1921), S.107-111, https://www. gutenberg.org/files/40373/40373-h/40373-h.htm.

135 https://tacanow.org/autism-prevalence/ (last accessed on 05.02.2025).

136 A. Zawadzka et al., "The Role of Maternal Immune Activation in the Pathogenesis of Autism: A Review of the Evidence, Proposed Mechanisms and Implications for Treatment," *Int J Mol Sci.* (2021), doi: 10.3390/ijms222111516.

137 M. J. D. Holland & Z. (Herausgeber) O'Toole, Anonymous (Autor): Schildkröten bis ganz nach unten: Wissenschaft und Mythos des Impfens. The Turtles Team 2023; Humphries S & Bystrianyk R: Die Impf-Illusion: Infektionskrankheiten, Impfungen und die unterdrückten Fakten. Kopp 2018; Hirte M: Impfen Pro & Contra: Das Handbuch für die individuelle Impfentscheidung. Mit aktualisiertem Corona-Kapitel. Knaur MensSana HC 2018;

138 S. Lee et al., "Acute central nervous system inflammation following COVID-19 vaccination: An observational cohort study," *Mult Scler.* (2023), 29:595-605. doi: 10.1177/13524585231154780.

139 J. H. Roh et al., "A potential association between COVID-19 vaccination and development of Alzheimer's disease," *QJM.* (2024), 117:709-716. doi: 10.1093/qjmed/hcae103.

140 https://www.cdc.gov/covid/vaccines/stay-up-to-date.html (1/7/2025, last ac-cessed on 3/13/2025).

141 J. Dahlgren et al., "Interleukin-6 in the maternal circulation reaches the rat fetus in mid-gestation," *Pediatr Res.* (2006), 60:147-151. doi: 10.1203/01.pdr.0000230026.74139.18.; A. Mouihate & H.Mehdawi, "Toll-like receptor 4-mediated immune stress in pregnant rats activates STAT3 in the fetal brain: role of interleukin-6," *Pediatr Res.* (2016), 79:781-787. doi: 10.1038/pr.2015.86.

142 D. Nikolopoulos et al., "Microglia activation in the presence of intact blood-brain barrier and disruption of hippocampal neurogenesis via IL-6 and IL-18 mediate early diffuse neuropsychiatric lupus," *Ann Rheum Dis.* (2023), 82:646-657. doi: 10.1136/ard2022-223506.; W. A. Banks et al., "Passage of cytokines across the blood-brain barrier," *Neuroimmunomodulation.* (1995), 2:241-248. doi: 10.1159/000097202.

143 A. Zawadzka et al., "The Role of Maternal Immune Activation in the Pathogenesis of Autism: A Review of the Evidence, Proposed Mechanisms and Implications for Treatment," *Int J Mol Sci.* (2021), doi: 10.3390/ijms222111516.

144 A. Mouihate, "Prenatal Activation of Toll-Like Receptor-4 Dampens Adult Hippocampal Neurogenesis in An IL-6 Dependent Manner," *Front Cell Neurosci.* (2016), doi: 10.3389/fncel.2016.00173.

145 Holland MJD & O'Toole Z (Herausgeber), Anonymous (Autor): Schildkröten bis ganz nach unten: Wissenschaft und Mythos des Impfens. The Turtles Team 2023, Kapitel 1: https://alschner-klartext.de/wp-content/uploads/2023/01/Schildkroetenbis-ganz-nach-unten-Kapitel1-Leseprobe.pdf; Referenzen: https://drive.google.com/file/d/16qHPe0odDOweuCDlwIEjarF4Aiu7oyM9/view Holland MJD & O'Toole Z (Publishers), Anonymous (author): Turtles All The Way Down: Vaccine Science and Myth. The Turtles Team, 2022, Chapter 1:

https://www.skepticalraptor.com/blog/wp-content/uploads/2023/05/Turtles-Book-English-Chapter-1.pdf; References: https://drive.google.com/file/d/16qHPe0odDOweuCDlwIEjarF4Aiu7oyM9/view.

146 Nehls, M., *Das Corona-Syndrom: Wie das Virus unsere Schwächen offenlegt – und wie wir uns nachhaltig schützen können*. Heyne 2021.

147 K. Weintraub, "The prevalence puzzle: Autism counts," *Nature*. (2011), 479:22-24. doi: 10.1038/479022a.

148 R. Lathe, "Microwave Electromagnetic Radiation and Autism," *EJAP*. (2009), 5:11-30. 10.7790/ejap.v5i1.144.

149 Y. Pu et al., "Maternal glyphosate exposure causes autism-like behaviors in offspring through increased expression of soluble epoxide hydrolase," *Proc Natl Acad Sci USA*. (2020), 117:11753-11759. doi: 10.1073/pnas.1922287117.; Y. Pu et al., "Autism-like Behaviors in Male Juvenile Offspring after Maternal Glyphosate Exposure," *Clin Psychopharmacol Neurosci*. (2021), 19:554-558. doi: 10.9758/cpn.2021.19.3.554.

150 J. Zaheer et al., "Pre/post-natal exposure to microplastic as a potential risk factor for autism spectrum disorder," *Environ Int. (*2022), doi: 10.1016/j.envint.2022.107121.

151 J. B. Adams et al., "Analyses of toxic metals and essential minerals in the hair of Arizona children with autism and associated conditions, and their mothers," *Biol Trace Elem Res*. (2006), 110:193-209. doi: 10.1385/BTER:110:3:193.

152 J. Wu et al., "Associations of essential element serum concentrations with autism spectrum disorder," *Environmental Science and Pollution Research*. (2022), doi: 10.1007/s11356-022-21978-1.

153 J. Wang et al., "Effects of different doses of lithium on the central nervous system in the rat valproic acid model of autism," *Chem Biol Interact*. (2023), doi: 10.1016/j.cbi.2022.110314.

154 M. Rizk et al., "Deciphering the roles of glycogen synthase kinase 3 (GSK3) in the treatment of autism spectrum disorder and related syndromes," *Mol Biol Rep*. (2021), 48:2669-2686. doi: 10.1007/s11033-021-06237-9.

155 Wang et al., "Effects of different doses of lithium."

156 Z. Liew et al., "Association Between Estimated Geocoded Residential Maternal Exposure to Lithium in Drinking Water and Risk for Autism Spectrum Disorder in Offspring in Denmark," *JAMA Pediatr*. (2023), 177:617-624. doi: 10.1001/jamapediatrics.2023.0346.

157 N. M. van der Lugt et al., "Fetal, neonatal and developmental outcomes of lithium-exposed pregnancies," *Early Hum Dev*. (2012), 88:375-378. doi: 10.1016/j.earlhumdev.2011.09.013.

158 E. M. P. Poels et al., "The effect of prenatal lithium exposure on the neuropsychological development of the child," *Bipolar Disord*. (2022), 24:310-319. doi: 10.1111/bdi.13133.

159 E. M. P. Poels et al., "Lithium exposure during pregnancy increases fetal growth," *J Psychopharmacol*. (2021), 35:178-183. doi: 10.1177/0269881120940914.

160 B. Lee, "Journal club: Does lithium in drinking water contribute to autism?" *The Transmitter*. (2023), doi: 10.53053/SCBZ3308.

161 M.Arora et al., "Fetal and postnatal metal dysregulation in autism," *Nat Commun*. (2017), doi: 10.1038/ncomms15493.

162 J. Christensen, "Study finds slightly higher risk of autism diagnosis in areas with more lithium in drinking water, but experts say more research is needed," https://edition. cnn.com/2023/04/03/health/autism-lithium-study/index .html; https://www.focus.de/gesundheit/news/neue-studie-forscher-warnen-vor -autismus-risiko-durch-leitungswasser_id_191457988.html; https://www.rtl.de /cms/hoeheres-autismus-risiko-forscher-machen-alarmierende-leitungswasser -entdeckung-5040157.html.

163 A. J. Russo et al., "Increased Glycogen Synthase Kinase 3 Alpha (GSK3A) in Children with Autism," *EC Paediatrics*. (2020), https://ecronicon.net/assets/ecpe /pdf/increasedglycogen-synthase-kinase-3-alpha-gsk3a-in-children-with-autism.pdf.

164 M. Siegel et al., "Preliminary investigation of lithium for mood disorder symptoms in children and adolescents with autism spectrum disorder," *J Child Adolesc Psychopharmacol.* (2014), 24:399-402. doi: 10.1089/cap.2014.0019.

165 J. Boi et al., "Medium-term efficacy data of medications in children and adolescents with autism spectrum disorder: an 18 months retrospective follow up study," *European Neuropsychopharmacology*. 2017, doi: 10.1016/S0924-977X(17)31928-4.

166 N. A. Spinazzi et al., "Co-occurring conditions in children with Down syndrome and autism: a retrospective study," *J Neurodev Disord.* (2023), doi: 10.1186 /s11689-023-09478-w.

167 A. Contestabile et al., "Lithium rescues synaptic plasticity and memory in Down syndrome mice," *J Clin Invest.* (2013), 123:348-361. doi: 10.1172/JCI64650.

168 J. Greenblatt, "Lithium for Irritability Related to Autism," *Psychiatry Redefined* (2023), https://www.psychiatryredefined.org/lithium-for-autism-irritability/.

169 J. F. Cade, "Lithium salts in the treatment of psychotic excitement," *Med J Aust.* (1949), 2:349-352. doi: 10.1080/j.1440-1614.1999.06241.x.

170 W. F. Parker et al., "Association Between Groundwater Lithium and the Diagnosis of Bipolar Disorder and Dementia in the United States," *JAMA Psychiatry*. (2018), 75:751754. doi: 10.1001/jamapsychiatry.2018.1020.

171 A. Muneer, "Wnt and GSK3 Signaling Pathways in Bipolar Disorder: Clinical and Therapeutic Implications," *Clin Psychopharmacol Neurosci*. 2017, 15:100-114. doi: 10.9758/cpn.2017.15.2.100.; A. Luca et al., "Gsk3 Signalling and Redox Status in Bipolar Disorder: Evidence from Lithium Efficacy," *Oxid Med Cell Longev*. (2016), doi: 10.1155/2016/3030547.

172 T. Hajek et al., "Smaller hippocampal volumes in patients with bipolar disorder are masked by exposure to lithium: a meta-analysis," *J Psychiatry Neurosci*. (2012), 37:333-343. doi: 10.1503/jpn.110143.

173 F. Benedetti et al., "Lithium and GSK3-β promoter gene variants influence white matter microstructure in bipolar disorder," *Neuropsychopharmacology*. (2013), 38:313-327. doi: 10.1038/npp.2012.172.

174 Chepenik LG et al., "Structure-function associations in hippocampus in bipolar disorder," *Biol Psychol.* 2012, 90:18-22. doi: 10.1016/j.biopsycho.2012.01.008.

175 S. Lucini-Paioni et al., "Lithium effects on Hippocampus volumes in patients with bipolar disorder," *J Affect Disord.* 2021, 294:521-526. doi: 10.1016/j.jad.2021.07.046.; L. C. Foland et al., "Increased volume of the amygdala and hippocampus in bi-polar patients treated with lithium," *Neuroreport.* (2008), 19:221-224. doi: 10.1097/WNR.0b013e3282f48108.

176 A. G. Pacholko, L. K. Bekar "Lithium orotate: A superior option for lithium therapy?" *Brain Behav.* (2021), doi: 10.1002/brb3.2262.

177 L. C. Foland et al., "Increased volume of the amygdala and hippocampus in bipolar patients treated with lithium," *Neuroreport.* (2008), 19:221-224. doi: 10.1097/WNR.0b013e3282f48108.; C. López-Jaramillo et al., "Increased hippocampal, thalamus and amygdala volume in long-term lithium-treated bipolar I disorder patients compared with unmedicated patients and healthy subjects," *Bipolar Disord.* (2017), 19:41-49. doi: 10.1111/bdi.12467.

178 S. Selek et al., "A longitudinal study of fronto-limbic brain structures in patients with bipolar I disorder during lithium treatment," *J Affect Disord.* (2013), 150:629-633. doi: 10.1016/j.jad.2013.04.020.

179 H. E. Sartori, "Lithium orotate in the treatment of alcoholism and related conditions," *Alcohol.* (1986), 3:97-100. doi: 10.1016/0741-8329(86)90018-2., https://www.sciencedirect.com/science/article/abs/pii/0741832986900182.

180 K. S. Anand & V. Dhikav, "Hippocampus in health and disease: An overview," *Ann Indian Acad Neurol.* (2012), 15:239-246. doi: 10.4103/0972-2327.104323.

181 R. M. Murray & S. W. Lewis, "Is schizophrenia a neurodevelopmental disorder?" *Br Med J* (Clin Res Ed.) (1988), doi: 10.1136/bmj.296.6614.63.

182 D. Wegrzyn et al., "Structural and Functional Deviations of the Hippocampus in Schizophrenia and Schizophrenia Animal Models," *Int J Mol Sci.* (2022), doi: 10.3390/ijms23105482.

183 S. Lovestone et al., "Schizophrenia as a GSK-3 dysregulation disorder," *Trends Neurosci.* (2007),: 142-149. doi: 10.1016/j.tins.2007.02.002.

184 R. Laxmi & S. Grover, "Unmasking of schizophrenia following COVID vaccination," *Indian J Psychiatry.* (2023), 65:385-386. doi: 10.4103/indianjpsychiatry.indianjpsychiatry_607_22.; M. Lazareva et al., "New-onset psychosis following COVID-19 vaccination: a systematic review," *Front Psychiatry.* (2024), doi: 10.3389/fpsyt.2024.1360338.

185 G. P. Amminger et al., "Long-chain omega-3 fatty acids for indicated prevention of psychotic disorders: a randomized, placebo-controlled trial," *Arch Gen Psychiatry.* (2010), 67:146-54. doi: 10.1001/archgenpsychiatry.2009.192.

186 G. P. Amminger et al., "Longer-term outcome in the prevention of psychotic disorders by the Vienna omega-3 study," *Nat Commun.* (2015), doi: 10.1038/ncomms8934; N. Mossaheb et al., "Predictors of longer-term outcome in the Vienna omega-3 high-risk study," *Schizophr Res.* (2018), 193:168-172. doi: 10.1016/j.schres.2017.08.010.

187 K. Zhongling et al., "Neuroinflammation in a Rat Model of Tourette Syndrome," *Front Behav Neurosci.* 2022, doi: 10.3389/fnbeh.2022.710116.

188 X. Wang et al., "The inflammatory injury in the striatal microglia-dopaminergic -neuron crosstalk involved in Tourette syndrome development," *Front Immunol.* (2023), doi: 10.3389/fimmu.2023.1178113.

189 M. Rapanelli et al., "Dysregulated intracellular signaling in the striatum in a pathophysiologically grounded model of Tourette syndrome," *Eur Neuropsychopharmacol.* (2014), 24:1896-1906. doi: 10.1016/j.euroneuro .2014.10.007.

190 H. M. Erickson Jr et al., "Comparison of lithium and haloperidol therapy in Gilles de la Tourette syndrome," *Adv Exp Med Biol.* (1977), 90:197-205. doi: 10.1007/978-1-46842511-6_11.; B. J. Hamra et al., "Remission of tics with lithium therapy: case report," *J Clin Psychiatry.* (1983), 44:73-74. https://pubmed .ncbi.nlm.nih.gov/6402503/.

191 P. M. Nunes et al., "Volumes of the hippocampus and amygdala in patients with borderline personality disorder: a meta-analysis," *J Pers Disord.* (2009), 23:333-345. doi: 10.1521/pedi.2009.23.4.333.

192 D. Carcone et al., "Disrupted Relationship between Hippocampal Activation and Subsequent Memory Performance in Borderline Personality Disorder," *J Affect Disord.* (2020), 274:1041-1048. doi: 10.1016/j.jad.2020.05.050.

193 T. Zetzsche et al., "Hippocampal volume reduction and history of aggressive behaviour in patients with borderline personality disorder," *Psychiatry Res.* (2007), 154:157-170. doi: 10.1016/j.pscychresns.2006.05.010.

194 A. Vallée et al., "Lithium: a potential therapeutic strategy in obsessive-compulsive disorder by targeting the canonical WNT/β pathway," *Transl Psychiatry.* (2021), doi: 10.1038/s41398-021-01329-3.

195 M. Leyton, P. Vezina, "Dopamine ups and downs in vulnerability to addictions: a neurodevelopmental model," *Trends Pharmacol Sci.* (2014), 35:268-276. doi: 10.1016/j. tips.2014.04.002.

196 https://www.aok.de/pk/magazin/koerper-psyche/sucht/von-dopamin-und-suchtver-halten-bis-zu-dopamin-detox/(8/1/2024, last accessed on 12/8/2024).

197 J. Beaulieu et al., "Lithium antagonizes dopamine-dependent behaviors mediated by an AKT/glycogen synthase kinase 3 signaling cascade," *Proc Natl Acad Sci USA.* (2004), 101:5099-5104, doi: 10.1073/pnas.0307921101.

198 S. Gadh, "Low-dose lithium impact in an addiction treatment setting," *Personalized Medicine in Psychiatry.* (2020), doi: 10.1016/j.pmip.2020.100059.

199 A. N. Edinoff et al., "Benzodiazepines: Uses, Dangers, and Clinical Considerations," *Neurol Int.* (2021), 13:594-607. doi: 10.3390/neurolint13040059.

200 O. Giotakos, "Is impulsivity in part a lithium deficiency state?" *Psychiatriki.* (2018), 29:264-270. doi: 10.22365/jpsych.2018.293.264.

201 M. Kessi et al., "Attention-deficit/hyperactive disorder updates," *Front Mol Neurosci.* (2022), doi: 10.3389/fnmol.2022.925049.

202 G. A. Dunn et al., "Neuroinflammation as a risk factor for attention deficit hyperactivity disorder," *Pharmacol Biochem Behav.* (2019), 182:22-34. doi: 10.1016/j. pbb.2019.05.005.

203 F. Turiaco et al., "Attention Deficit Hyperactivity Disorder (ADHD) and Polyphenols: A Systematic Review," *Int J Mol Sci.* (2024), doi: 10.3390/ijms25031536.

204 V. Dhikav et al., "Hippocampal Volume in Children with Attention Deficit Hyperactivity Disorder and Speech and Language Delay," *Ann Indian Acad Neurol.* (2023), 26:431-434. doi: 10.4103/aian.aian_77_23.

205 M. A. Mines, "Hyperactivity: glycogen synthase kinase-3 as a therapeutic target," *Eur J Pharmacol.* (2013), 708:56-59. doi: 10.1016/j.ejphar.2013.02.055.

206 Y. Cheng et al., "Stress-induced neuroinflammation is mediated by GSK3-dependent TLR4 signaling that promotes susceptibility to depression-like behavior," *Brain Behav Immun.* (2016), 53:207-222. doi: 10.1016/j.bbi.2015.12.012.

207 H. O. Kalkman, "Inhibition of Microglial GSK3β Activity Is Common to Different Kinds of Antidepressants: A Proposal for an In Vitro Screen to Detect Novel Antidepressant Principles," *Biomedicines.* (2023), doi: 10.3390/biomedicines11030806.

208 J. Leschik et al., "Stress-Related Dysfunction of Adult Hippocampal Neurogenesis-An Attempt for Understanding Resilience?" *Int J Mol Sci.* (2021), doi: 10.3390/ijms22147339.

209 H. Eliwa et al., "Adult hippocampal neurogenesis: Is it the alpha and omega of antidepressant action?" *Biochem Pharmacol.* (2017), 141:86-99. doi: 10.1016/j.bcp.2017.08.005.

210 J. E. Malberg et al., "Adult Neurogenesis and Antidepressant Treatment: The Surprise Finding by Ron Duman and the Field 20 Years Later," *Biol Psychiatry.* (2021), 90:96-101. doi: 10.1016/j.biopsych.2021.01.010.

211 P. Duda et al., "GSK3β: A Master Player in Depressive Disorder Pathogenesis and Treatment Responsiveness," *Cells.* (2020), doi: 10.3390/cells9030727.

212 X. Li & R. S. Jope, "Is glycogen synthase kinase-3 a central modulator in mood regulation?" *Neuropsychopharmacology.* (2010), 35:2143-2154. doi: 10.1038/npp.2010.105.

213 E. Jiménez et al., "Genetic variability at IMPA2, INPP1 and GSK3β increases the risk of suicidal behavior in bipolar patients," *Eur Neuropsychopharmacol.* (2013), 23:14521462. doi: 10.1016/j.euroneuro.2013.01.007.; H. O. Kalkman, "The GSK3-NRF2 Axis in Suicide," *Psychiatry Int.* (2021), 2:108-119. https://doi.org/10.3390/psychiatryint2010008.

214 Kalkman, Hans O. 2021. "The GSK3-NRF2 Axis in Suicide" *Psychiatry International* 2, no. 1: 108-119. https://doi.org/10.3390/psychiatryint2010008.

215 N. Zakharova et al., "Telomere Length as a Marker of Suicidal Risk in Schizophrenia," *Consort Psychiatr.* (2022), 03:37-47. doi: 10.17816/CP171.; D. Martinez et al., "Shorter telomere length and suicidal ideation in familial bipolar disorder," *PLoS One.* (2022), doi: 10.1371/journal.pone.0275999.

216 A. Wu & J. J. Zhang, "Neuroinflammation, memory, and depression: new approaches to hippocampal neurogenesis," *J Neuroinflammation*. (2023), doi: 10.1186/s12974-02302964-x.

217 Giotakos O., "Is impulsivity in part a lithium deficiency state? *Psychiatriki*. 2018 Jul-Sep;29(3):264-270. doi: 10.22365/jpsych.2018.293.264.

218 T. Huebner et al., "Morphometric brain abnormalities in boys with conduct disorder," *J Am Acad Child Adolesc Psychiatry*. (2008), 47:540-547. doi: 10.1097 /CHI.0b013e3181676545.

219 M. H. Teicher et al., "Childhood maltreatment is associated with reduced volume in the hippocampal subfields CA3, dentate gyrus, and subiculum," *Proc Natl Acad Sci USA*. (2012), 109:563-572. doi: 10.1073/pnas.1115396109.

220 A. Abdolalizadeh et al., "Larger left hippocampal presubiculum is associated with lower risk of antisocial behavior in healthy adults with childhood conduct history," *Sci Rep*. (2023), doi: 10.1038/s41598-023-33198-9.

221 D. Chatterjee & J. M. Beaulieu, "Inhibition of glycogen synthase kinase 3 by lithium, a mechanism in search of specificity," *Front Mol Neurosci*. (2022), doi: 10.3389/fnmol.2022.1028963.

222 A. Perlmutter, "How Inflammation Changes Our Thinking," *Psychology Today*. (11.02.2020), https://www.psychologytoday.com/intl/blog/the-modern -brain/202002/how-inflammation-changes-our-thinking.

223 E. Won & Y. K. Kim, "Neuroinflammation-Associated Alterations of the Brain as Potential Neural Biomarkers in Anxiety Disorders," *Int J Mol Sci*. (2020), doi: 10.3390/ijms21186546.

224 R. R. Uchida et al., 'Decreased left temporal lobe volume of panic patients measured by magnetic resonance imaging," *Braz J Med Biol Res. (*2003), 36:925-929. doi: 10.1590/s0100-879x2003000700014.

225 M. Chennaoui et al., "Effects of exercise on brain and peripheral inflammatory biomarkers induced by total sleep deprivation in rats," *J Inflamm* (Lond.) (2015), doi: 10.1186/s12950-015-0102-3.

226 D. Cutuli et al., "Pre-reproductive Parental Enriching Experiences Influence Progeny's Developmental Trajectories," *Front Behav Neurosci*. (2018), doi: 10.3389 /fnbeh.2018.00254.

227 J. Madrigal et al., "The Increase in TNF-α Levels Is Implicated in NF-κB Activation and Inducible Nitric Oxide Synthase Expression in Brain Cortex after Immobilization Stress," *Neuropsychopharmacol*. (2002), 26:155-163. doi: 10.1016/S0893133X(01)00292-5

228 O. Kofman, "The role of prenatal stress in the etiology of developmental behavioural disorders," *Neurosci Biobehav*. (2002), 26, 457-470. https://pubmed.ncbi.nlm .nih.gov/12204192/; C. L. Howerton & T. L. Bale, "Prenatal programing: at the intersection of maternal stress and immune activation," *Horm Behav*. (2012), 62:237-242. doi: 10.1016/j.yhbeh.2012.03.007.

229 B. Zupan et al., "Maternal Brain TNF-α Programs Innate Fear in the Offspring," *Curr Biol*. (2017), 27:3859-3863.e3. doi: 10.1016/j.cub.2017.10.071.

230 A. Dellarole et al., "Neuropathic pain-induced depressive-like behavior and hippocampal neurogenesis and plasticity are dependent on TNFR1 signaling." *Brain Behav Immun.* (2014), 41:65-81. doi: 10.1016/j.bbi.2014.04.003.

231 I. Lladó et al., "Varicella zoster virus reactivation and mRNA vaccines as a trigger," *JAAD Case Rep. (*2021), 15:62-63. doi: 10.1016/j.jdcr.2021.07.011.

232 L. Leung & C. M. Cahill, "TNF-alpha and neuropathic pain—a review," *J Neuroinflammation.* 2010, doi: 10.1186/1742-2094-7-27.

233 G. J. Norman et al., "Stress and IL-1beta contribute to the development of depressive-like behavior following peripheral nerve injury," *Mol Psychiatry.* (2010), 15:404-414. doi: 10.1038/mp.2009.91.; Q. Li et al., "Hippocampal PKR/NLRP1 Inflammasome Pathway Is Required for the Depression-Like Behaviors in Rats with Neuropathic Pain," *Neuroscience.* (2019), 412:16-28. doi: 10.1016 /j.neuroscience.2019.05.025.

234 W. J. Ren et al., "Peripheral nerve injury leads to working memory deficits and dysfunction of the hippocampus by upregulation of TNF-α in rodents," *Neuropsychopharmacology.* (2011), 36:979-992. doi: 10.1038/npp.2010.236.

235 A. Ezzati et al., "The relationship between hippocampal volume, chronic pain, and depressive symptoms in older adults," *Psychiatry Res Neuroimaging.* (2019), 289: 10-12. doi: 10.1016/j.pscychresns.2019.05.003.

236 M. Nehls, "Unified theory of Alzheimer's disease (UTAD): implications for prevention and curative therapy," *J Mol Psychiatry. (*2016), doi: 10.1186 /s40303-016-0018-8.

237 D. W. Maixner & H. R. Weng, "The Role of Glycogen Synthase Kinase 3 Beta in Neuroinflammation and Pain," *J Pharm Pharmacol* (Los Angel.) (2013), doi: 10.13188/2327204X.1000001.

238 L. Mazzardo-Martins et al., "Glycogen synthase kinase 3-specific inhibitor ARA014418 decreases neuropathic pain in mice: evidence for the mechanisms of action," *Neuroscience.* (2012), 226:411-420. doi: 10.1016 /j.neuroscience.2012.09.020.

239 X. Dong et al., "Function of GSK-3 signaling in spinal cord injury (Review.)," *Exp Ther Med. (*2023), doi: 10.3892/etm.2023.12240.

240 I. Weinsanto et al., "Lithium reverses mechanical allodynia through a mu opioid-dependent mechanism," *Mol Pain.* (2018), doi: 10.1177/1744806917754142.

241 S. Nezamoleslami et al., "Lithium reverses the effect of opioids on eNOS/nitric oxide pathway in human umbilical vein endothelial cells," *Mol Biol Rep.* (2020), 47:68296840. doi: 10.1007/s11033-020-05740-9.

242 D. C. Buse et al., "Sex differences in the prevalence, symptoms, and associated features of migraine, probable migraine and other severe headache: results of the American Migraine Prevalence and Prevention (AMPP) Study," *Headache.* (2013), 53:1278-1299. doi: 10.1111/head.12150.

243 A. Chakravarty & A. Sen, "Migraine, neuropathic pain and nociceptive pain: towards a unifying concept," *Med Hypotheses.* (2010), 74:225-231. doi: 10.1016 /j.mehy.2009.08.034.

244 L. Biscetti et al., "The putative role of neuroinflammation in the complex pathophysiology of migraine: From bench to bedside," *Neurobiol Dis.* (2023), doi: 10.1016/j. nbd.2023.106072.

245 O. Kursun et al., "Migraine and neuroinflammation: the inflammasome perspective," *J Headache Pain.* (2021), doi: 10.1186/s10194-021-01271-1.

246 https://www.charite.de/service/pressemitteilung/artikel/detail/warum_migrae-ne_haeufig_waehrend_der_menstruation_auftritt (2/23/2023, last accessed 3/6/2023).

247 G. Yao et al., "NO up-regulates migraine-related CGRP via activation of an Akt/GSK3β/NF-κB signaling cascade in trigeminal ganglion neurons," *Aging* (Albany NY.) (2020), 12:6370-6384. doi: 10.18632/aging.103031.

248 B. Raffaelli et al., "Sex Hormones and Calcitonin Gene-Related Peptide in Women With Migraine: A Cross-sectional, Matched Cohort Study," *Neurology.* (2023), 100:1825-1835. doi: 10.1212/WNL.0000000000207114.

249 G. Yao et al., "NO up-regulates migraine-related CGRP via activation."

250 J. L. Medina, "Cyclic migraine: a disorder responsive to lithium carbonate," *Psychosomatics.* (1982), 23:625-37. doi: 10.1016/S0033-3182(82)73362-6.

251 B. Benkli et al., "Circadian Features of Cluster Headache and Migraine: A Systematic Review, Meta-analysis, and Genetic Analysis," *Neurology.* (2023), 100:2224-2236. doi: 10.1212/WNL.0000000000207240.

252 M. Leone et al., "Management of chronic cluster headache," *Curr Treat Options Neurol.* (2011), 13:56-70. doi: 10.1007/s11940-010-0106-5.

253 R. P. Silva Neto & K. J. Almeida, "Lithium-responsive Headaches," *Headache Medicine.* (2010), 01:25-28. doi: 10.48208/HeadacheMed.2010.7.

254 M. M. Mehndiratta & S. A. Wadhai, "International Epilepsy Day: A day noted for global public education & awareness," *Indian J Med Res.* (2015), 141:143-144. doi: 10.4103/0971-5916.155531.

255 O. K. Steinlein, "Genetics and epilepsy," *Dialogues Clin Neurosci.* (2008), 10:29-38. doi: 10.31887/DCNS.2008.10.1/oksteinlein.

256 E. Magiorkinis et al., "Hallmarks in the history of epilepsy: epilepsy in antiquity," *Epilepsy Behav.* (2010), 17:103-108. doi: 10.1016/j.yebeh.2009.10.023.

257 S. Fordington & M. Manford, "A review of seizures and epilepsy following traumatic brain injury," *J Neurol.* (2020), 267:3105-3111. doi: 10.1007/s00415-020-09926-w.

258 Y. Chen et al., "Neuroinflammatory mediators in acquired epilepsy: an update," *Inflamm Res.* (2023), 72:683-701. doi: 10.1007/s00011-023-01700-8.

259 T. Bast, "The hippocampal learning-behavior translation and the functional significance of hippocampal dysfunction in schizophrenia," *Curr Opin Neurobiol.* (2011), 21:492-501. doi: 10.1016/j.conb.2011.01.003.

260 A. R. Mawson & J. Binu, "Vaccination and Neurodevelopmental Disorders: A Study of Nine-Year-Old Children Enrolled in Medicaid," *Science, Public Health Policy, and the Law.* (2025), 6:2019-2025. https://publichealthpolicyjournal.com/vaccination-and-neurodevelopmental-disorders-a-study-of-nine-year-old-children-enrolled-in-medicaid/.

261 K. S. Anand & V. Dhikav, "Hippocampus in health and disease: An overview," *Ann Indian Acad Neurol.* (2012), 15:239-246. doi: 10.4103/0972-2327.104323.

262 H. Zheng et al., "Evidence for Accelerated Hippocampal Volume Loss with Age in Patients with Left Hemisphere Focal Seizures," *Structural Imaging.* (2022), https://aesnet.org/abstractslisting/evidence-for-accelerated-hippocampal-volume-loss-with-age in-patients-with-left-hemisphere-focal-seizures.

263 A. Vezzani et al., "Neuroinflammatory pathways as treatment targets and biomarkers in epilepsy," *Nat Rev Neurol.* 2019, 15:459-472. doi: 10.1038/s41582-019-0217-x.; Y. C. Lai et al., "Anakinra usage in febrile infection related epilepsy syndrome: an international cohort," *Ann Clin Transl Neurol.* (2020), 7:2467-2474. doi: 10.1002/acn3.51229.

264 S. L. Bojja et al., "What is the Role of Lithium in Epilepsy?" *Curr Neuropharmacol.* (2022), 20:1850-1864. doi: 10.2174/1570159X20666220411081728.

265 G. Jiang et al., "Lithium affects rat hippocampal electrophysiology and epileptic seizures in a dose dependent manner, *Epilepsy Res.* (2018), 146:112-120. doi: 10.1016/j.eplepsyres.2018.07.021. Erratum doi: 10.1016/j.eplepsyres.2018.12.002.

266 H. E. Sartori, "Lithium orotate in the treatment of alcoholism and related conditions," *Alcohol.* (1986), 3:97-100. doi: 10.1016/0741-8329(86)90018-2., https://www.sciencedirect.com/science/article/abs/pii/0741832986900182.

267 S. R. Shri et al., "Role of GSK-3β Inhibitors: New Promises and Opportunities for Alzheimer's Disease," *Adv Pharm Bull.* (2023), 13:688-700. doi: 10.34172/apb.2023.071.; A. Fiorentini et al., "Lithium improves hippocampal neurogenesis, neuropathology and cognitive functions in APP mutant mice," *PLoS One.* (2010), doi: 10.1371/journal.pone.0014382.

268 "Alzheimer's disease facts and figures," *Alzheimer's Dement.* (2024), 20:3708-3821. doi: 10.1002/alz.13809.

269 M. Nehls, "Unified theory of Alzheimer's disease (UTAD): implications for prevention and curative therapy," *J Mol Psychiatry.* (2016), doi: 10.1186/s40303-016-0018-8.

270 R. K. Fenech et al., "Low-Dose Lithium Supplementation Influences GSK3β Activity in a Brain Region Specific Manner in C57BL6 Male Mice," *J Alzheimers Dis.* (2023), 91:615-626. doi: 10.3233/JAD-220813.

271 C. Gherardelli et al., "Lithium Enhances Hippocampal Glucose Metabolism in an In Vitro Mice Model of Alzheimer's Disease," *Int J Mol Sci.* (2022), doi: 10.3390/ijms23158733.

272 P. V. Nunes et al., "Lithium and risk for Alzheimer's disease in elderly patients with bipolar disorder," *Br J Psychiatry.* (2007), 190:359-360. doi: 10.1192/bjp.bp.106.029868.

273 S. Chen et al., "Association between lithium use and the incidence of dementia and its subtypes: A retrospective cohort study," *PLoS Med.* (2022), doi: 10.1371/journal.pmed.1003941.

274 K. Yucel et al., "Bilateral Hippocampal Volume Increase in Patients with Bipolar Disorder and Short-term Lithium Treatment," *Neuropsychopharmacol.* (2008), 33:361-367, https://doi.org/10.1038/sj.npp.1301405.

275 L. G. Chepenik et al., "Structure-function associations in hippocampus in bipolar disorder," *Biol Psychol.* (2012), 90:18-22. doi: 10.1016/j.biopsycho.2012.01.008.

276 C. Morrison et al., "The use of hippocampal grading as a biomarker for preclinical and prodromal Alzheimer's disease," *Hum Brain Mapp.* (2023), 44:3147-3157. doi: 10.1002/hbm.26269.

277 O. V. Forlenza et al., "Disease-modifying properties of long-term lithium treatment for amnestic mild cognitive impairment: randomised controlled trial," *Br J Psychiatry.* (2011), 198:351-356. doi: 10.1192/bjp.bp.110.080044. PMID: 21525519.; O. V. Forlenza et al., "Does lithium prevent Alzheimer's disease?" *Drugs Aging.* (2012), 29:335-342. doi: 10.2165/11599180-000000000-00000.

278 M. A. Nunes et al., "Microdose lithium treatment stabilized cognitive impairment in patients with Alzheimer's disease," *Curr Alzheimer Res.* (2013), 10:104-107. doi: 10.2174/1567205011310010014.

279 M. A. Nunes et al., "Chronic Microdose Lithium Treatment Prevented Memory Loss and Neurohistopathological Changes in a Transgenic Mouse Model of Alzheimer's Disease," *PLoS One.* (2015), doi: 10.1371/journal.pone.0142267.

280 V. A. Fajardo et al., "Examining the Relationship between Trace Lithium in Drinking Water and the Rising Rates of Age-Adjusted Alzheimer's Disease Mortality in Texas,"*J Alzheimers Dis.* (2018), 61:425-434. doi: 10.3233/JAD-170744.

281 R. Koike et al., "Memory formation in old age requires GSK-3β," *Aging Brain.* (2021), doi: 10.1016/j.nbas.2021.100022.

282 J. Fraiha-Pegado et al., "Trace lithium levels in drinking water and risk of dementia: a systematic review," *Int J Bipolar Disord.* (2024), doi: 10.1186/s40345-024-00348-5.

283 Müller-Jung J: Alzheimer-Antikörper: Ein großer Hoffnungsträger floppt. FAZ, 11/23/2016, https://www.faz.net/aktuell/wissen/medizin-ernaehrung/ein-schmerzhafter-flop-fuer-die-alzheimer-forschung-14541369.html (last accessed on 3/19/2025).

284 M. Nehls, "Unified theory of Alzheimer's disease (UTAD): implications for prevention and curative therapy," *J Mol Psychiatry.* (2016), doi: 10.1186/s40303-016-0018-8.

285 G. R. Dawson et al., "Age-related cognitive deficits, impaired long-term potentiation and reduction in synaptic marker density in mice lacking the beta-amyloid precursor protein," *Neuroscience.* 1999, 90:1-13. doi: 10.1016/s0306-4522(98)00410-2.

286 H. Zheng et al., "beta-Amyloid precursor protein-deficient mice show reactive gliosis and decreased locomotor activity," *Cell.* (1995), 81:525-531. doi: 10.1016/00928674(95)90073-x.; Y. Senechal et al., "Amyloid precursor protein knockout mice show age-dependent deficits in passive avoidance learning," *Behav Brain Res.* (2008), 186:126-132. doi: 10.1016/j.bbr.2007.08.003.

287 https://www.mayoclinic.org/drugs-supplements/aducanumab-avwa-intravenous-Route/description/drg-20516708 (2/1/2025, last accessed on 3/9/2025).

288 F. Alves et al., "Accelerated Brain Volume Loss Caused by Anti-β-Amyloid Drugs: A Systematic Review and Meta-analysis," *Neurology.* (2023), 100:2114-2124. doi: 10.1212/WNL.0000000000207156.

289 P. Whitehouse et al., "Making the Case for Accelerated Withdrawal of Aducanumab," *J Alzheimers Dis.* (2022), 87:1003-1007. doi: 10.3233/JAD-220262.

290 Nehls M: Die Alzheimer-Lüge: Die Wahrheit über eine vermeidbare Krankheit. Heyne 2014, S. 160-163; neu überarbeitete Auflage erscheint im Herbst 2025.

291 J. M. Orgogozo et al., "Subacute meningoencephalitis in a subset of patients with AD after Abeta42 immunization," *Neurology.* 2003, 61:46-54. doi: 10.1212/01.wnl.0000073623.84147.a8.

292 J. A. Nicoll et al., "Neuropathology of human Alzheimer disease after immunization with amyloid-beta peptide: a case report," *Nat Med.* (2003), 9:448-52. doi: 10.1038/nm840.

293 P. Vollmar et al., "Active immunization with amyloid-beta 1-42 impairs memory performance through TLR2/4-dependent activation of the innate immune system," *J Immunol.* (2010), 185:6338-6347. doi: 10.4049/jimmunol.1001765.

294 C. Piller, "Blots on a field?" *Science.* (2022), 377:358-363. doi: 10.1126/science.add9993.; C. Piller, "Picture imperfect. Scores of papers by Eliezer Masliah, prominent neuroscientist and top NIH official, fall under suspicion," *Science.* (2024), https://www. science.org/content/article/research-misconduct-finding-neuroscientist-eliezer-masliah-papers-under-suspicion, doi: 10.1126/science.z2o7c3k.

295 Nehls, M., *Die Alzheimer-Lüge: Die Wahrheit über eine vermeidbare Krankheit.* Heyne 2014; neu überarbeitete Auflage erscheint im Herbst 2025.

296 M. Nehls, "Unified theory of Alzheimer's disease (UTAD): implications for prevention and curative therapy," *J Mol Psychiatry.* (2016), doi: 10.1186/s40303-016-0018-8.

297 https://www.agm-online.de/demenz-und-alzheimer/wissensportal-einblickde-menz/01-2015-ist-alzheimer-eine-luege-eine-stellungnahme (January 2015, last accessed on 3/19/2025).

298 C. Piller, "Picture imperfect."

299 I. Terao et al., "Comparative efficacy of lithium and aducanumab for cognitive decline in patients with mild cognitive impairment or Alzheimer's disease: A systematic review and network meta-analysis," *Ageing Res Rev.* (2022), doi: 10.1016/j.arr.2022.101709.

300 I. Terao & W. Kodama, "Comparative efficacy, tolerability and acceptability of donanemab, lecanemab, aducanumab and lithium on cognitive function in mild cognitive impairment and Alzheimer's disease: A systematic review and network meta-analysis," *Ageing Res Rev.* doi: 10.1016/j.arr.2024.102203.

301 https://www.ema.europa.eu/en/news/leqembi-recommended-treatment-early-alz-heimers-disease (last accessed 3/12/2025).
302 Terao & Kodama, "Comparative Efficacy."
303 J. M. Kang & A. P. Tanna, "Glaucoma," *Med Clin North Am.* (2021), 105:493-510. doi: 10.1016/j.mcna.2021.01.004.
304 S. M. DiCesare et al., "GSK3 inhibition reduces ECM production and prevents age-related macular degeneration-like pathology," *JCI Insight.* (2024), doi: 10.1172/jci.insight.178050.
305 A. Singh et al., "The Actions of Lithium on Glaucoma and Other Senile Neurodegenerative Diseases Through GSK-3 Inhibition: A Narrative Review," *Cureus.* (2022), doi: 10.7759/cureus.28265.
306 J. Gallagher et al., "Long-Term Dementia Risk in Parkinson Disease," *Neurology.* (2024), doi: 10.1212/WNL.0000000000209699.
307 S. S. Khan et al., "GSK-3β: An exuberating neuroinflammatory mediator in Parkinson's disease," *Biochemical Pharmacology* (2023), doi: 10.1016/j.bcp.2023.115496.
308 A. Turkistani et al., "Therapeutic Potential Effect of Glycogen Synthase Kinase 3 Beta (GSK-3β) Inhibitors in Parkinson Disease: Exploring an Overlooked Avenue," *Mol Neurobiol.* (2024), 61:7092-7108. doi: 10.1007/s12035-024-04003-z.
309 C. A. Lazzara & Y. H. Kim, "Potential application of lithium in Parkinson's and other neurodegenerative diseases," *Front Neurosci.* (2015), doi: 10.3389/fnins.2015.00403.
310 F. Bright et al., "Neuroinflammation in frontotemporal dementia," *Nat Rev Neurol. (*2019), 15:540-555. doi: 10.1038/s41582-019-0231-z.
311 N. Magrath Guimet et al., "Advances in Treatment of Frontotemporal Dementia," *J Neuropsychiatry Clin Neurosci.* (2022), 34:316-327. doi: 10.1176/appi.neuropsych.21060166.
312 F. Bright et al., "Neuroinflammation in frontotemporal dementia," *Nat Rev Neurol.* 2019, 15:540-555. doi: 10.1038/s41582-019-0231-z.
313 W. R. Bevan-Jones et al., "Neuroinflammation and protein aggregation co-localize across the frontotemporal dementia spectrum," *Brain.* (2020), 143:1010-1026. doi: 10.1093/brain/awaa033.
314 T. Gianferrara et al., "Glycogen Synthase Kinase 3β Involvement in Neuroinflammation and Neurodegenerative Diseases," *Curr Med Chem.* 2022, 29:4631-4697. doi: 10. 2174/0929867329666220216113517.
315 V. Llorca-Bofí et al., "Lithium management of periodic mood fluctuations in behavioural frontotemporal dementia: a case report," *Front Psychiatry.* (2024), doi: 10.3389/fpsyt.2023.1325145.; D. P. Devanand et al., "Low-dose Lithium Treatment for Agitation and Psychosis in Alzheimer Disease and Frontotemporal Dementia: A Case Series," *Alzheimer Dis Assoc Disord.* (2017), 31:73-75. doi: 10.1097/WAD.0000000000000161.

316 N. Ramesh & U. B. Pandey, "Autophagy Dysregulation in ALS: When Protein Aggregates Get Out of Hand," *Front Mol Neurosci.* (2017), doi: 10.3389/fnmol.2017.00263.; F. Fornai et al., "Autophagy and amyotrophic lateral sclerosis: The multiple roles of lithium," *Autophagy.* (2008), 4:527-530. doi: 10.4161/auto.5923.

317 M. Manchia et al., "Lithium and its effects: does dose matter?" *Int J Bipolar Disord.* (2024), doi: 10.1186/s40345-024-00345-8.

318 V. Danivas et al., "Off label use of lithium in the treatment of Huntington's disease: A case series," *Indian J Psychiatry.* (2013), 55:81-83. doi: 10.4103/0019-5545.105522.

319 L. Riedl et al., "Frontotemporal lobar degeneration: current perspectives," *Neuropsychiatr Dis Treat.* (2014), 10:297-310. doi: 10.2147/NDT.S38706.

320 C. Munteanu et al., "Lithium Biological Action Mechanisms after Ischemic Stroke," *Life* (Basel.) 2022, doi: 10.3390/life12111680.

321 Q. Zhou et al., "Discovery of novel phosphodiesterase-1 inhibitors for curing vascular dementia: Suppression of neuroinflammation by blocking NF-κB transcription regulation and activating cAMP/CREB axis," *Acta Pharm Sin B.* (2023), 13:1180-1191. doi: 10.1016/j.apsb.2022.09.023.

CHAPTER 5

1 Kieselbach J: Behandlung von Covid-19: Gegen den Sturm. Der Spiegel. https://www.spiegel.de/wissenschaft/medizin/coronavirus-forscher-testen-medikament-gegen-zytokinsturm-a-301dcdb8-85c5-440a-86cf-4a3df22f57b5, (last accessed on 3/23/2025).

2 J. L. Ferrara et al., "Cytokine storm of graft-versus-host disease: a critical effector role for interleukin-1," *Transplant Proc.* (1993), 25:1216-1217. https://pubmed.ncbi.nlm.nih.gov/8442093/.

3 K. Y. Yuen & S. S. Wong, "Human infection by avian influenza A H5N1," *Hong Kong Med J.* (2005), 11:189-199.

4 J. R. Tisoncik et al., "Into the eye of the cytokine storm," *Microbiol Mol Biol Rev.* (2012), 76:16-32. doi: 10.1128/MMBR.05015-11.

5 K. J. Huang et al., "An interferon-gamma-related cytokine storm in SARS patients," *J Med Virol.* (2005), 75:185-194. doi: 10.1002/jmv.20255.

6 P. Schreyer, Lockdown aus Angst vor der Biowaffe? https://multipolar-magazin.de/artikel/lockdown-aus-angst-vor-der-biowaffe(3/13/2025, last accessed on 3/14/2025); Nehls M: Das indoktrinierte Gehirn: Wie wir den globalen Angriff auf unsere mentale Freiheit erfolgreich abwehren. Mental Enterprises 2023.

7 S. Bindoli et al., "The amount of cytokine-release defines different shades of Sars-Cov2 infection," *Exp Biol Med* (Maywood.) (2020), 245:970-976. doi: 10.1177/1535370220928964.

8 P. Miossec, "Understanding the cytokine storm during COVID-19: Contribution of preexisting chronic inflammation," *Eur J Rheumatol.* (2020), 7:97-98. doi: 10.5152/eurjrheum.2020.2062.

9 Nehls, M., *Herdengesundheit: Der Weg aus der Corona-Krise und die natürliche Alternative zum globalen Impfprogramm.* Mental Enterprises 2022.

10 A. Giustina & A. M. Formenti, "Preventing a covid-19 pandemic," *BMJ.* (2020), doi: 10.1136/bmj.m810.

11 B. McCall, "Vitamin D Deficiency in COVID-19 Quadrupled Death Rate," www.medscape.com/viewarticle/942497(last accessed on 3/14/2025).

12 L. Borsche et al., "COVID-19 Mortality Risk Correlates Inversely with Vitamin D3 Status, and a Mortality Rate Close to Zero Could Theoretically Be Achieved at 50 ng/mL 25(OH)D3: Results of a Systematic Review and Meta-Analysis," *Nutrients.* (2021), doi: 10.3390/nu13103596.

13 Nehls, M., *Das Corona-Syndrom: Wie das Virus unsere Schwächen offenlegt – und wie wir uns nachhaltig schützen können.* Heyne 2021.

14 M. Entrenas Castillo et al., "Effect of calcifediol treatment and best available therapy versus best available therapy on intensive care unit admission and mortality among patients hospitalized for COVID-19: A pilot randomized clinical study," *J Steroid Biochem Mol Biol.* (2020), doi: 10.1016/j.jsbmb.2020.105751.; J. F. Alcala-Diaz et al., "Calcifediol Treatment and Hospital Mortality Due to COVID-19: A Cohort Study," *Nutrients.* (2021), doi: 10.3390/nu13061760.

15 C. Spuch et al., "Does Lithium Deserve a Place in the Treatment Against COVID-19? A Preliminary Observational Study in Six Patients, Case Report," *Front Pharmacol (*2020), doi: 10.3389/fphar.2020.557629.

16 W. Bartens, Corona: Hilflos im Unwetter. Süddeutsche Zeitung. https://www.Sueddeutsche.de/gesundheit/corona-covid-19-behandlung-1.5110786, (11/12/2020, last accessed on 3/23/2025).

17 C. Spuch et al., "Efficacy and Safety of Lithium Treatment in SARS-CoV-2 Infected Patients," *Front Pharmacol* (2022), doi: 10.3389/fphar.2022.850583.

18 G. Gómez-Bernal, "Lithium for the 2019 novel coronavirus," *Medical Hypotheses* (2020), doi: 10.1016/j.mehy.2020.109822.; A. B. Qaswal et al., "The Potential Role of Lithium as an Antiviral Agent against SARS-CoV-2 via Membrane Depolarization: Review and Hypothesis," *Sci Pharm.* 2021, doi: 10.3390/scipharm89010011.; M. E. Snitow et al., "Lithium and Therapeutic Targeting of GSK-3," *Cells* (2021), doi: 10.3390/cells10020255.

19 A. Murru et al., "Lithium's antiviral effects: a potential drug for CoViD-19 disease?" *Int J Bipolar Disord* (2020), doi: 10.1186/s40345-020-00191-4.

20 Amsterdam JD et al., "Suppression of herpes simplex virus infections with oral lithium carbonate–a possible antiviral activity. *Pharmacotherapy.* 1996, 16:1070-1075. https://pubmed.ncbi.nlm.nih.gov/8947995/.

21 S. M. Harrison et al., "Lithium chloride inhibits the coronavirus infectious bronchitis virus in cell culture," *Avian Pathol* (2007), 36:109-114, https://www.tandfonline.com/doi/full/10.1080/03079450601156083.

22 Liu X. et al., "Targeting the coronavirus nucleocapsid protein through GSK-3 inhibition. *Proc Natl Acad Sci USA.* 2021, doi: 10.1073/pnas.2113401118.

23 A. Borsini et al., "Neurogenesis is disrupted in human hippocampal progenitor cells upon exposure to serum samples from hospitalized COVID-19 patients with neurological symptoms," *Mol Psychiatry.* (2022), 27:5049-5061. doi: 10.1038 /s41380-02201741-1.

24 F. L. Fontes-Dantas et al., "SARS-CoV-2 Spike protein induces TLR4-mediated longterm cognitive dysfunction recapitulating post-COVID-19 syndrome in mice," *Cell Rep* (2023), doi: 10.1016/j.celrep.2023.112189.; K. Shirato & T. Kizaki, "SARS-CoV-2 spike protein S1 subunit induces pro-inflammatory responses via toll-like receptor 4 signaling in murine and human macrophages," *Heliyon.* (2021), doi: 10.1016/j.heliyon.2021.e06187.

25 P. I. Parry et al., "'Spikeopathy': COVID-19 Spike Protein Is Pathogenic, from Both Virus and Vaccine mRNA," *Biomedicines* (2023), https://www.mdpi .com/2227-9059/11/8/2287.

26 C. Spuch et al., "Efficacy and Safety of Lithium Treatment in SARS-CoV-2 Infected Patients," *Front Pharmacol* (2022), doi: 10.3389/fphar.2022.850583.

27 T. Guttuso Jr et al., "Lithium Aspartate for Long COVID Fatigue and Cognitive Dysfunction: A Randomized Clinical Trial," *JAMA Netw Open.* (2024), doi: 10.1001/jamanetworkopen.2024.36874.

28 P. Mehrbod et al., "The roles of apoptosis, autophagy and unfolded protein response in arbovirus, influenza virus, and HIV infections," *Virulence.* (2019), 10:376-413. doi: 10.1080/21505594.2019.1605803.

29 H. Zhou et al., "Bidirectional interplay between SARS-CoV-2 and autophagy," *mBio.* (2023), doi: 10.1128/mbio.01020-23.

30 C. Vidoni et al., "Targeting autophagy with natural products to prevent SARSCoV-2 infection." *J Tradit Complement Med.* (2022), 12:55-68. doi: 10.1016 /j.jtcme.2021.10.003.

31 T. J. Matthew et al., "Exploring autophagy in treating SARS-CoV-2 spike protein-related pathology," *Endocrine and Metabolic Science.* (2024), https://doi .org/10.1016/j. endmts.2024.100163.

32 M. Singer et al., "The Third International Consensus Definitions for Sepsis and Septic Shock (Sepsis-3)," *JAMA.* (2016), 315:801-810. doi: 10.1001 /jama.2016.0287.

33 C. Fleischmann-Struzek & K. Rudd, "Challenges of assessing the burden of sepsis," *Med Klin Intensivmed Notfmed.* (2023), 118:68-74. doi: 10.1007 /s00063-023-01088-7.

34 K. E. Rudd et al., "Global, regional, and national sepsis incidence and mortality, 19902017: analysis for the Global Burden of Disease Study," *Lancet.* (2020), 395:200-211. doi: 10.1016/S0140-6736(19)32989-7.

35 H. Dong et al., "Lithium ameliorates lipopolysaccharide-induced microglial activation via inhibition of toll-like receptor 4 expression by activating the PI3K/Akt/FoxO1 pathway," *J Neuroinflammation.* (2014), doi: 10.1186 /s12974-014-0140-4.

36 A. Nassar & A. N. Azab, "Effects of lithium on inflammation," *ACS Chem Neurosci.* (2014), 5:451-458. doi: 10.1021/cn500038f.

37 A. Albayrak et al., "Protective effects of lithium: A new look at an old drug with potential antioxidative and anti-inflammatory effects in an animal model of sepsis," *International Immunopharmacology.* (2013), 16:35-40. https://doi.org/10.1016/j.intimp.2013.03.018.; J. H. Lee et al., "Lithium Chloride Protects against Sepsis-Induced Skeletal Muscle Atrophy and Cancer Cachexia," *Cells.* (2021), doi: 10.3390/cells10051017.

38 J. Y. Kim et al., "Mortality and incidence rate of acute severe trauma patients in the emergency department: a report from the National Emergency Department Information System (NEDIS) of Korea, 2018-2022," *Clin Exp Emerg Med.* (2023), 10:55-62. doi: 10.15441/ceem.23.147.

39 R. Li et al., "Traumatic inflammatory response: pathophysiological role and clinical value of cytokines," *Eur J Trauma Emerg Surg.* (2024), 50:1313-1330. doi: 10.1007/s00068-023-02388-5.

40 M. Z. Xiao et al., "Postoperative delirium, neuroinflammation, and influencing factors of postoperative delirium: A review," *Medicine* (Baltimore.) (2023), doi: 10.1097/MD.0000000000032991.

41 Alam et al., "Surgery, neuroinflammation and cognitive impairment," *EBioMedicine.* (2018), 37:547-556. doi: 10.1016/j.ebiom.2018.10.021.

42 P. R. Leeds et al., "A new avenue for lithium: intervention in traumatic brain injury," *ACS Chem Neurosci.* (2014), 5:422-433. doi: 10.1021/cn500040g.

43 P. Dohare et al., "GSK3β Inhibition Restores Impaired Neurogenesis in Preterm Neonates With Intraventricular Hemorrhage," *Cereb Cortex.* (2019), 29:3482-3495. doi: 10.1093/cercor/bhy217.

44 A. Marcuzzi et al., "Autoinflammatory Diseases and Cytokine Storms-Imbalances of Innate and Adaptative Immunity," *Int J Mol Sci.* (2021), doi: 10.3390/ijms222011241.

45 M. Mobasseri et al., "Prevalence and incidence of type 1 diabetes in the world: a systematic review and meta-analysis," *Health Promot Perspect.* (2020), 10:98-115. doi: 10.34172/hpp.2020.18.; J. M. Norris et al., "Type 1 diabetes-early life origins and changing epidemiology," *Lancet Diabetes Endocrinol.* (2020), 8:226-238. doi: 10.1016/S2213-8587(19)30412-7.

46 F. W. Miller, "The increasing prevalence of autoimmunity and autoimmune diseases: an urgent call to action for improved understanding, diagnosis, treatment, and prevention," *Curr Opin Immunol.* (2023), doi: 10.1016/j.coi.2022.102266.

47 A. O. Adebajo, "Low frequency of autoimmune disease in tropical Africa," *Lancet.* (1997), 349:361-362. doi: 10.1016/s0140-6736(05)62867-x.

48 J. B. Classen & D. C. Classen, "Association between type 1 diabetes and hib vaccine. Causal relation is likely," *BMJ.* (1999), https://pmc.ncbi.nlm.nih.gov/articles/PMC1116914/; D. O. Ricke, "Immediate onset signatures of autoimmune diseases after vaccination," *Global Translational Medicine* (2023), doi: 10.36922/gtm.1455.

49 T. Noori et al., "The role of glycogen synthase kinase 3 beta in multiple sclerosis," *Biomed Pharmacother.* (2020), doi: 10.1016/j.biopha.2020.110874.

50 H. Link, "The cytokine storm in multiple sclerosis," *Mult Scler.* (1998), 04:12-15. doi: 10.1177/135245859800400104.

51 H. P. Eugster et al., "IL-6-deficient mice resist myelin oligodendrocyte glycoprotein-induced autoimmune encephalomyelitis," *Eur. J. Immunol.* (1998), 28:2178-2187.

52 M. Stampanoni Bassi et al., "Interleukin-6 Disrupts Synaptic Plasticity and Impairs Tissue Damage Compensation in Multiple Sclerosis," *Neurorehabil Neural Repair.* (2019), 33:825-835. doi: 10.1177/1545968319868713.

53 P. De Sarno et al., "Lithium prevents and ameliorates experimental autoimmune encephalomyelitis," *J Immunol.* (2008), 181:338-345. doi: 10.4049/ jimmunol.181.1.338.

54 M. Ahn et al., "Potential involvement of glycogen synthase kinase (GSK)-3β in a rat model of multiple sclerosis: evidenced by lithium treatment," *Anat Cell Biol.* (2017), 50:48-59. doi: 10.5115/acb.2017.50.1.48.

55 A. L. Rowse et al., controls central nervous system autoimmunity through modulation of IFN-γ signaling," PLoS One. (2012), doi: 10.1371/journal .pone.0052658.

56 S. Ghosouri et al., "Evaluation of in vivo lithium chloride effects as a GSK3-β inhibitor on human adipose derived stem cells differentiation into oligodendrocytes and remyelination in an animal model of multiple sclerosis," *Mol Biol Rep.* (2023), 50:16171625. doi: 10.1007/s11033-022-08181-8.

57 A. Turkistani et al., "Therapeutic Potential Effect of Glycogen Synthase Kinase 3 Beta (GSK-3β) Inhibitors in Parkinson Disease: Exploring an Overlooked Avenue," *Mol Neurobiol.* (2024), 61:7092-7108. doi: 10.1007/s12035-024-04003-z.

58 M. Silberstein, "Correlation between premorbid IL-6 levels and COVID-19 mortality: potential role for Vitamin D. Int.," *Immunopharm.* (2020);88:106995. doi: 10.1016/j. intimp.2020.106995.; M. Silberstein, "Vitamin D: a simpler alternative to tocilizumab for trial in COVID-19?" *Med. Hypotheses.* (2020), doi: 10.1016/j.mehy.2020.109767.

59 A. Miclea et al., "A Brief Review of the Effects of Vitamin D on Multiple Sclerosis," *Front Immunol.* (2020), doi: 10.3389/fimmu.2020.00781.; D. Lemke et al., "Vitamin D Resistance as a Possible Cause of Autoimmune Diseases: A Hypothesis Confirmed by a Therapeutic High-Dose Vitamin D Protocol," *Front Immunol.* (2021), doi: 10.3389/fimmu.2021.655739.

60 C. B. Tauil et al., "Suicidal ideation, anxiety, and depression in patients with multiple sclerosis," *Arq Neuropsiquiatr.* (2018), 76:296-301. doi: 10.1590/0004-282X20180036.

61 S. Rossi et al., "Neuroinflammation drives anxiety and depression in relapsing-remitting multiple sclerosis," *Neurology.* (2017) Sep 26;89(13):1338-1347. doi: 10.1212/WNL.0000000000004411.

62 J. R. Rinker 2nd et al., "Randomized feasibility trial to assess tolerance and clinical effects of lithium in progressive multiple sclerosis," *Heliyon.* (2020), doi: 10.1016/j.heliyon.2020.e04528.

63 R. Strawbridge et al., "Identifying the neuropsychiatric health effects of low-dose lithium interventions: A systematic review," *Neuroscience & Biobehavioral Reviews.* (2023), doi: 10.1016/j.neubiorev.2022.104975.

64 F. Cano-Cano et al., "IL-1β Implications in Type 1 Diabetes Mellitus Progression: Systematic Review and Meta-Analysis," *J Clin Med.* (2022), doi: 10.3390/jcm11051303.

65 M. Alibashe-Ahmed et al., "Toll-like receptor 4 inhibition prevents autoimmune diabetes in NOD mice," *Sci Rep.* (2019), doi: 10.1038/s41598-019-55521-z.

66 S. R. Jung et al., "Lithium and exercise ameliorate insulin-deficient hyperglycemia by independently attenuating pancreatic α-cell mass and hepatic gluconeogenesis," *Korean J Physiol Pharmacol.* (2024), 28:31-38. doi: 10.4196/kjpp.2024.28.1.31.

67 K. Kiran. et al., "Autoimmune activation and cytokine storm: biochemical and physiological mechanisms in rheumatoid arthritis pathology," *J Pop Therap Clin Pharm.* doi: 10.53555/2vhzx593.

68 C. L. Thompson et al., "Lithium chloride prevents interleukin-1β induced cartilage degradation and loss of mechanical properties," *J Orthop Res.* (2015), 33:1552-1559. doi: 10.1002/jor.22913.; T. Minashima et al., "Lithium protects against cartilage degradation in osteoarthritis," *Arthritis Rheumatol.* (2014), 66:1228-1236. doi: 10.1002/art.38373.

69 J. Zhang et al., "Microdose lithium protects against pancreatic islet destruction and renal impairment in streptozotocin-elicited diabetes (Type1)," *Antioxidants*, (2021), doi: 10.3390/antiox10010138.

70 D. Nikolopoulos et al., "Microglia activation in the presence of intact blood-brain barrier and disruption of hippocampal neurogenesis via IL-6 and IL-18 mediate early diffuse neuropsychiatric lupus," *Ann Rheum Dis.* (2023), 82:646-657. doi: 10.1136/ard2022-223506.

71 S. P. Lenz et al., "Lithium chloride enhances survival of NZB/W lupus mice: influence of melatonin and timing of treatment," *Int J Immunopharmacol.* (1995), 17:581-592. doi: 10.1016/0192-0561(95)00032-w.

CHAPTER 6

1 K. Zarse et al., "Low-dose lithium uptake promotes longevity in humans and metazoans," *Eur J Nutr.* (2011), 50:387-389. doi: 10.1007/s00394-011-0171-x.; V. A. Fajardo et al., "Trace lithium in Texas tap water is negatively associated with all-cause mortality and premature death," *Appl Physiol Nutr Metab.* (2018), 43:412-414. doi: 10.1139/apnm-2017-0653.

2 X. Liu & Z. Yao, "Chronic over-nutrition and dysregulation of GSK3 in diseases," *Nutr Metab* (Lond.) (2016), doi: 10.1186/s12986-016-0108-8.

3 L. F. M. Rocha et al., "Gene Editing for Treatment and Prevention of Human Diseases: A Global Survey of Gene Editing-Related Researchers," *Hum Gene Ther.* (2020), 31:852862. doi: 10.1089/hum.2020.136.; J. van Haasteren et al., "The delivery challenge: fulfilling the promise of therapeutic genome editing," *Nat Biotechnol.* (2020), 38:845-855. doi: 10.1038/s41587-020-0565-5.

4 O. Sofola-Adesakin et al., "Lithium suppresses Aβ pathology by inhibiting translation in an adult Drosophila model of Alzheimer's disease," *Front Aging Neurosci.* (2014), doi: 10.3389/fnagi.2014.00190.

5 J. I. Castillo-Quan et al., "Lithium Promotes Longevity through GSK3/NRF2-Dependent Hormesis," *Cell Rep.* (2016) Apr 19;15(3):638-650. doi: 10.1016 /j.celrep.2016.03.041.

6 Blackburn E. H. et al., "Human telomere biology: A contributory and interactive factor in aging, disease risks, and protection," *Science.* 2015, 350:1193-1198. doi: 10.1126/science.aab3389.

7 R. M. Post, "The New News about Lithium: An Underutilized Treatment in the United States," *Neuropsychopharmacology.* (2018), 43:1174-1179. doi: 10.1038 /npp.2017.238.

8 R. P. Barnes et al., "The impact of oxidative DNA damage and stress on telomere homeostasis," *Mech Ageing Dev.* (2019), 177:37-45. doi: 10.1016 /j.mad.2018.03.013.

9 R. Khairova et al., "Effects of lithium on oxidative stress parameters in healthy subjects," *Mol Med Rep.* (2012), 5:680-682. doi: 10.3892/mmr.2011.732.

10 F. Coutts et al., "The polygenic nature of telomere length and the anti-ageing properties of lithium," *Neuropsychopharmacology.* (2019), 44:757-765. doi: 10.1038 /s41386018-0289-0.

11 J. I. Castillo-Quan et al., "A triple drug combination targeting components of the nutrient-sensing network maximizes longevity," *Proc Natl Acad Sci USA.* (2019) Oct 15;116(42):20817-20819. doi: 10.1073/pnas.1913212116.

12 ps://www.pharmazeutische-zeitung.de/kombinationstherapie-verlaengert-das-le-ben /(10/29/2019, last accessed on 3/20/2025).

13 R. Zhao et al., "Fasting promotes acute hypoxic adaptation by suppressing mTOR-mediated pathways," *Cell Death Dis.* (2021), doi: 10.1038/s41419-021-04351-x.

14 P. S. Rejeki et al., "Combined Aerobic Exercise with Intermittent Fasting Is Effective for Reducing mTOR and Bcl-2 Levels in Obese Females," *Sports* (Basel.) doi: 10.3390/sports12050116.

15 S. H. Baik et al., "Intermittent fasting increases adult hippocampal neurogenesis," *Brain Behav.* (2020) Jan;10(1):e01444. doi: 10.1002/brb3.1444.

16 T. Viel et al., "Microdose lithium reduces cellular senescence in human astrocytes – a potential pharmacotherapy for COVID-19?" *Aging* (Albany NY.) (2020), 12:1003510040. doi: 10.18632/aging.103449.

17 M. Toricelli et al., "Microdose Lithium Treatment Reduced Inflammatory Factors and Neurodegeneration in Organotypic Hippocampal Culture of Old

SAMP-8 Mice," *Cell Mol Neurobiol.* (2021), 41:1509-1520. doi: 10.1007/s10571-020-00916-0.

18 F. B. Ahmad & R. N. Anderson, "The leading causes of death in the US for 2020," *JAMA.* (2021), 325:1829–1830. doi: 10.1001/jama.2021.5469.

19 A. W. Voors, "Minerals in the municipal water and atherosclerotic heart death," *Am J Epidemiol.* (1971), 93:259-266. https://pubmed.ncbi.nlm.nih.gov/5550342/; Voors AW: "Lithium depletion and atherosclerotic heart-disease," *Lancet.* 1970, 2:670. doi: 10.1016/s0140-6736(70)91446-7.; A. W. Voors, "Minerals in the municipal water and atherosclerotic heart death," *Am J Epidemiol.* (1971), 93:259-266. https://pubmed.ncbi.

20 World Health Organization, "Trace elements in human nutrition and health," (1996), https://www.who.int/publications/i/item/9241561734, S. 225.

21 W. F. Parker et al., "Association Between Groundwater Lithium and the Diagnosis of Bipolar Disorder and Dementia in the United States," *JAMA Psychiatry.* (2018), 75:751754. doi: 10.1001/jamapsychiatry.2018.1020.

22 P. W. Cheng et al., "Wnt Signaling Regulates Blood Pressure by Downregulating a GSK3β-Mediated Pathway to Enhance Insulin Signaling in the Central Nervous System," *Diabetes.* (2015), 64:3413-3424. doi: 10.2337/db14-1439.

23 B. Bosche et al., "A differential impact of lithium on endothelium-dependent but not on endothelium-independent vessel relaxation," *Prog. Neuropsychopharmacol. Biol. Psychiatry.* (2016), 67:98-106. doi: 10.1016/j.pnpbp.2016.02.004.; B. Bosche et al., "Lowdose lithium stabilizes human endothelial barrier by decreasing MLC phosphorylation and universally augments cholinergic vasorelaxation capacity in a direct manner," *Front. Physiol.* doi: 10.3389/fphys.2016.00593.

24 S. I. Hamstra et al., "Low-dose lithium feeding increases the SERCA2a-to-phospholamban ratio, improving SERCA function in murine left ventricles," *Exp. Physiol.* (2020), 105:666-675. doi: 10.1113/EP088061.

25 N. Mehta & R. Vannozzi, "Lithium-induced electrocardiographic changes: A complete review," *Clin. Cardiol.* (2017), 40:1363-1367. doi: 10.1002/clc.22822.; S. Moradi et al., "Cardiac chronotropic hypo-responsiveness and atrial fibrosis in rats chro-nically treated with lithium," *Auton. Neurosci.* (2019), 216:46-50. doi: 10.1016/j.autneu.2018.09.002.; V. Bisogni et al., "Antihypertensive therapy in patients on chronic lithium treatment for bipolar disorders," *J Hypertens.* (2016), 34:20-28. doi: 10.1097/HJH.0000000000000758.

26 S. I. Hamstra et al., "Beyond its Psychiatric Use: The Benefits of Low-dose Lithium Supplementation," *Curr Neuropharmacol.* (2023), 21:891-910. doi: 10.2174/1570159X2 0666220302151224.

27 S. Wang et al., "Glycogen synthase kinase-3β inhibition alleviates activation of the NLRP3 inflammasome in myocardial infarction," *J Mol Cell Cardiol.* (2020), 149:82-94. doi: 10.1016/j.yjmcc.2020.09.009.

28 N. Ates et al., "Phosphorylation of PI3K/Akt at Thr308, but not phosphorylation of MAPK kinase, mediates lithium-induced neuroprotection against cerebral ischemia in mice," *Exp Neurol.* (2022), doi: 10.1016/j.expneurol.2022.113996.

29 B. Chen et al., "The neuroprotective mechanism of lithium after ischaemic stroke," *Commun Biol.* (2022), doi: 10.1038/s42003-022-03051-2.

30 O. P. Almeida et al., "Lithium and Stroke Recovery: A Systematic Review and MetaAnalysis of Stroke Models in Rodents and Human Data," *Stroke*. (2022), 53:2935-2944. doi: 10.1161/STROKEAHA.122.039203.; W. Wang et al., "Exploring the Neuroprotective Effects of Lithium in Ischemic Stroke: A literature review," *Int J Med Sci.* (2024), 21:284-298. doi: 10.7150/ijms.88195.

31 S. E. Mohammadianinejad et al., "The effect of lithium in post-stroke motor recovery: a double-blind, placebo-controlled, randomized clinical trial," *Clin Neuropharmacol.* (2014), 37:73–78. doi: 10.1097/WNF.0000000000000028.

32 W. Cai et al., "Post-stroke DHA Treatment Protects Against Acute Ischemic Brain Injury by Skewing Macrophage Polarity Toward the M2 Phenotype," *Transl Stroke Res.* (2018), 9:669-680. doi: 10.1007/s12975-018-0662-7.

33 J. H. O'Keefe et al., "Omega-3 Blood Levels and Stroke Risk: A Pooled and Harmonized Analysis of 183 291 Participants From 29 Prospective Studies," *Stroke.* (2024), 55:50-58. doi: 10.1161/STROKEAHA.123.044281.

34 K. U. Eckardt et al., "Evolving importance of kidney disease: from subspecialty to global health burden," *Lancet.* (2013), 382:158-169. doi: 10.1016/S0140 -6736(13)60439-0.

35 J. A. Hurcombe et al., "Podocyte GSK3 is an evolutionarily conserved critical regulator of kidney function," *Nat Commun.* 2019, doi: 10.1038 /s41467-018-08235-1.

36 M. Alsady et al., "Lithium in the Kidney: Friend and Foe?" *J Am Soc Nephrol.* (2016), 27:1587-1595. doi: 10.1681/ASN.2015080907.

37 A. Post et al., "Dietary lithium intake, graft failure and mortality in kidney transplant recipients," *Nephrol Dial Transplant.* (2023), 38:1867-1879. doi: 10.1093/ndt/gfac340.

38 M. Blüher, "Obesity: global epidemiology and pathogenesis," *Nat Rev Endocrinol.* (2019), 15:288-298. doi: 10.1038/s41574-019-0176-8.

39 E. D. Peselow et al., "Lithium carbonate and weight gain," *J. Affect. Disord.* (1980), 2:303–310. doi: 10.1016/0165-0327(80)90031-2.; H. Mangge et al., "Weight gain during treatment of bipolar disorder (BD)-facts and therapeutic options," *Front. Nutr.* (2019), doi: 10.3389/fnut.2019.00076.

40 V. A. Fajardo et al., "Examining the relationship between trace lithium in drinking water and the rising rates of age-adjusted Alzheimer's disease mortality in Texas," *J. Alzheimers Dis.* (2017), 61:425-434. doi: 10.3233/JAD-170744.

41 S. E. Choi et al., "Atherosclerosis induced by a high-fat diet is alleviated by lithium chloride via reduction of VCAM expression in ApoE-deficient mice," *Vascul. Pharmacol.* (2010), 53:264–272. doi: 10.1016/j.vph.2010.09.004.

42 S. Jung et al., "Effect of lithium on the mechanism of glucose transport in skeletal muscles," *J. Nutr. Sci. Vitaminol.* (Tokyo) 2017, 63:365–371. doi: 10.3177 /jnsv.63.365.

43 R. A. Ryan, "Low-Dose Lithium as a Therapy for High-Fat Diet Induced Obesity: A Burning Topic in Metabolic Research and Adipose Tissue Browning," Brock University. (2022), http://hdl.handle.net/10464/15745; M. S. Finch et al., "Creatine and low-dose lithium supplementation separately alter energy expenditure, body mass, and adipose metabolism for the promotion of thermogenesis," *iScience.* (2024), doi: 10.1016/j. isci.2024.109468.

44 S. E. Choi et al., "Atherosclerosis induced by a high-fat diet is alleviated by lithium chloride via reduction of VCAM expression in ApoE-deficient mice," *Vascul. Pharmacol.* (2010), 53:264–272. doi: 10.1016/j.vph.2010.09.004.

45 S. I. Hamstra et al., "Beyond its Psychiatric Use: The Benefits of Low-dose Lithium Supplementation," *Curr Neuropharmacol.* (2023), 21:891-910. doi: 10.2174/1570159X2 0666220302151224.

46 O. Köhler-Forsberg et al., "Association of Lithium Treatment with the Risk of Osteoporosis in Patients with Bipolar Disorder," *JAMA Psychiatry.* (2022), 79:454-463. doi: 10.1001/jamapsychiatry.2022.0337.; L. J. Williams et al., "Lithium use and bone health in women with bipolar disorder: A cross-sectional study," *Acta Psychiatr Scand.* (2024) Apr;149(4):332-339. doi: 10.1111/acps.13660.; V. W. S. Ng et al., "Lithium and the risk of fractures in patients with bipolar disorder: A population-based cohort study," *Psychiatry Res.* (2024), doi: 10.1016 /j.psychres.2024.116075.

47 T. J. Margetts et al., "From the Mind to the Spine: The Intersecting World of Alzheimer's and Osteoporosis," *Curr Osteoporos Rep.* (2024), 22:152-164. doi: 10.1007/s11914-02300848-w.

48 N. Kurgan et al., "Low dose lithium supplementation activates Wnt/β-catenin signalling and increases bone OPG/RANKL ratio in mice," *Biochem Biophys Res Commun.* (2019), 511:394-397. doi: 10.1016/j.bbrc.2019.02.066.

49 L. Li et al., "Acceleration of bone regeneration by activating Wnt/β-catenin signalling pathway via lithium released from lithium chloride/calcium phosphate cement in osteoporosis. *Sci Rep.* (2017), doi: 10.1038/srep45204.

50 P. Clément-Lacroix et al., "Lrp5-independent activation of Wnt signaling by lithium chloride increases bone formation and bone mass in mice," *Proc Natl Acad Sci USA.* (2005), 102:17406-17411. doi: 10.1073/pnas.0505259102.

51 K. Vachhani et al., "Low-dose lithium regimen enhances endochondral fracture healing in osteoporotic rodent bone," *J Orthop Res.* (2018), 36:1783-1789. doi: 10.1002/jor.23799.

52 J. Woo, "Sarcopenia," *Clin. Geriatr. Med.* (2017), 33:305-314. doi: 10.1016 /j.cger.2017.02.003.

53 K. J. P. Verhees et al., "Glycogen synthase kinase-3β is required for the induction of skeletal muscle atrophy," *Am. J. Physiol. Cell Physiol.* (2011), 301:995-1007. doi: 10.1152/ajpcell.00520.2010.; T. M. Mirzoev et al., "The role of GSK-3β in the regulation of protein turnover, myosin phenotype, and oxidative capacity in skeletal muscle under disuse conditions," *Int. J. Mol. Sci.* (2021), doi: 10.3390 /ijms22105081.

54 J. Park et al., "Effects of Long-Term Endurance Exercise and Lithium Treatment on Neuroprotective Factors in Hippocampus of Obese Rats," *Int J Environ Res Public Health.* (2020), doi: 10.3390/ijerph17093317.
55 R. W. Baranowski et al., "Exploring the Effects of Greek Yogurt Supplementation and Exercise Training on Serum Lithium and Its Relationship With Musculoskeletal Outcomes in Men," *Front Nutr.* (2021), doi: 10.3389/fnut.2021.798036.
56 R. Sender & R. Milo, "The distribution of cellular turnover in the human body," *Nat Med.* (2021), 27:45-48. doi: 10.1038/s41591-020-01182-9.
57 C. Paddock, "Immune system kills spontaneous blood cancer cells every day," https://www.medicalnewstoday.com/articles/272092 (2/3/2014): S. Afshar-Sterle et al., "Fast ligand-mediated immune surveillance by T cells is essential for the control of spontaneous B cell lymphomas," *Nat Med.* (2014), 20:283-290. doi: 10.1038/nm.3442.
58 E. Y. Villegas-Vázquez et al., "Lithium: A Promising Anticancer Agent," *Life* (Basel.) (2023), doi: 10.3390/life13020537.
59 R. Y. Huang et al., "Use of lithium and cancer risk in patients with bipolar disorder: population-based cohort study," *Br J Psychiatry.* (2016), 209:393-399. doi: 10.1192 /bjp.bp.116.181362.
60 W. F. Parker et al., "Association Between Groundwater Lithium and the Diagnosis of Bipolar Disorder and Dementia in the United States," *JAMA Psychiatry.* (2018), 75:751754. doi: 10.1001/jamapsychiatry.2018.1020.
61 T. Domoto et al., "Glycogen Synthase Kinase 3β in Cancer Biology and Treatment," *Cells.* (2020), doi: 10.3390/cells9061388.
62 E. Y. Villegas-Vázquez et al., "Lithium: A Promising Anticancer Agent," *Life* (Basel.) (2023), doi: 10.3390/life13020537.
63 H. A. Bischoff-Ferrari et al., DO-HEALTH Research Group, "Effect of Vitamin D Supplementation, Omega-3 Fatty Acid Supplementation, or a Strength-Training Exercise Program on Clinical Outcomes in Older Adults: The DO-HEALTH Randomized Clinical Trial," *JAMA.* (2020) Nov 10;324(18):1855-1868. doi: 10.1001/jama.2020.16909. PMID: 33170239; PMCID: PMC7656284.
64 T. Domoto et al., "Glycogen Synthase Kinase 3β in Cancer Biology and Treatment," *Cells.* (2020), doi: 10.3390/cells9061388.
65 M. Khasraw et al., "Using lithium as a neuroprotective agent in patients with cancer," *BMC Med.* (2012), doi: 10.1186/1741-7015-10-131.
66 J. Mutz et al., "The duration of lithium use and biological ageing: telomere length, frailty, metabolomic age and all-cause mortality," *Geroscience.* (2024), 46:5981-5994. doi: 10.1007/s11357-024-01142-y.

CHAPTER 7

1 Brecht, B., *Leben des Galilei: Schauspiel*, Suhrkamp 1998.
2 J. F. Cade, "Lithium salts in the treatment of psychotic excitement," *Med J Aust.* (1949), 2:349-352. doi: 10.1080/j.1440-1614.1999.06241.x.
3 Schrauzer, "Lithium: occurrence, dietary intakes, nutritional essentiality."

4 D. A. Hart, "Lithium Ions as Modulators of Complex Biological Processes: The Conundrum of Multiple Targets, Responsiveness and Non-Responsiveness, and the Potential to Prevent or Correct Dysregulation of Systems during Aging and in Disease," *Biomolecules*. (2024), doi: 10.3390/biom14080905.

5 E. J. Calabrese et al., "Lithium and hormesis: Enhancement of adaptive responses and biological performance via hermetic mechanisms," *J Trace Elem Med Biol.* (2023), doi: 10.1016/j.jtemb.2023.127156.

6 J. I. Castillo-Quan et al., "Lithium Promotes Longevity through GSK3/NRF2 -Dependent Hormesis" *Cell Rep.* (2016), 15:638-650. doi: 10.1016 /j.celrep.2016.03.041.

7 J. S. Hayward & J.D. Eckerson, "Physiological responses and survival time prediction for humans in ice-water," *Aviat Space Environ Med.* (1984), 55:206-211. https://pubmed.ncbi.nlm. nih.gov/6721816/; B. Knechtle et al., "Water Swimming-Benefits and Risks: A Narrative Review," *Int J Environ Res Public Health.* (2020), doi: 10.3390/ijerph17238984. PMC7730683.

8 J. I. Castillo-Quan et al., "Lithium Promotes Longevity through GSK3 /NRF2-Dependent Hormesis," *Cell Rep.* (2016), 15:638-650. doi: 10.1016 /j.celrep.2016.03.041.

9 M. Medina & F. Wandosell, "Deconstructing GSK-3: The Fine Regulation of Its Activity," *Int J Alzheimers Dis.* (2011), doi: 10.4061/2011/479249.

10 K. C. Duong-Ly & J. R. Peterson, "The human kinome and kinase inhibition," *Curr Protoc Pharmacol.* (2013), doi: 10.1002/0471141755.ph0209s60.

11 E. Beurel et al., "Glycogen synthase kinase-3 (GSK3): regulation, actions, and diseases," *Pharmacol Ther.* (2015), 148:114-131. doi: 10.1016/j.pharmthera .2014.11.016.

12 R. S. Jope et al., "Stressed and Inflamed, Can GSK3 Be Blamed?" *Trends Biochem Sci.* (2017), 42:180-192. doi: 10.1016/j.tibs.2016.10.009.; X. Liu & Z. Yao, "Chronic over-nutrition and dysregulation of GSK3 in diseases," *Nutr Metab* (Lond.) 2016, https://doi. org/10.1186/s12986-016-0108-8; H. O. Kalkman, "The GSK3-NRF2 Axis in Suicide," *Psychiatry Int.* (2021), 2:108-119. https://doi .org/10.3390/psychiatryint2010008.

13 Verbraucherzentrale: Lithiumorotat: Welche Wirkung hat dieses Nahrungsergänzungsmittel und weshalb ist es in Deutschland so schwer und teuer zu bekommen? https://www.verbraucherzentrale.de/faq/projekt-klartext-nem /lithiumorotat-40582 October 16, 2024, last accessed on December 30, 2024).

14 https://report24.news/ist-lithium-gut-fuers-gehirn-schadet-fluor-wirklich -bestsellerautor-dr-nehls-im-gespraech/ (March 28, 2024, last accessed February 10, 2025); In conversation with Tucker Carlson: How the Government Uses Fear -Mongering to Alter Your Brain, https://www.youtube.com/watch?v=GhFr7WeIeQc (April 2, 2024, last accessed on February 10,.2025).

15 Policy 2002/46/EC of the European Parliament and of the Council of 10 June 2002 on the approximation of the laws of the Member States relating to

food supplements. https://eur-lex.europa.eu/legal-content/DE/TXT /PDF/?uri=CELEX:32002L0046.

16 www.kanzlei-roehrig.de; https://www.youtube.com/watch?v=1yS5FWgNdV4 (November 1, 2024).

17 A. M. Pérez-Granados & M. P. Vaquero, "Silicon, aluminium, arsenic and lithium: essentiality and human health implications," *J Nutr Health Aging.* (2002), 6:154-162. https://pubmed.ncbi.nlm.nih.gov/12166372/.

18 Grimm H-U: Die Ernährungsfalle: Wie die Lebensmittelindustrie unser Essen manipuliert – Das Lexikon. Heyne 2010, S. 24ff.

19 Aluminiumhaltige Lebensmittel, Lebensmittelzusatzstoffe, Kosmetika, Lebensmittelkontaktmaterialien und Spielzeuge. https://www.bundestag.de/resource /blob/928354/6a673927f71e5e93755e0a5987b42d23/WD-5-137-22-pdf.pdf (November 25, 2022, last accessed August 29, 2024).

20 https://www.lgl.bayern.de/lebensmittel/chemie/schwermetalle/aluminium /ue_2015_2016_aluminium.htm (last accessed on August 29, 2024).

21 https://www.efsa.europa.eu/de/glossary/tolerable-weekly-intake.

22 Ł. Bryliński et al., "Aluminium in the Human Brain: Routes of Penetration, Toxicity, and Resulting Complications," *Int J Mol Sci.* (2023), doi: 10.3390/ijms24087228.

23 Evaluations of the Joint FAO/WHO Expert Committee on Food Additives (JECFA): ALUMINIUM-containing food additives. 2011, https://apps.who.int /food-additivescontaminants-jecfa-database/Home/Chemical/6179.

24 https://www.efsa.europa.eu/en/efsajournal/pub/1086 (June 3, 2009, last accessed on February 8, 2025).

25 Schrauzer & de Vroey, "Effects of nutritional lithium supplementation on mood."

26 D. Szklarska & P. Rzymski, "Is Lithium a Micronutrient? From Biological Activity and Epidemiological Observation to Food Fortification," *Biol Trace Elem Res. (*2019), 189:18-27. doi: 10.1007/s12011-018-1455-2.

27 M. Richardson, "A Search for Genius in Weimar Germany: The Abraham Lincoln Stiftung and American Philanthropy," In: Bulletin of the GHI Washington, Issue 26 (Spring 2000.) (= Bulletin of the German Historical Institute Washington, DC.) Online unter: https://perspectivia.net/receive/ploneimport3_mods_00003232.

28 Müller-Hill B: Das Blut von Auschwitz und das Schweigen der Gelehrten. In: Doris Kaufmann (Hrsg.): Geschichte der Kaiser-Wilhelm-Gesellschaft im Nationalsozialismus. Wallstein 2020, S. 189-227.

29 Schönherr H-P: Erben der Nazi-Forscher. 17.03.2024, https://taz. de/75-Jahre-MaxPlanck-Gesellschaft/!5991583/.

30 S. Müller-Wille, "Eugenics: Then and now," *Metascience.* (2010), 20:347-349. doi:10.1007/s11016-010-9477-1.

31 https://www.cshl.edu/archives/institutional-collections/eugenics-record-office / (zuletzt abgerufen am 17.02.2025).

32 http://archive.carnegiefoundation.org/publications/pdfs/elibrary/Carnegie_Flex -ner_Report.pdf, Introduction, S. 10.

33 https://corbettreport.com/bigoil/ (06.10.2017, zuletzt aufgerufen am 07.01.2025).
34 D. Starr, "Ethics. Revisiting a 1930s scandal, AACR to rename a prize," *Science.* (2003), 300:573-574. doi: 10.1126/science.300.5619.573.
35 https://web.archive.org/web/20210118011339/https://www.reuters.com/article/us-maryland-lawsuit-infections-idUSKCN1OY1N3 (2019); https://journals.library.columbia.edu/index.php/bioethics/article/download/12795/6359/34794 (2024).
36 Nehls M: Das indoktrinierte Gehirn: Wie wir den globalen Angriff auf unsere mentale Freiheit erfolgreich abwehren. Mental Enterprises 2023.
37 https://www.who.int/about/funding/contributors/the-rockefeller-foundation.
38 P. Weindling, "Philanthropy and World Health: The Rockefeller Foundation and the League of Nations Health Organisation," *Minerva.*(1997), 35:269-281. https://www. jstor.org/stable/41821072.
39 L. Tournès, "The Rockefeller Foundation and the Transition from the League of Nations to the UN (1939–1946.)," *Journal of Modern European History.* (2014), 12:323-341. doi: 10.17104/1611-8944_2014_3_323.
40 https://www.who.int/about/funding/contributors/the-rockefeller-foundation (April 20, 2022, last accessed January 07, 2025).
41 T. Niccoli & L. Partridge, "Ageing as a risk factor for disease," *Curr Biol.* (2012), 22:741752. doi: 10.1016/j.cub.2012.07.024.
42 Handel S: Volkskrankheit Demenz:Allianz gegen das Vergessen. 13.12.2015, https://www.sueddeutsche.de/muenchen/volkskrankheit-demenz-allianz-gegen-das-verges-sen-1.2780522.
43 M. B. Demay et al., "Vitamin D for the Prevention of Disease: An Endocrine Society Clinical Practice Guideline," *JCEM.* (2024), 109:1907-1947. doi: 10.1210/clinem/dgae290.
44 G. Agyralides, "The future of medicine: an outline attempt using state-of-the-art business and scientific trends," *Front Med (*Lausanne.) 2024, doi: 10.3389/fmed.2024.1391727.
45 https://mx.pinterest.com/pin/394769342186797421/.
46 A. M. Martelli et al., "Pathobiology and Therapeutic Relevance of GSK-3 in Chronic Hematological Malignancies," *Cells.* (2022), doi: 10.3390/cells11111812.
47 X. Li & R. S. Jope, "Is glycogen synthase kinase-3 a central modulator in mood regulation?" *Neuropsychopharmacology.* (2010), 35:2143-2154. doi: 10.1038/npp.2010.105.
48 R. M. Arciniegas Ruiz, H. Eldar-Finkelman, "Glycogen Synthase Kinase-3 Inhibitors: Preclinical and Clinical Focus on CNS-A Decade Onward," *Front Mol Neurosci.* (2022), doi: 10.3389/fnmol.2021.792364.
49 A. Memon et al., "Association between naturally occurring lithium in drinking water and suicide rates: systematic review and meta-analysis of ecological studies," *Br J Psychiatry.* (2020), 217:667-678. doi: 10.1192/bjp.2020.128.
50 https://www.periodpaper.com/products/1903-ad-londonderry-lithia-spring-drin-king-water-health-original-advertising-070234-tin1-041.

51 Londonderry Lithia Spring Water Co (1891), https://archive.org/details/39002086471951.med.yale.edu/page/2/mode/2up.

52 R. S. El-Mallakh, "Prethymoleptic uses of lithium in the United States," *Louisville Med* (1996), 56:461-463. doi: 10.1176/ajp.156.1.129.; S. A. Hedya et al., *Lithium Toxicity*, (StatPearls Publishing 2023), https://www.ncbi.nlm.nih.gov/books/NBK499992/.

53 J. F. Aita et al., "7-Up anti-acid lithiated lemon soda or early medicinal use of lithium," *Nebr Med J.* (1990), 75:277-279, https://pubmed.ncbi.nlm.nih.gov/2074904/; U. Seidel et al., Lithium Content of 160 Beverages and Its Impact on Lithium Status in Drosophila melanogaster," *Foods.* (2020), doi: 10.3390/foods9060795.

54 A. C. Kaufman, "The Original 7-Up Was A Mind-Altering Substance," *Huffpost*, https://www. huffpost.com/entry/7up-history_n_5836322 (06.12.2017, zuletzt abgerufen am 24.11.2024).

55 L. W. Hanlon et al., "Lithium chloride as a substitute for sodium chloride in the diet; observations on its toxicity," *J Am Med Assoc.* (1949), 139:688-692. doi: 10.1001/jama.1949.02900280004002.

56 Time: Medicine: Case of trie Substitute Sal. 28.02.1949, https://time.com/archive/6602227/medicine-case-of-trie-substitute-salt/.

57 Hanlon et al., "Lithium chloride as a substitute for sodium chloride."

58 A. C. Corcoran et al., "Lithium poisoning from the use of salt substitutes," *J Am Med Assoc. (*1949), 139:685-688. doi: 10.1001/jama.1949.02900280001001.

59 R. L. Stern, "Severe lithium chloride poisoning with complete recovery; report of case," *J Am Med Assoc.* (1949), 139:710-711. doi: 10.1001/jama.1949.72900280001006.

60 D. A. Hart, "Lithium Ions as Modulators of Complex Biological Processes: The Conundrum of Multiple Targets, Responsiveness and Non-Responsiveness, and the Potential to Prevent or Correct Dysregulation of Systems during Aging and in Disease," *Biomolecules.* (2024), doi: 10.3390/biom14080905.

61 J. F. Cade, "Lithium salts in the treatment of psychotic excitement."

62 T. M. Sissung et al., "The Dietary Supplement Health And Education Act: are we healthier and better informed after 27 years?" *Lancet Oncol.* (2021), 22:915-916. doi: 10.1016/S1470-2045(21)00084-X.

63 D. K. Pauzé & D. E. Brooks, "Lithium toxicity from an Internet dietary supplement," *J Med Toxicol.* (2007), 3:61-62. doi: 10.1007/BF03160910.

64 Dörner, K., *Gesundheitssystem: In der Fortschrittsfalle.* Dtsch Arztebl 2002, 38:24622466. https://www.aerzteblatt.de/archiv/32976/Gesundheitssystem-In-der-Fort-schrittsfalle.

65 Peter C. Gøtzsche, "Prescription Drugs Are the Leading Cause of Death And psychiatric drugs are the third leading cause of death," *Mad in America – Science, Psychiatry and Social Justice.* (April 16, 2024), https://www.madinamerica.com/2024/04/prescription-drugs-are-the-leading-cause-of-death/.

66 Holland, M. & O'Toole, Z., (Publishers), Anonymous (author): Turtles All The Way Down: Vaccine Science and Myth. The Turtles Team, 2022, Chapter 1: https://www.skepticalraptor.com/blog/wp-content/uploads/2023/05/Turtles-Book-English-Chapter-1.pdf; References: https://drive.google.com/file/d/16qHPe0odDOweuCDlwIEjarF4Aiu7oyM9/view.

67 A. R. Mawson & J. Binu, "Vaccination and Neurodevelopmental Disorders: A Study of Nine-Year-Old Children Enrolled in Medicaid," *Science, Public Health Policy, and the Law*. (2025), 6:2019-2025. https://publichealthpolicyjournal.com/vaccination-and-neurodevelopmental-disorders-a-study-of-nine-year-old-children-enrolled-in-medicaid/.

68 https://michael-nehls.de/infos/reformprogramm-gesundheitssystem/ (17.12.2024).

69 M. Nehls, "What does the brain need to think peace?" IC Forum 2024, Switzerland: https://www.youtube.com/watch?v=4wtJ22zn2Kg&t=0s; German version of a public lecture on 04.05.2024: https://www.youtube.com/watch?v=3dmIF-wwcNik.

70 Nehls M: Das indoktrinierte Gehirn: Wie wir den globalen Angriff auf unsere mentale Freiheit erfolgreich abwehren. Mental Enterprises 2023.

71 https://michael-nehls.de/infos/vitamin-d/ (November 2023); https://michael-nehls.de/infos/biowaffe-gegen-das-ungeborene-kind/ (August 2024); https://michaelnehls.de/infos/rsv-amp-anti-vitamin-d-kampagne/ (August 2024).

72 F. Deruelle, "The pharmaceutical industry is dangerous to health. Further proof with COVID-19," *Surg Neurol Int*. (2022), doi: 10.25259/SNI_377_2022.

73 F. Bergman, "Dutch Government Official Admits Covid Pandemic Was 'Military Operation': 'Ministry of Health Obeys NATO'" https://slaynews.com/news/dutch-government-official-admits-covid-pandemic-military-operation-ministry-health-obeys-nato/ (09.11.2024, zuletzt abgerufen am 25.02.2025).

74 F. Deruelle, "Microwave radiofrequencies, 5G, 6G, graphene nanomaterials: Technologies used in neurological warfare," *Surg Neurol Int.* (2024), doi: 10.25259/SNI_731_2024.

75 J. F. Kennedy: Address "The President and the Press" Before the American Newspaper Publishers Association, New York City. 27.04.1961, https://www.presidency.ucsb.edu/documents/address-the-president-and-the-press-before-the-american-newspaper-publishers-association.

76 https://www.epochtimes.de/politik/ausland/us-senat-robert-f-kennedy-jr-als-gesundheitsminister-bestaetigt-a5039241.html (13.02.2025, zuletzt abgerufen am 19.02.2025).

77 *https://www.cdc.gov/covid/vaccines/stay-up-to-date.html.*

78 M. R. Goldstein & L. Mascitelli, "Is violence in part a lithium deficiency state?" *Med Hypotheses*. (2016), 89:40-42. doi: 10.1016/j.mehy.2016.02.002.

INDEX

Page numbers followed by f indicate figure

J